Decentralized Charging Coordination of Large-scale Plug-in Electric Vehicles in Power Systems

Zhongjing Ma

Decentralized Charging Coordination of Large-scale Plug-in Electric Vehicles in Power Systems

Springer

Zhongjing Ma
School of Automation
Beijing Institute of Technology
Beijing, China

ISBN 978-981-13-7651-1 ISBN 978-981-13-7652-8 (eBook)
https://doi.org/10.1007/978-981-13-7652-8

This Springer imprint is published by the registered company Springer Nature Singapore Pte Ltd.
The registered company address is: 152 Beach Road, #21-01/04 Gateway East, Singapore 189721, Singapore

Preface

Plug-in electric vehicles (PEVs) that obtain their energy from the electricity grid have been achieving significant market penetration in recent years. Such vehicles will reduce consumption of exhaustible petroleum resources and could reduce pollutant emissions including greenhouse gases. As their population grows, however, the disordered or improperly coordinated charging behaviors of the high-penetration PEVs may have significant negative impacts on the grid system. For example, a large number of PEVs began charging around the time most people finish their evening commute, a new demand peak could result, possibly requiring substantial new generation capacity capability. Due to the coupling constraints over the multi-time period among the PEV population, it is challenging to implement the optimal charging behaviors for large-population PEVs in practical ways.

This book focuses on the designs of decentralized optimization methods to implement optimal charging strategies for large-scale PEVs in electrical power systems. This book studies several classes of charging coordination problems of large-scale PEVs by considering distinct characteristics of PEV populations and electrical power systems. It designs decentralized methods via distinct optimization schemes, such as non-cooperative games, mean-field games, bidding in electricity market, to implement the optimal/nearly optimal charging strategies in effective ways. The performances of proposed algorithms, such as convergence, computational complexity and optimality, etc., are verified, respectively, in rigorous ways. Numerical simulation cases are also studied and analyzed in detail to verify the developed results.

This book may benefit the researchers and engineers in universities and electrical power industry who design and develop PEV charging strategies, utilization of energy resources, and economic operation & control of electrical power systems, and those researchers in universities and research institutes who are interested in the areas of systems and control, optimization, game theory, auction mechanism design, and especially their applications in efficient allocation of divisible resources, like electricity. This book may be useful to senior undergraduate and postgraduate students who are in a range of electrical engineering, computer sciences, and operation research as well.

Beijing, China
February 2019

Zhongjing Ma
mazhongjing@bit.edu.cn

Acknowledgements

This monograph would not be possible without those people who have encouraged and helped me during my research career.

First, I would like to express my sincere gratitude to Prof. Duncan Callaway and Prof. Ian Hiskens. I had started my research in the research direction of charging coordination of large-population PEVs in smart grid from January 2009 at the University of Michigan, Ann Arbor, as a post-doctoral research fellow advised by Prof. Duncan Callaway and Prof. Ian Hiskens. During that period of time, I had been dedicated to this topic which had been one of the hottest research directions in recent years. I also thank Prof. Peter Caines and Prof. Roland Malhame who were my advisors during my graduate study in the field of systems and control at the McGill University, Montreal, Canada. All of them provided me many insightful comments and contributed significantly to my research work.

I really appreciate valuable contributions from Suli Zou, my former student and close collaborator. We have worked together for about 9 years since she was still in her undergraduate study at School of Automation, Beijing Institute of Technology.

I also thank some of my other graduate students including Xiaokun Yin, Long Ran, Xingyu Shi, Yongjian Cai, Nan Yang, Peng Wang, Xu Zhou, Fei Yang, etc. Besides, I would like to thank my team members, Dr. Zhigang Gao, Dr. Hongwei Ma, and Dr. Liang Wang, and other colleagues at the research institute of Electrical Engineering who have not only provided me with their valuable suggestions for my research but also enriched my research life in Beijing Institute Technology for our warm team environment throughout my research activities.

In addition, I also would like to thank the editor, Jasmine Dou, and reviewers and other staffs for their assistances.

I also thank the financial supports from National Natural and Science Foundation, China (NNSFC), State Grid, China.

Last not least, I would like to express my deepest appreciations to my wife, Lijuan Wu, and our parents, and our dear daughter, Cathy Ma. They have always provided me with their endless encouragement and supports.

Contents

List of Figures

List of Tables

Chapter 1
Introduction

1.1 Backgrounds and Motivations of the Book

Along with the rapid consumption of exhaustible nonrenewable petroleum energy resources and high emission of green gas, the plug-in electric vehicles (PEVs) achieve a high-speed development. These vehicles may reduce consumption of petroleum resources and may reduce pollutant emissions including greenhouse gases. However the improper or disordered charging behaviors of high-penetration PEVs may have significant negative impacts on power grid, like more investments on grid system, higher energy costs and increased peak load etc, see [1–15]. For example, if a large number of PEVs began charging around the time most people finish their evening commute, a new demand peak could result, possibly requiring substantial new generation capacity and ramping capability [9].

In the United States, the average vehicle is driven about 30 miles per day, and the average one-way commute time is around 25 min [16]. This implies PEVs usually spend a significant amount of time parked and available to charge. On the other hand, charging times for electric vehicles are likely to be 8 h or less, well under the time that most PEVs will actually be available for charging. This suggests that there could be useful flexibility with respect to when and how fast vehicles charge. If vehicle charging can be coordinated, it could be possible to construct aggregated charge profiles that avoid detrimental power system impacts and minimize system-wide costs. Hence in order to support the accommodation of high-penetration PEVs in power grid, it is important to study how to coordinate the PEV charging behaviors.

Usually, in those researches which explored the potential impacts of high penetrations of PEVs on the power grid [1–3, 7, 17], it assumes that PEV charging patterns "fill the valley" of night-time demand, see Fig. 1.1. Subject to the valley-filling behaviors, the overnight valley of the base demand would be replaced by a total demand that remained relatively constant during the charging period. These studies do not address the issue of how to coordinate PEV charging patterns. Actually, though the valley-filling charging behavior is conceptually straightforward, the

Z. Ma, *Decentralized Charging Coordination of Large-scale Plug-in Electric Vehicles in Power Systems*, https://doi.org/10.1007/978-981-13-7652-8_1

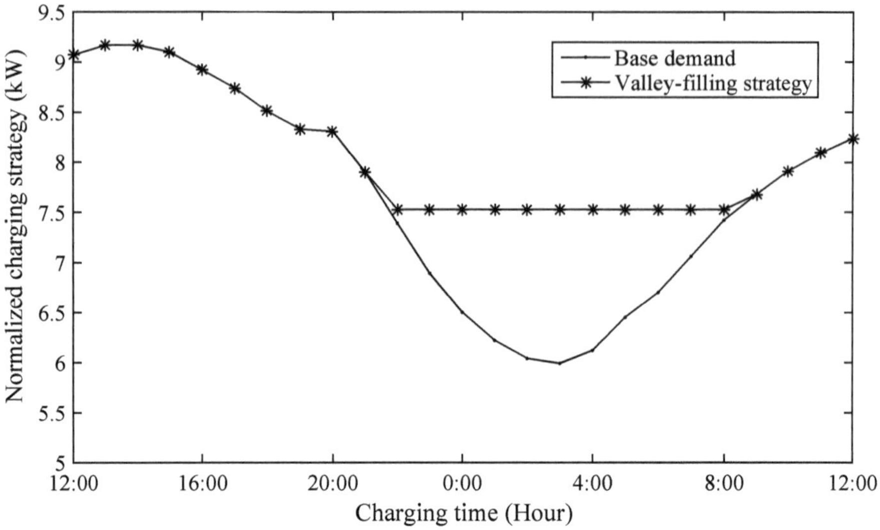

Fig. 1.1 Typical (non-PEV) base demand in summer for the region managed by the Midwest Independent System Operator (MISO). An ideal valley-filling profile is also shown

process of achieving it in a practical manner with a large population of independent PEVs presents a lot of challenges.

Recently lots of research work has been dedicated to studying how to properly coordinate the PEV charging behaviors to mitigate the involved negative impacts, see [18–30] and references therein.

Possible charging strategies could be implemented in centralized ways to mitigate their negative effects on the grid in centralized ways, e.g. [31–36] and the references therein.

In these centralized strategies, a central operator dictates precisely when and at what rate every individual PEV will be charged. Decisions could be made on the basis of system-level considerations only, or they could factor vehicle-level preferences, for example desired charging interval, final state of charge, and budget. These strategies could be further distinguished by whether they attempt to identify charge patterns that are in some way optimal, or instead follow rules-of-thumb that seek to achieve aggregate charging patterns that are reasonably close to optimal.

However, in practice, a centralized approach may not be permitted to consumers, who are accustomed to having complete decision-making authority over their electricity consumption patterns. Also the PEVs may not be willing to share their private information with others, the transmission of complete information may create heavy communication signals, and optimizing over a large population of PEVs will have high computational complexity.

In the literature, quite a few researches have been dedicated to proposing decentralized methods to achieve an optimal or near-optimal outcome for large-scale PEVs to preserve individual authority and distribute the computational burden [37–44]. However, as it will be discussed below, the challenge of coordinating many autonomous agents to achieve an optimal or near-optimal outcome is non-trivial.

Game-Based Large-Population PEV Charging Coordination Problems

As discussed in [18], it is difficult to effectively fill the night-time valley based upon the time-based or fixed price; and in [45] the authors adopt a real-time price model which has been widely applied in the literature, e.g., [46, 47] for demand response management, and [48, 49, 148] for electric vehicle charging/discharging coordination. More specifically, the electricity price at an instant is determined by the total demand at that instant and represents the generation marginal cost. Decentralized strategies allow individual PEVs to determine their own charging pattern respectively. Vehicle charging decisions could, for example, be made on the basis of time-of-day or electricity price. The outcome of a decentralized approach may or may not be optimal, depending on the information and methods used to determine local charging patterns. Care must be taken to ensure charging strategies cannot inadvertently synchronize the responses of large numbers of PEVs, as the resulting abrupt changes in aggregate demand could potentially destabilize grid operations [18].

In this book, it explores the potential of decentralized strategies by applying the method which utilizes real-time marginal electricity price information. It allows each PEV to choose and implement its own local charging strategy, with the aim of minimizing its individual charging cost.

The decentralized PEV charging coordination problem studied in this book is a form of non-cooperative game, where a large number of selfish PEVs share electricity resources on a finite collection of charging intervals.

This charging game falls within the class of *potential games* identified by Monderer and Shapley [50]. It is mathematically equivalent to routing and flow control games in telecommunications, where networks of parallel links are congested [51]. From this point of view, PEV charging games are conceptually similar to network games [52].

Substantial work has been presented in the literature on the computation of Nash equilibria, or ε-Nash equilibria, for potential games, especially in relation to network games. Research on *centralized mechanisms* includes Christoudoulou et al. [53], who consider two classes of potential games, selfish routing games and cut games, and Even-Dar et al. [54] who study the number of steps required to reach a Nash equilibrium in load balancing games. Research on *decentralized or distributed mechanisms* includes Berenbrink [55], who propose a strongly distributed setting for load balancing games such that all agents update their strategy simultaneously. Also, Even-Dar et al. [56] present convergence results for an approximate ε-Nash equilibrium under a non-centralized setting in routing games, and Fischer et al. [57, 58] propose a distributed and concurrent process for convergence to Wardrop equilibria [59] in adaptive routing problems. The computation time of each of these methods is related to the number of agents, the size of network, and the accuracy parameter ε if applicable. In [60] Chien and Sinclair studied convergence to approximate NE in congestion games, In [61] W. Sandholm studied the computation convergence of NE of a class of potential games with (infinite) continuous player sets along evolutionary dynamics.

Actually non-cooperative game theory has been widely used to study the supply side of electricity markets, especially in the context of imperfect competition, see [62–65]. Some game theoretic work has been done to understand demand-side behavior in the face of dynamic pricing tariffs, primarily in the context of demand aggregators [66, 67]. Though the work also applies game-theoretic principles to understand outcomes in electricity markets, it considers multi-period demand-side behavior with a local energy constraint applied to the total energy consumed across all periods, and analyzes the problem from a decentralized perspective.

PEVs are supposed to be coupled with each other through a common price signal which is determined by the average charging strategy of the PEV population. Therefore each PEV effectively interacts with the average charging strategy of the rest of the PEV population. As the population grows substantially, the influence of each individual PEV on that average charging strategy becomes negligible. Accordingly, in the infinite population limit, all PEVs will observe the same average strategy as they calculate their optimal local strategy. In this situation, a collection of local charging strategies is a Nash equilibrium, in case: Each PEV's charging strategy is optimal with respect to a single commonly observed charging trajectory, and the average of all the local optimal charging strategies is equal to that common trajectory.

The proposed algorithm converges to an ε-Nash equilibrium for a finite population, and ε tends to zero as the population size approaches infinity. In this limiting case, the Nash equilibrium corresponds to a Wardrop equilibrium [59]. This method also has connections with the Nash certainty equivalence principle (or mean-field games), proposed by Huang et al. [68, 69] in the context of large-scale games for sets of weakly coupled linear stochastic control systems.

Charging Coordination Problems Considering Battery Degradation Cost

Besides the charging pricing cost considered earlier which is an important factor for each individual PEV, it has been widely studied, e.g., [70–77], that some key characteristics of chemical batteries, like the state of health, the growth of resistance, the cycle life, etc., are effected by the charging behaviors as well. As a consequence, it is proper to address the need for a charging coordination scheme which considers the tradeoffs between system-wide economic efficiency discussed in the above and battery degradation concerns.

Charging behavior affects key battery characteristics, including the state of health, the resistance impedance growth and the cycle life, which are all strongly related to the energy capacity of a battery [70, 71]. Intermittent charging may also shorten the battery lifespan [21]. Optimal charging strategies that take into account both the total energy cost and the battery state of health have been studied for single PEVs [70, 78]. These ideas form the basis for the extension, undertaken in this book, to large-scale coordination. It analyzes the battery degradation cost of the $LiFePO_4$ battery with respect to the charging power.

The system considers the tradeoff between the total generation cost and the local costs associated with overloading and battery degradation. Furthermore, it considers the charging flexibility of PEVs. PEVs can independently determine their total charging capacity according to their own characteristics and the environment in which they are charged. The optimal charging strategies are analyzed as the parameter varies. some research work, e.g., [70, 78], studied the optimal charging behavior for a single PEV by taking into account both total the energy cost and the state of health of batteries.

Alternatively in this book, a decentralized charging coordination method is proposed for large-scale PEVs dealing with the total generation cost and the accumulated battery degradation cost over multi-time intervals.

Decentralized PEV Charging Coordination Under a Hierarchical Structure

The decentralized charging methods described above and other methods studied in the literature, e.g., [79], have several benefits such that the individual authority remains and only limited individual charging information is required.

However the system under decentralized methods mentioned above has to directly deal with the transactions with all of the individuals. As a consequence, it is still challenging and impractical to implement the decentralized charging behaviors for large-population PEVs, since it requires complicated management, computation and heavy communications between the system operator and PEVs. In the realistic power market, the electricity is always traded under hierarchical structure, from the generator individual end-users and intermediaries, say utilities.

Actually much research work, studied in [80–83], has presented hierarchical charging systems for PEVs by organizing the PEVs into several groups, and setting a collection of intermediaries to manage the charging behaviors of the PEVs belonging to them. More specifically, in [80], the power grid is divided into multiple areas with a charging management center setting up the management of PEVs in a certain geographic area, providing communication interfaces between the PEVs and the dispatching centers. [81] focused on the PEVs charging coordination scheme for multiple parking decks in multifamily dwellings accounting for every user's satisfaction in a centralized way and distributed method, while in [82], it presents a hierarchical and zonal architecture for dispatching PEVs which aims to optimize the objective composed of electricity generation cost, battery degradation cost and penalty cost between the upper and lower layers. [83] introduced a hierarchical network resource allocation model. In the above work, the agents need to collect the charging characteristics of individual PEVs and coordinate the charging behaviors of these PEVs. As a result, agents take implement complicated optimization computing and heavy information transmission missions.

In this book, it considers that each of the individual PEVs deals with the tradeoff between the battery degradation cost and the charging pricing cost; then the system can improve the longevity of the batteries and may mitigate the oscillation caused by the greedy behavior for the cheap electricity resource by individual PEVs. As mentioned before, this kind of cost plays the same role as the congestion pricing which has been widely used in Internet traffic control, see [84], and has been adopted in [48] to schedule PEV charging strategies in the power grid.

To deal with the charging coordination of large-population of PEVs, it formulates a class of decentralized PEV charging coordination problems under a hierarchical structure such that the electricity is traded from the generators to load units via (intermediate) agents.

PEV Charging & Discharging Coordination Problems

PEVs could also serve as energy storage systems (ESSs), which can promote the integration of renewable energy sources (RES) and maintain the stability of the grid against the intermittence and randomness of the RES. Due to this, it has emerged to offer benefits to the power grid by considering both charging and discharging behaviors for individual PEVs, [23, 85–88], e.g., shaving peak load [89] and providing ancillary services [90, 91], and some recent work has studied the interaction between the PEVs and the RES, such as [92, 93]. In this book, it is focused on the optimal bidirectional, say charging and discharging, coordination of PEVs in order to minimize total system costs and reduce the effects of the RES on the power grid.

Compared with the work on how to effectively coordinate PEVs' charging behaviors, the PEV management algorithms under bidirectional, say including both charging and discharging behaviors, scheme can achieve smaller load variance through cost minimization or benefit maximization [94–98]. Specifically, in [94], the authors firstly analyzed the bidirectional coordination of PEVs via a global scheme, and then proposed a local scheme in which PEVs are divided into groups to reduce the communicational and computational burden. It can achieve a near-optimal performance under the proposed hierarchical coordination method. In [95], a coordinated charging and discharging algorithm was proposed to minimize PEVs' charging cost, under the consideration of network limits. In [96], a real-time scheduling algorithm was applied to estimate the capacity of V2G and the availability of the RES within a building energy management system. Stochastic characteristics of PEV driving behaviors were taken into account in [97, 98]. The aim of [97] is to minimize the overall load variance of power grid in a centralized way, whilst [98] aims at evaluating the V2G capacity using aggregate PEV model.

Such work [94–98] deals with the charging and discharging coordination of PEVs in different scenarios under centralized schemes, which usually require complex and expensive communication infrastructure, powerful central controller with computation modular, and full information of all the PEVs. Also, decisions of all the PEVs are controlled by the central controller. In contrast, charging/discharging decisions are made and executed by the PEVs themselves in decentralized ways. The communication and computation burden can be largely reduced in the meanwhile, especially for large scale systems. A couple of decentralized methods have been designed alternatively to achieve coordination solutions. For example, in PEV charging coordination, real-time pricing [45, 99] and noncooperative games [100, 148] are leveraged. However, these developed decentralized methods cannot be directly applied to solve bidirectional coordination, since the evolution of the SOC of PEVs' batteries is no longer non-decreasing as in the charging-only scenarios.

Considering how to effectively coordinate the charging and discharging behaviors of PEVs, [101–105] have dedicated to developing decentralized schemes. The performance of these methods is highly related to the shape of the inelastic demand trajectory, which means the resulted coordination strategy may be far away from the social optimum in case the shape of the inelastic demand trajectory doesn't fulfill the shapes considered.

Charging Coordination Considering Dynamic Electricity Prices

Pressures of rapid consumption of exhaustible energy, increasing market competition and reliability have led to an increasing push for demand-side participation, especially in the form of increased price responsiveness of electricity demand, in competitive power markets [106, 107]. Meanwhile, consumers are getting exposed to a more volatile electricity price and they may decide to modify the profile of their demand to reduce their electricity costs [108]. Demand Response (DR) appears as a very promising opportunity for consumers and brings several advantages for the whole system [109–111]. DR can be defined as the changes in electricity usage by end-use customers in response to changes in the price of electricity over time, or the incentive payments designed to induce lower electricity use at times of high electricity prices or when system reliability is jeopardized [112, 113].

PEVs may migrate from one time period to another dependent on the price prevailing at other time periods, and shiftable demands are said to be lossless if the total energy use over the full time horizon is fixed [106, 108]. This brings attention to inter-temporal factors of price elasticity or customer utility. In many researches, the demand-price interaction is characterized by price elasticity matrices [114, 115], which represent the sensitivity of electricity demand with respect to the change of price. Moreover, the concept of "cross-elasticity" is developed in [116], to characterize the inter-temporal factors of DR systems, such as [110, 117–121]. However, there may be approximation in obtaining the values of the elasticity, e.g. a simplified linearized price demand curve in [122] and a stepwise linear price demand curve in [123]. In addition, the elasticity values are fixed in the above researches, whereas in actual they are uncertain due to a variety of reasons including latency in communication and change in consumption behavior. Therefore, consumer loads have different response patterns to the electricity prices under different scenarios. In [124], the authors try to consider this uncertainty and inexact DR information by allowing the price-elastic demand curve to vary within a certain range, and propose an efficient robust unit commitment approach for the solution. To fix the related issues of elasticity values, it originally applies an equality energy constraint of total electricity use in this work to reflect the inter-temporal load characteristic. Further, this constraint is transformed as a deviation cost that is added into the valuation function of end-users.

In this work, it considers large-scale incorporation of inter-temporal shifting load units in electricity markets, like PEVs studied in this book. Centrally administered demand management methods was an important feature of many regulated utilities, like Lagrangian/Newton based methods [108], interior point method [118], linear programming [125] and mixed integer programming [124, 126]. However, the valuation function and the constraint sets of users are private, and it will be burden for the market operator to solve an optimization problem directly coordinating a significantly large number of customers. Seasoned to cope with these issues, hierarchical DR management framework is proposed via DR aggregators between the market operators and the end-users in [107, 127, 128], such that the resultant large optimization problem can be decoupled into separate subproblems. In [129], an efficient pricing method based on VCG mechanism is proposed to seek social benefit. In [106, 121], DR is incorporated with auction based mechanisms where loads simply respond to real-time or forecasted prices.

Since auctions in electricity markets considering demand response may have greater potential to increase efficiency than relying solely on a more passive demand response, in this book, it designs a pricing mechanism to induce users behave efficiently and obtain the maximum social welfare following the iterative update strategies for individual PEVs. It also specifies certain sufficient conditions under which the system may converge to the efficient solution.

Charging Coordination Problems in Constrained Systems

Besides high electricity generation costs, the uncoordinated charging behaviors of PEV populations in the distribution networks may also induce localized overloading, large power losses and voltage problems [7, 14, 32–34, 130–133]. Hence how to coordinate the PEV charging behaviors in distribution networks has been widely studied in the literature, to minimize the power losses [32], to maximize the profits of the utility considering voltage constraints [134], to minimize the battery degradation cost and deviation cost from the referred SOC value under the constraints of transformers [33], and to shave the peak demand and minimize the power losses considering voltage regulation [135].

Some research work devotes to avoiding overloading in distribution networks. In [136], the PEV charging was coordinated to minimize individual costs in response to the time-varying prices with additional incentives provided to mitigate potential overloads in the network. In [137], the energy costs of the distribution networks were minimized with overload limitation by adopting Lyapunov optimization and Lagrange dual decomposition techniques, while in [138], the variance of the aggregate load was minimized with feeder overload constraints relaxed by adding an overload cost function to the objective function. A decentralized PEV coordinating method was proposed in [139] by solving Lagrangian dual problem to avoid the feeder overloading, while it didn't consider the total amount of the energy charged by each PEV and the optimal solution was implemented on a single time slot. The above work decouples the overload constraints by implementing Lagrangian multipliers with slack parameters, which may violate the feeder capacity. In addition, the application of slack parameters may result in suboptimal solutions.

Due to the coupling relationship among the charging behaviors of individual PEVs introduced by feeder overload constraints, it is usually challenging to design a method to coordinate the charging behaviors of PEV populations in decentralized ways. In this chapter, a class of charging coordination problems of PEV populations is formulated in a distribution network by considering the hard constraint of the feeder capacity. It further develops a gradient-projection based decentralized method for the underlying constrained optimization problems.

V2G Coordination Problems: Voltage and Frequency Regulations with PEVs

To operate safely, power grid has to be ensured with the amplitude of frequency and voltage close to their nominal values by balancing the production and consumption of both active and reactive power, see [140–143]. As discussed earlier, PEVs can play as distributed electricity storages for power grid, see [144–147], since PEVs can deliver electricity to power grid and consume electricity from power grid. Nowadays quite a few researches have been dedicated to studying how to apply vehicle-to-grid (V2G) structure for frequency regulations in smart grid, e.g. [148–152]. In practice, most of reactive power loads are inductive, so power compensation for voltage regulation mainly involves adjusting capacitors, e.g. [153–156]. More recently [157] proposed a V2G regulation model such that PEVs are applied to provide ancillary services of frequency and voltage regulations for power grid by coordinating their active and reactive power rates, and further formulated the underlying regulation problems as two joint optimization problems under different pricing and contract scenarios.

An auction-based decentralized method is proposed for V2G coordination problems initialized by [157] under the progressive second price (PSP) auction mechanism which was firstly proposed by [158, 159] to efficiently allocate a single network divisible resource.

Under the PSP auction mechanism, the incentive compatibility holds, i.e. the truth-telling bid profile is the best response of agent. Hence the PSP auction can be considered as the extension of the second price auction proposed in [160–162] to allocate indivisible items with truth-telling bid profiles in an efficient way. It is worth to note that the incentive compatibility does not hold in general for other sealed auction mechanisms, like uniform market clearing price and pay as bid auction mechanisms etc., see [163, 164], which have been widely adapted to economically dispatch generations in day-ahead deregulated electricity markets, see [165]. In the allocation problems of a single divisible resource under the PSP auction mechanism, each agent only reports a two-dimension bid profile which is composed of a maximum amount of demand and an associated buying price, and is used to replace his complete (private) utility function. In [159], the authors verified that the efficient incentive compatible bid profile is a Nash equilibrium for PSP network auction games. More recently many progresses have been developed in PSP auction games. For example, [166] extended the results from single-side auction games to double-side ones in the network resource allocations; [167–169] proposed a decentralized algorithm, under quantized-PSP auction mechanism for power electricity sharing games, following which the auction game system converges to the efficient solution.

In the V2G regulation problems in smart grid, suppose that each of autonomous PEVs deals with the tradeoff between costs and benefits with respect to supplied active and reactive power service rates; while power grid deals with the tradeoff between the benefits with the aggregated service rates provided by the PEV populations and the costs he needs to pay for the services. The players participating in the underlying auction games, including power grid as a single buyer and autonomous PEVs as sellers, do not report a full cost or valuation function but only a multi-dimensional bid signal, say the maximum active and reactive power quantities that power grid wants and the maximum per-unit prices he is willing to pay, the maximum active and reactive power quantities that an autonomous PEV can provide and the minimum per-unit prices he asks. It is worth to analyze the existence of the efficient NE for the underlying auction game of the V2G coordination problems and design an update algorithm to implement the efficient solution as well.

1.2 Contributions

- *Decentralized charging coordination method for large-population PEVs*
  In this book, it presents a novel procedure that identifies a Nash equilibrium by simultaneously updating each PEV's best individual strategy with respect to the average charging strategy of the whole population. It is proposed that this procedure would be undertaken prior to the actual charging interval. The computation time of this algorithm is unrelated to the number of PEVs, since they simultaneously and independently update their charging strategy. Under certain mild conditions, the proposed decentralized charging coordination procedure drives the system asymptotically to a unique Nash equilibrium that is nearly globally optimal (valley filling). In the case of homogeneous populations, where all vehicles have identical parameters, this unique Nash equilibrium becomes a perfect valley-filling charging strategy.
  The proposed algorithm converges to an ε-Nash equilibrium for a PEV charging system with a finite population, and ε tends to zero as the population size approaches infinity. In this limiting case, the Nash equilibrium corresponds to a Wardrop equilibrium [59]. This method also has connections with the Nash certainty equivalence principle (or mean-field games), proposed by Huang et al. [68, 69] in the context of large-scale games for sets of weakly coupled linear stochastic control systems.
  However due to the heterogeneity among individual PEVs, by adapting proper deviation cost and following the simultaneous best strategy update mechanism, the game systems converge to an equilibrium which is not a valley-filling in general. In order to improve the performance, in this work, it presents an improved charging behavior update mechanism. More specifically, all individuals in a homogeneous subpopulation simultaneously update their best strategies considering the tradeoff between the low charging price and a low deviation cost penalizing against the deviation of the individual strategy from the average behavior in this subpopulation. This procedure continues until no individuals update their best strategies any longer. The proposed strategy update procedure is applied to all sub-populations successively.
  By adapting the proposed update mechanism, at the equilibrium, the deviation cost vanishes, that is to say, the deviation cost is only introduced to enforce individuals to update their behavior towards the average behavior and vanishes at the equilibrium since all individuals in a subpopulation possess an identical strategy. Hence in case the update procedure converges, the system reaches the socially optimal valley-

filling equilibrium. It also specifies technical sufficient condition under which the update procedure converges.

- *Decentralized method for PEV charging coordination problems considering battery degradation cost*
  In the decentralized approach to charging coordination proposed in this part, participating PEVs simultaneously determine their optimal charging strategy with respect to an energy price forecast. These proposed charging strategies are used to estimate the total demand over the charging horizon. An updated price forecast is obtained as a weighted average of the previous price forecast and the generation marginal cost evaluated at this latest demand forecast. The revised price is (re)broadcast to the PEVs, and the process repeats. It is shown that convergence is guaranteed under mild conditions. Upon convergence, the price profile is coincident with the generation marginal cost over the charging horizon. As a consequence, the resulting collection of PEV charging strategies is efficient (socially optimal). Moreover, convergence is obtained without the need for artificial deviation costs to damp oscillations, as in Chap. 2 and some research work in the literature, e.g. [38]. Cost terms introduced to mitigate the effects of local demand peaks and battery degradation play the same role as congestion pricing used for traffic control in communication networks [84], which has been adopted in [48] to schedule PEV charging.
  As the $LiFePO_4$ battery, a lithium-ion type battery, has been widely applied in the PEV market, e.g., GM Chevrolet Spark, Nissan Leaf, BYD (e6, F3DM, F6DM), Renault Clio, etc., the health model for $LiFePO_4$ cell units was specified in [170] based upon the evolution of battery cell characteristics developed in [171, 172]. By adopting the analysis given in [170], it analyzes the battery degradation cost, which is in a quadratic form with respect to the charging strategy, for the $LiFePO_4$ battery with respect to the charging power in this book. In [94], a quadratic form of $f_n(\cdot)$ is adopted as well.
  In the proposed decentralized framework, each of the individual PEVs deals with the tradeoff between the battery degradation cost and the electricity cost; then the system can improve the longevity of the batteries and may mitigate the oscillation caused by the greedy behavior for the cheap electricity resource by individual PEVs. As a result, the deviation cost artificially introduced to mitigate the oscillation in Chap. 2 and some research work in the literature, e.g. [38], is no longer required. The battery degradation cost plays the same role as the congestion pricing which has been widely used in Internet traffic control, see [84], and has been adopted in [48] to schedule PEV charging strategies in the power grid.
- *Decentralized PEV charging coordination under a hierarchical structure*
  In this book, by adapting the exact individual cost model described earlier, say each of the individual PEVs deals with the tradeoff between the battery degradation cost and the charging pricing cost; then the system can improve the longevity of the batteries and may mitigate the oscillation caused by the greedy behavior for the cheap electricity resource by individual PEVs.
  To deal with the charging coordination of large-population of PEVs here, it formulates a class of decentralized PEV charging coordination problems under a

hierarchical structure such that the electricity is traded from the generators to load units via (intermediate) agents. To reimburse the transaction operation cost on agent, an agent operation price is charged to all of the individual vehicles under this agent. Each of the agents sets a charging price which is composed of the generation price and the operation price, and broadcasts it to the collection of individuals under its supervision. Each of the PEVs under an agent implements the best charging behavior with respect to the charging price by minimizing individual charging cost. The agents collect the implemented charging behaviors, updates the operation price and submits the aggregated charging behaviors to the central system operator; the system operator updates the marginal generation cost with respect to the collected updated aggregated charging strategies from collection of agents. It is further shown that the hierarchical charging system can converge to the efficient (or socially optimal) solution by applying the proposed decentralized method under certain mild conditions. The proposed decentralized method under the hierarchical structure presents the advantage of the autonomy of the individual PEVs and the low computation and communication capability requirements on the system.

- *Decentralized optimization method for PEV charging & discharging coordination problems with uncertainties*
  In this book, firstly a class of charging & discharging coordination problems for PEVs are formulated which deal with the tradeoff between total generation cost and the accumulated battery degradation cost of PEV populations [45, 99]. Then a decentralized algorithm is proposed by applying the gradient projection method and by following this modeling issue, the coordination behaviors of all the individual PEVs can be updated locally and simultaneously. The convergence and optimality of the proposed algorithm are verified in case the step-size parameter of the update procedure is set in a certain region. It is worth to mention that, our proposed method doesn't rely on the shape of the inelastic demand. The system is guaranteed to converge to the social optimum no matter what the shape of inelastic demand is. Moreover, to address the forecast errors appeared in the predictions of PEV populations, RES productions and inelastic load profiles, a receding horizon based algorithm is constructed such that the prediction proceeds at each instant. As a result, the system can obtain a coordination strategy against those forecast errors.
- *Decentralized charging coordination method considering dynamic prices*
  In the auction based pricing mechanism, each user submits a multi dimension bid related to the inter-temporal factor to the system. The system implements an optimal allocation with respect to the submitted bid profile, and broadcasts an associated system price to all the users. Due to the cross-elastic correlation among the intervals, the best updated response of users, at an instant, is determined by both the demand at this instant and the total demand over the whole horizon. In addition, it develops a dynamic dispatch algorithm for distributing the energy where all the individual users update their best bidding responses simultaneously with respect to the system price which is determined by the previous collection of bid profile. The expected system price dictates users' best response; on the other

hand, the resulting load pattern will alter this price. As a result, the algorithm continues until the system price is fixed. It can show that, following the update procedure of the individual best response behaviors, the system converges to the efficient coordination solution under certain mild sufficient conditions.

- *Decentralized method for PEV charging coordination problems with coupling constraints*
  In case the capacity constraints of feeder lines are not considered, then by the decoupling relationship among the admissible sets of coordination behaviors of all the PEVs, the gradient projection on an individual admissible (charging) set is independent upon the admissible sets of all the others. Following this modeling characteristic, the coordination behaviors of all the PEVs can be updated locally and simultaneously. However, as demonstrated via numerical simulations, the updated behaviors in the decentralized form may not belong to the admissible space under capacity constraints. Consequently, in order to avoid the overloading on the feeder lines during the dynamic updates, the step size for each update step is properly adjusted to satisfy the capacity constraints. More specifically, if the updated behaviors are not satisfied with the capacity constraints at some iteration step; it re-implements the decentralized gradient projection with an adjusted smaller valued step-size parameters, and this process continues until the updated behaviors satisfy the constraints. Furthermore, suppose that the updated behaviors reach the exact capacity constraints on some feeder line at some instant; then it considers that the charging behaviors of all the PEVs on these feeder lines shall be fixed and won't be updated any longer.
- *Game-based decentralized methods for V2G coordination problems*
  It verifies the existence of the efficient NE for the V2G auction game, though there may exist other inefficient NEs. Moreover in order to deal with the V2G coordination problems with large-scale PEVs, it designs a class of V2G coordination auction games with aggregator players each of which represents a collection of individual PEVs. To implement the efficient NE, in this part, the PSP auction mechanism, proposed in [167], is adapted and extended to the underlying V2G auction games with constrained multi-dimension bid strategies. As illustrated with the numerical examples, the game system converges to some Nash equilibria which is near to the efficient NE.

1.3 Organization

By exploiting both spatial and temporal coordinations, it introduces an online charging/discharging strategy considering range anxieties for mobile PEVs. Specifically, to collect the real-time information required by the proposed strategy, a heterogeneous wireless infrastructure is proposed by combining wide-coverage cellular networks with economic high-rate vehicular ad hoc networks (VANETs).

- This chapter gives the overall and the brief introduction of the main works in this monograph.
- In Chap. 2, it formulates a class of PEV charging coordination problems in smart grid and develops a decentralized method such that it implements the so-called valley-filling charging strategy in case each individual PEV updates its own charging behavior with respect to others.
- Considering the battery degradation cost analyzed in Chap. 3, the decentralized method developed in Chap. 2 is updated and analyzed.
- In Chap. 4, PEV charging coordination problems are analyzed under a hierarchical structure and solved in decentralized ways.
- Due to the energy storage property of PEV batteries, in Chap. 5, PEVs are considered as an energy storage units. It further designs a decentralized optimization method, for the underlying charging and discharging coordination problems, by applying which the optimal solution could be implemented.
- In Chap. 6, it designs a decentralized dynamic pricing mechanism for demand response of inter-temporal shifting loads, say PEVs as specific case in this work, based on auction rules, to induce users to behave efficiently and obtain system wide benefit. Moreover, it proposes a dynamic algorithm to implement the efficient results under certain sufficient conditions.
- It studies in Chap. 7 the charging coordination of PEVs in distribution networks with capacity-constrained feeder lines. A gradient-projection based decentralized method is proposed such that the system converges a nearly optimal solution which satisfies the capacity coupling constraints of feeder lines.
- In Chap. 8, it formulates frequency and voltage regulations problems of PEV populations as a class of VCG-style auction games, and shows the existence of the efficient NE for the underlying auction games. In order to deal with large-scale PEVs, it proposes games with aggregator players, and extends the quantized-PSP mechanism to implement the efficient NE.

References

1. P. Denholm, W. Short, An evaluation of utility system impacts and benefits of optimally dispatched plug-in hybrid electric vehicles. Technical Report NREL/TP-620-40293 (National Renewable Energy Laboratory, 2006)
2. S. Rahman, G.B. Shrestha, An investigation into the impact of electric vehicle load on the electric utility distribution system. IEEE Trans. Power Deliv. **8**(2), 591–597 (1993)
3. F. Koyanagi, Y. Uriu, Modeling power consumption by electric vehicles and its impact on power demand. Electr. Eng. Jpn. **120**(4), 40–47 (1997)
4. F. Koyanagi, T. Inuzuka, Y. Uriu, R. Yokoyama, Monte Carlo simulation on the demand impact by quick chargers for electric vehicles, in *IEEE Power Engineering Society Summer Meeting*, vol. 2, Istanbul, Turkey, 18–22 July 1999, pp. 1031–1036
5. X. Yu, Impact assessment of PHEV charge profiles on generation expansion using energy modeling system, in *IEEE Power and Energy Society General Meeting*, Pittsburgh, PA (2008)
6. S.W. Hadley, A. Tsvetkova, Potential impacts of plug-in hybrid electric vehicles on regional power generation. Technical Report ORNL/TM-2007/150 (ORNL, 2008)

7. L. Fernandez, T. Roman, R. Cossent, C. Domingo, Assessment of the impact of plug-in electric vehicles on distribution networks. IEEE Trans. Power Syst **26**(1), 206–213 (2011)
8. A. Maitra, J. Taylor, D. Brooks, M. Alexander, M. Duvall, Integrating plug-in-electric vehicles with the distribution system, in *20th International Conference and Exhibition on Electricity Distribution – Part 1*, 8–11 June 2009
9. D.M. Lemoine, D.M. Kammen, A.E. Farrell, An innovation and policy agenda for commercially competitive plug-in hybrid electric vehicles. Environ. Res. Lett. **3**, 014003 (2008)
10. D. Vallejo, J. Albusac, C. Glez-Morcillo, J.J. Castro-Schez, L. Jimenez, A multi-agent approach to intelligent monitoring in smart grids. Int. J. Syst. Sci. (2013). https://doi.org/10.1080/00207721.2013.783644
11. L. Kelly, A. Rowe, P. Wild, Analyzing the impacts of plug-in electric vehicles on distribution networks in british columbia, in *Electrical Power & Energy Conference* (2010)
12. E. Ungar, K. Fell, Plug in, turn on, and load up. IEEE Power Energy Mag. **8**(3), 33–35 (2010)
13. L. Wang, Potential impacts of plug-in hybrid electric vehicles on locational marginal prices, in *IEEE Energy 2030*, Atlanta, Georgia, 17–18 November 2008
14. P. Richardson, D. Flynn, A. Keane, Impact assessment of varying penetrations of electric vehicles on low voltage distribution systems, in *IEEE Power and Energy Society General Meeting* (2010), pp. 1–6
15. M. Kintner-Meyer, K. Schneider, R. Pratt, Impacts assessment of plug-in hybrid vehicles on electric utilities and regional US power grids, part 1: Technical analysis. Technical Report (Pacific Northwest National Laboratory, 2007)
16. S.C. Davis, S.W. Diegel, R.G. Boundy, Transportation energy data book. Technical Report ORNL-6985 (Oak Ridge National Laboratory, 2010)
17. F. Koyanagi, T. Inuzuka, Y. Uriu, R. Yokoyama, Monte Carlo simulation on the demand impact by quick chargers for electric vehicles, in *Proceedings of the IEEE Power Engineering Society Summer Meeting*, vol. 2 (1999), pp. 1031–1036
18. D.S. Callaway, I.A. Hiskens, Achieving controllability of electric loads. Proc. IEEE **99**(1), 184–199 (2011)
19. M. Vaya, G. Andersson, Centralized and decentralized approaches to smart charging of plug-in vehicles, in *2012 IEEE Power and Energy Society General Meeting* (2012), pp. 1–8
20. C. Ahn, C.-T. Li, H. Peng, Optimal decentralized charging control algorithm for electrified vehicles connected to smart grid. J. Power Sources **196**, 10369–10379 (2011)
21. L. Gan, U. Topcu, S.H. Low, Stochastic distributed protocol for electric vehicle charging with discrete charging rate, in *IEEE Power & Energy Society General Meeting* (2012), pp. 1–8
22. Y. Ota, H. Taniguchi, Autonomous distributed v2g (vehicle-to-grid) satisfying scheduled charging. IEEE Trans. Smart Grid **3**(1), 559–564 (2012)
23. W. Kempton, J. Tomic, Vehicle-to-grid power fundamentals: calculating capacity and net revenue. J. Power Sources **144**(1), 268–279 (2005)
24. W. Kempton, J. Tomic, Vehicle-to-grid power implementation: From stabilizing the grid to supporting large-scale renewable energy. J. Power Sources **144**(1), 280–294 (2005)
25. M.D. Galus, M. Zima, G. Andersson, On integration of plug-in hybrid electric vehicles into existing power system structures. Energy Policy **38**(11), 6736–6745 (2010)
26. H. Lund, W. Kempton, Integration of renewable energy into the transport and electricity sectors through V2G. Energy Policy **36**, 3578–3587 (2008)
27. S.L. Andersson, A.K. Elofsson, M.D. Galus, L. Göoransson, S. Karlsson, F. Johnsson, G. Andersson, Plug-in hybrid electric vehicles as regulating power providers: case studies of sweden and germany. Energy policy **38**(6), 2751–2762 (2010)
28. J.R. Pillai, B. Bak-Jensen, Integration of vehicle-to-grid in the western Danish power system. IEEE Trans. Sustain. Energy **2**(1), 12–19 (2011)
29. C.D. White, K.M. Zhang, Using vehicle-to-grid technology for frequency regulation and peak-load reduction. J. Power Sources **196**(8), 3972–3980 (2011)
30. P. Mitra, G.K. Venayagamoorthy, Wide area control for improving stability of a power system with plug-in electric vehicles. IET Gener. Transm. Distrib. **4**(10), 1151–1163 (2010)

31. O. Sundstrom, C. Binding, Planning electric-drive vehicle charging under constrained grid conditions. Technical Report (IBM - Zurich, Switzerland, 2010)
32. K. Clement-Nyns, E. Haesen, J. Driesen, The impact of charging plug-in hybrid electric vehicles on a residential distribution grid. IEEE Trans. Power Syst. **25**(1), 371–380 (2010)
33. R. Hermans, M. Almassalkhi, I.A. Hiskens, Incentive-based coordinated charging control of plug-in electric vehicles at the distribution-transformer level, in *American Control Conference (ACC)*, Montreal, Canada (2012), pp. 264–269
34. M.D. Galus, G. Andersson, Demand management of grid connected plug-in hybrid electric vehicles (PHEV), in *IEEE Energy 2030*, Atlanta, Georgia, 17–18 November 2008
35. N. Rotering, M. Ilic, Optimal charge control of plug-in hybrid electric vehicles in deregulated electricity markets. IEEE Trans. Power Syst. **26**(3), 1021–1029 (2011)
36. S. Cha, T. Reen, N. Hah, Optimal charging strategies of electric vehicles in the UK power market, in *The first Conference on Innovative Smart Grid Technologies*, Gaithersburg, Maryland, 19–21 January 2010, pp. 1–8
37. Z. Ma, D.S. Callaway, I.A. Hiskens, Decentralized charging control of large populations of plug-in electric vehicles. IEEE Trans. Control Syst. Technol. **21**(1), 67–78 (2013)
38. L. Gan, U. Topcu, S. Low, Optimal decentralized protocol for electric vehicle charging. IEEE Trans. Power Syst. **28**(2), 940–951 (2013)
39. Q. Li, T. Cui, R. Negi, F. Franchetti, M. Ilic, On-line decentralized charging of plug-in electric vehicles in power systems. Technical Report arXiv:1106.5063v2 [math.OC], 2011
40. S. Stüdli, E. Crisostomi, R. Middleton, R. Shorten, A flexible distributed framework for realising electric and plug-in hybrid vehicle charging policies. Int. J. Control. **85**(8), 1130–1145 (2012)
41. D. Wu, D. Aliprantis, L. Ying, Load scheduling and dispatch for aggregators of plug-in electric vehicles. IEEE Trans. Smart Grid **3**(1), 368–376 (2012)
42. G. Binetti, A. Davoudi, D. Naso, B. Turchiano, F.L. Lewis, Scalable real-time electric vehicles charging with discrete charging rates. IEEE Trans. Smart Grid **6**(5), 2211–2220 (2015)
43. C.K. Wen, J.C. Chen, J.H. Teng, P. Ting, Decentralized plug-in electric vehicle charging selection algorithm in power systems. IEEE Trans. Smart Grid **3**(4), 1779–1789 (2012)
44. C. Ahn, C.-T. Li, H. Peng, Optimal decentralized charging algorithm for electrified vehicles connected to smart grid. J. Power Sources **196**(23), 10369–10379 (2011)
45. Z. Ma, S. Zou, L. Ran, X. Shi, I. Hiskens, Efficient decentralized coordination of large-scale plug-in electric vehicle charging. Automatica **69**, 35–47 (2016)
46. A.-H. Mohsenian-Rad, A. Leon-Garcia, Optimal residential load control with price prediction in real-time electricity pricing environments. IEEE Trans. Smart Grid **1**(2), 120–133 (2010)
47. P. Samadi, A. Mohsenian-Rad, R. Schober, V. Wong, J. Jatskevich, Optimal real-time pricing algorithm based on utility maximization for smart grid, in *Proceedings of 1st IEEE International Conference on Smart Grid Communications*, Gaithersburg, 4–6 October 2010
48. Z. Fan, A distributed demand response algorithm and its application to PHEV charging in smart grids. IEEE Trans. Smart Grid **3**(3), 1280–1290 (2012)
49. L. Gan, N. Chen, A. Wierman, U. Topcu, S.H. Low, Real-time deferrable load control: handling the uncertainties of renewable generation, in *Fourth International Conference on Future Energy Systems* (ACM, 2013)
50. D. Monderer, L.S. Shapley, Potential games. Games Econ. Behav. **14**, 124–143 (1996)
51. E. Altman, T. Başar, T. Jiménez, N. Shimkin, Routing into two parallel links: game-theoretic distributed algorithms. J. Parallel Distrib. Comput. **61**(9), 1367–1381 (2001)
52. E. Altman, T. Boulogne, R. El-Azouzi, T. Jimenez, L. Wynter, A survey on networking games in telecommunications. Comput. Oper. Res. **33**, 286–311 (2006)
53. G. Christodoulou, V. Mirrokni, A. Sidiropoulos, Convergence and approximation in potential games, in *Proceedings of the 23rd Symposium on Theoretical Aspects of Computer Science (STACS)* (2006), pp. 349–360
54. E. Even-Dar, A. Kesselman, Y. Mansour, Convergence time to Nash equilibirum in load balancing. ACM Trans. Comput. Log. **2**(3), 111–132 (2001)

55. P. Berenbrink, T. Friedetzky, L. Goldberg, P. Goldberg, Z. Hu, R. Martin, Distributed selfish load balancing, in *Proceedings of the 17th Annual ACM-SIAM Symposium on Discrete Algorithms*, Miami, Florida, 22–26 January 2006, pp. 354–363
56. A. Blum, E. Even-Dar, K. Ligett, Routing without regret: On convergence to Nash equilibria of regret-minimizing algorithms in routing games, in *Proceedings of the 25th ACM Symposium on Principles of Distributed Computing* (2006), pp. 45–52
57. S. Fischer, L. Olbrich, B. Vocking, Approximating Wardrop equilibria with finitely many agents. Distrib. Comput. **21**(2), 129–139 (2008)
58. S. Fischer, H. Racke, B. Vocking, Fast convergence to Wardrop equilibria by adaptive sampling methods, in *Proceedings of the 38th Annual ACM Symposium on Theory of Computing*, Seattle, WA, 21–23 May 2006, pp. 653–662
59. J.G. Wardrop, Some theoretical aspects of road traffic research. Proc. Inst. Civ. Eng., Part 2, pages 1: 325–378 (1952)
60. S. Chien, A. Sinclair, Convergence to approximate Nash equilibria in congestion games. Games Econ. Behav. **71**(2), 315–327 (2009)
61. W. Sandholm, Potential games with continuous player sets. J. Econ. Theory **97**, 81–108 (2001)
62. S. Borenstein, J.B. Bushnell, F.A. Wolak, Measuring market inefficiencies in California's restructured wholesale electricity market. Am. Econ. Rev. **92**(5), 1376–1405 (2002)
63. J.B. Bushnell, E.T. Mansur, C. Saravia, Vertical arrangements, market structure, and competition: an analysis of restructured US electricity markets. Am. Econ. Rev. **98**(1), 237–266 (2008)
64. B.E. Hobbs, Linear complementarity models of Nash-Cournot competition in bilateral and POOLCO power markets. IEEE Trans. Power Syst. **16**(2), 194–202 (2002)
65. L.B. Cunningham, R. Baldick, M.L. Baughman, An empirical study of applied game theory: transmission constrained Cournot behavior. IEEE Trans. Power Syst. **17**(1), 166–172 (2002)
66. E. Pettersen, *Managing end-user flexibility in electricity markets* (Fakultet for samfunnsvitenskap og teknologiledelse, Institutt for industriell økonomi og teknologiledelse, NTNU, 2004)
67. A.B. Philpott, E. Pettersen, Optimizing demand-side bids in day-ahead electricity markets. IEEE Trans. Power Syst. **21**(2), 488–498 (2006)
68. M. Huang, P.E. Caines, R.P. Malhamé, Individual and mass behaviour in large population stochastic wireless power control problems: centralized and nash equilibrium solutions, in *Proceedings of the 42th IEEE International Conference on Decision and Control*, Maui, Hawaii (2003), pp. 98–103
69. M. Huang, P.E. Caines, R.P. Malhamé, Large-population cost-coupled LQG problems with non-uniform agents: individual-mass behaviour and decentralized epsilon-nash equilibria. IEEE Trans. Autom. Control. **52**(9), 1560–1571 (2007)
70. S. Bashash, S.J. Moura, J.C. Forman, H.K. Fathy, Plug-in hybrid electric vehicle charge pattern optimization for energy cost and battery longevity. J. Power Sources **196**, 541–549 (2011)
71. J. Wang, P. Liu, J. Hicks-Garner, E. Sherman, S. Soukiazian, M. Verbrugge, H. Tataria, J. Musser, P. Finamore, Cycle-life model for graphite-LiFePO4 cells. J. Power Sources **194**(8), 3942–3948 (2011)
72. B.Y. Liaw, R.G. Jungst, G. Nagasubramanian, H.L. Case, D.H. Doughty, Modeling capacity fade in lithium-ion cells. J. Power Sources **140**(1), 157–161 (2005)
73. S. Santhanagopalan, Q. Guo, P. Ramadass, R.E. White, Review of models for predicting the cycling performance of lithium ion batteries. J. Power Sources **156**(2), 620–628 (2006)
74. M. Safari, M. Morcrette, A. Teyssot, C. Delacourt, Multimodal physics-based aging model for life prediction of Li-ion batteries. J. Electrochem. Soc. **156**(3), A145–A153 (2009)
75. J. Christensen, J. Newman, Stress generation and fracture in lithium insertion materials. J. Solid State Electrochem. **10**(5), 293–319 (2006)
76. G. Ning, R.E. White, B.N. Popov, A generalized cycle life model of rechargeable Li-ion batteries. Electrochim. Acta **51**(10), 2012–2022 (2006)
77. J.C. Forman, S.J. Moura, J.L. Stein, H.K. Fathy, Optimal experimental design for modeling battery degradation, in *ASME 2012 5th Annual Dynamic Systems and Control Conference joint with the JSME 2012 11th Motion and Vibration Conference*, Fort Lauderdale, Florida, USA, 17–19 October 2012, pp. 309–318

78. K.W.E. Cheng, B.P. Divakar, H. Wu, K. Ding, H.F. Ho, Battery-management system (BMS) and SOC development for electrical vehicles. IEEE Trans. Veh. Technol. **60**(1), 76–88 (2011)
79. Z. Li, Q. Guo, H. Sun, S. Xin, J. Wang, A new real-time smart-charging method considering expected electric vehicle fleet connections. IEEE Trans. Power Syst. **pp**(99), 1–2 (2014)
80. S. Wang, L. Han, D. Wang, M. Shahidehpour, Z. Li, Hierarchical charging management strategy of plug-in hybrid electric vehicles to provide regulation service, in *2012 3rd IEEE PES International Conference and Exhibition on Innovative Smart Grid Technologies (ISGT Europe)* (IEEE, Berlin, 2012), pp. 1–6
81. W. Qi, Z. Xu, Z.J.M. Shen, Z. Hu, Y. Song, Hierarchical coordinated control of plug-in electric vehicles charging in multifamily dwellings. IEEE Trans. Smart Grid **5**(3), 1465–1474 (2014)
82. W. Yao, J. Zhao, F. Wen, Y. Xue, G. Ledwich, A hierarchical decomposition approach for coordinated dispatch of plug-in electric vehicles. IEEE Trans. Power Syst. **28**(3), 2768–2778 (2013)
83. W. Tang, R. Jain, Hierarchical auction mechanisms for network resource allocation. IEEE J. Sel. Areas Commun. **30**(11), 2117–2125 (2012)
84. F. Kelly, A. Maulloo, D. Tan, Rate control for communication networks: Shadow prices, proportional fairness and stability. J. Oper. Res. Soc. **49**(3), 237–252 (1998)
85. Y. Ma, T. Houghton, A. Cruden, D. Infield, Modeling the benefits of vehicle-to-grid technology to a power system. IEEE Trans. Power Syst. **27**(2), 1012–1020 (2012)
86. M. Zeng, S. Leng, Y. Zhang, Power charging and discharging scheduling for V2G networks in the smart grid, in *IEEE International Conference on Communications Workshops*, Budapest, 9–13 June 2013, pp. 1052–1056
87. J. Lin, K.C. Leung, V.O.K. Li, Optimal scheduling with vehicle-to-grid regulation service. IEEE Internet Things J. **1**(6), 556–569 (2014)
88. H. Liang, B.J. Choi, W. Zhuang, X. Shen, Optimizing the energy delivery via V2G systems based on stochastic inventory theory. IEEE Trans. Smart Grid **4**(4), 2230–2243 (2013)
89. M. Esmaili, M. Rajabi, Optimal charging of plug-in electric vehicles observing power grid constraints. IET Gener. Transm. Distrib. **8**(4), 583–590 (2014)
90. S. Falahati, S.A. Taher, M. Shahidehpour, Grid frequency control with electric vehicles by using of an optimized fuzzy controller. Appl. Energy **178**, 918–928 (2016)
91. T.N. Pham, H. Trinh, V.H. Le, Load frequency control of power systems with electric vehicles and diverse transmission links using distributed functional observers. IEEE Trans. Smart Grid **7**(1), 238–252 (2016)
92. X. DaQian, G. Joós, M. Lévesque, M. Maier, Integrated V2G, G2V, and renewable energy sources coordination over a converged fiber-wireless broadband access network. IEEE Trans. Smart Grid **4**(3), 1381–1390 (2013)
93. L. Igualada, C. Corchero, M. Cruz-Zambrano, F.-J. Heredia, Optimal energy management for a residential microgrid including a vehicle-to-grid system. IEEE Trans. Smart Grid **5**(4), 2163–2172 (2014)
94. Y. He, B. Venkatesh, L. Guan, Optimal scheduling for charging and discharging of electric vehicles. IEEE Trans. Smart Grid **3**(3), 1095–1105 (2012)
95. K. Clement-Nyns, E. Haesen, J. Driesen, The impact of vehicle-to-grid on the distribution grid. Electr. Power Syst. Res. **81**(1), 185–192 (2011)
96. K.N. Kumar, B. Sivaneasan, P.H. Cheah, P.L. So, V2G capacity estimation using dynamic EV scheduling. IEEE Trans. Smart Grid **5**(2), 1051–1060 (2014)
97. L. Jian, Y. Zheng, X. Xiao, C.C. Chan, Optimal scheduling for vehicle-to-grid operation with stochastic connection of plug-in electric vehicles to smart grid. Appl. Energy **146**, 150–161 (2015)
98. H. Zhang, H. Zechun, X. Zhiwei, Y. Song, Evaluation of achievable vehicle-to-grid capacity using aggregate PEV model. IEEE Trans. Power Syst. **32**(1), 784–794 (2017)
99. Z. Ma, S. Zou, X. Liu, A distributed charging coordination for large-scale plug-in electric vehicles considering battery degradation cost. IEEE Trans. Control. Syst. Technol. **23**(5), 2044–2052 (2015)

100. S. Zou, Z. Ma, X. Liu, I. Hiskens, An efficient game for coordinating electric vehicle charging. IEEE Trans. Autom. Control. **62**(5), 2374–2389 (2017)
101. M.A. Ortega-Vazquez, Optimal scheduling of electric vehicle charging and vehicle-to-grid services at household level including battery degradation and price uncertainty. IET Gener. Transm. Distrib. **8**(6), 1007–1016 (2014)
102. D.T. Hoang, P. Wang, D. Niyato, E. Hossain, Charging and discharging of plug-in electric vehicles (PEVs) in vehicle-to-grid (V2G) systems: a cyber insurance-based model. IEEE Access **5**, 732–754 (2017)
103. Z. Chen, Z. Hu, Optimal coordination of charging and discharging control of large-scale plug-in electric vehicles, in *Transportation Electrification Asia-Pacific* (2014), pp. 1–6
104. H. Xing, F. Minyue, Z. Lin, Y. Mou, Decentralized optimal scheduling for charging and discharging of plug-in electric vehicles in smart grids. IEEE Trans. Power Syst. **31**(5), 4118–4127 (2016)
105. H. Li, Z. Ma, A distributed charging and discharging coordination for large-population plug-in electric vehicles, in *Proceedings of the 2015 Chinese Intelligent Systems Conference*, 8 November 2016, pp. 415–424
106. J. Wang, S. Kennedy, J. Kirtley, A new wholesale bidding mechanism for enhanced demand response in smart grids, in *Innovative Smart Grid Technologies (ISGT)*, Gaithersburg, MD, 19–21 January 2010, pp. 1–8
107. M. Parvania, M. Fotuhi-Firuzabad, M. Shahidehpour, Optimal demand response aggregation in wholesale electricity markets. IEEE Trans. Smart Grid **4**(4), 1957–1965 (2013)
108. L. Goel, Q. Wu, P. Wang, Y. Ding, Reliability enhancement of deregulated power systems considering demand-price elasticity, in *7th IEEE International Power Engineering Conference (IPEC)* (2005), pp. 934–938
109. R. Walawalkar, S. Fernands, N. Thakur, K.R. Chevva, Evolution and current status of demand response (DR) in electricity markets: Insights from PJM and NYISO. Energy **35**(4), 1553–1560 (2010)
110. H.A. Aalami, M.P. Moghaddam, G.R. Yousefi, Demand response modeling considering interruptible/curtailable loads and capacity market programs. Appl. Energy **87**(1), 243–250 (2010)
111. G. Liu, K. Tomsovic, A full demand response model in co-optimized energy and reservemarket. Electr. Power Syst. Res. **111**, 62–70 (2014)
112. M.H. Albadi, E.F. El-Saadany, A summary of demand response in electricity markets. Electr. Power Syst. Res. **78**(11), 1989–1996 (2008)
113. J. Aghaei, M.I. Alizadeh, Demand response in smart electricity grids equipped with renewable energy sources: a review. Renew. Sustain. Energy Rev. **18**, 64–72 (2013)
114. R. Rajaraman, J.V. Sarlashkar, F.L. Alvarado, The effect of demand elasticity on security prices for the poolco and multi-lateral contract models. IEEE Trans. Power Syst. **12**(3), 1177–1184 (1997)
115. M.G. Lijesen, The real-time price elasticity of electricity. Energy Econ. **29**(2), 249–258 (2007)
116. D. Kirschen, G. Strbac, P. Cumperayot, D. de Paiva Mendes, Factoring the elasticity of demand in electricity prices. IEEE Trans. Power Syst. **15**(2), 612–617 (2000)
117. A.K. David and Y.Z. Li. Effect of inter-temporal factors on the real time pricing of electricity. IEEE Trans. Power Syst., 8(1) (1993) (Institute of Electrical and Electronics Engineers)
118. W. Uturbey, A.S. Costa, Optimal power flow with inter-temporal constraints as an aiding tool for demand-side management. IEE Proc.-Gener. Transm. Distrib. **149**(1), 37–43 (2002)
119. P. Faria, Z. Vale, Demand response in electrical energy supply: An optimal real time pricing approach. Energy **36**(8), 5374–5384 (2011)
120. J. Aghaei, M.I. Alizadeh, Multi-objective self-scheduling of CHP (combined heat and power)-based microgrids considering demand response programs and ESSs (energy storage systems). Energy **55**, 1044–1054 (2013)
121. L. Goel, Q. Wu, P. Wang, Nodal price volatility reduction and reliability enhancement of restructured power systems considering demand-price elasticity. Electr. Power Syst. Res. **78**(10), 1655–1663 (2008)

122. C.L. Su, D. Kirschen, Quantifying the effect of demand response on electricity markets. IEEE Trans. Power Syst. **24**(3), 1199–1207 (2009)
123. A. Khodaei, M. Shahidehpour, S. Bahramirad, SCUC with hourly demand response considering intertemporal load characteristics. IEEE Trans. Smart Grid **2**(3), 564–571 (2011)
124. C. Zhao, J. Wang, J.P. Watson, Y. Guan, Multi-stage robust unit commitment considering wind and demand response uncertainties. IEEE Trans. Power Syst. **28**(3), 2708–2717 (2013)
125. M. Ali, J. Jokisalo, K. Siren, M. Lehtonen, Combining the demand response of direct electric space heating and partial thermal storage using LP optimization. Electr. Power Syst. Res. **106**, 160–167 (2014)
126. D. Setlhaolo, X. Xia, J. Zhang, Optimal scheduling of household appliances for demand response. Electr. Power Syst. Res. **116**, 24–28 (2014)
127. N. Gatsis, G.B. Giannakis, Decomposition algorithms for market clearing with large-scale demand response. IEEE Trans. Smart Grid **4**(4), 1976–1987 (2013)
128. J. Medina, N. Muller, I. Roytelman, Demand response and distribution grid operations: Opportunities and challenges. IEEE Trans. Smart Grid **1**(2), 193–198 (2010)
129. P. Samadi, H. Mohsenian-Rad, R. Schober, V.W.S. Wong, Advanced demand side management for the future smart grid using mechanism design. IEEE Trans. Smart Grid **3**(3), 1170–1180 (2012)
130. L. Kelly, A. Rowe, P. Wild, Analyzing the impacts of plug-in electric vehicles on distribution networks in British Columbia, in *IEEE Electrical Power and Energy Conference*, Montreal, Canada (2009)
131. O. Beaude, S. Lasaulce, M. Hennebel, J. Daafouz, Minimizing the impact of ev charging on the electricity distribution network, in *European Control Conference* (2015), pp. 648–653
132. Q. Gong, S. Midlam-Mohler, V. Marano, G. Rizzoni, Study of PEV charging on residential distribution transformer LiFe. IEEE Trans. Smart Grid **3**(1), 404–412 (2012)
133. S. Argade, V. Aravinthan, W. Jewell, Probabilistic modeling of ev charging and its impact on distribution transformer loss of life, in *Electric Vehicle Conference* (2012)
134. K. Zhan, Z. Hu, Y. Song, Z. Luo, Z. Xu, Long Jia. Coordinated electric vehicle charging strategy for optimal operation of distribution network, in *3rd IEEE PES International Conference and Exhibition on Innovative Smart Grid Technologies (ISGT Europe)* (2012), pp. 1–6
135. A.S. Masoum, S. Deilami, P.S. Moses, M.A. Masoum, A. Abu-Siada, Smart load management of plug-in electric vehicles in distribution and residential networks with charging stations for peak shaving and loss minimization considering voltage regulation. IET Gener. Transm. Distrib. **5**(8), 877–888 (2011)
136. M.R. Sarker, M. Ortega-Vazquez, D.S. Kirschen, Optimal coordination and scheduling of demand response via monetary incentives. IEEE Trans. Smart Grid **6**(3), 1341–1352 (2015)
137. B. Yang, J. Li, Q. Han, T. He, Distributed control for charging multiple electric vehicles with overload limitation. IEEE Trans. Parallel Distrib. Syst. **27**(12), 3441–3454 (2016)
138. A. Ghavami, K. Kar, A. Gupta, Decentralized charging of plug-in electric vehicles with distribution feeder overload control. IEEE Trans. Autom. Control. **61**(11), 3527–3532 (2016)
139. O. Ardakanian, C. Rosenberg, S. Keshav, Distributed control of electric vehicle charging, in *International Conference on Future Energy Systems* (2013), pp. 101–112
140. I. Cvetkovic, T. Thacker, Future home uninterruptible renewable energy system with vehicle-to-grid technology, *IEEE Energy Conversion Congress and Exposition, 2009. ECCE 2009* (2009), pp. 2675–2681
141. S. Weckx, R. D'Hulst, B. Claessens, J. Driesensam, Multiagent charging of electric vehicles respecting distribution transformer loading and voltage limits. IEEE Trans. Smart Grid **5**(6), 2857–2867 (2014)
142. M. Liu, P.K. Phanivong, S. Yang, D.S. Callaway, Decentralized charging control of electric vehicles in residential distribution networks. IEEE Trans. Control. Syst. Technol. **27**(1), 266–281 (2019)
143. Q.R. Hamid, J.A. Barria, Distributed recharging rate control for energy demand management of electric vehicles. IEEE Trans. Power Syst. **28**(3), 2688–2699 (2013)

144. M. Erol-Kantarci, H. Mouftah, Management of PHEV batteries in the smart grid: Towards a cyber-physical power infrastructure, in *7th International Wireless Communications and Mobile Computing Conference (IWCMC)*, (2011), pp. 795–800
145. C. Wu, H. Mohsenian-Rad, J. Huang, Vehicle-to-aggregator interaction game. IEEE Trans. Smart Grid **3**(1), 434–442 (2012)
146. M.H.K. Tushar, C. Assi, M. Maier, M.F. Uddin, Smart microgrids: optimal joint scheduling for electric vehicles and home appliances. IEEE Trans. Smart Grid **5**(1), 239–250 (2014)
147. J. Lin, K.C. Leung, V.O.K. Li, Optimal scheduling with vehicle-to-grid regulation service. IEEE Internet Things J. **1**(6), 556–569 (2014)
148. C. Wu, H. Mohsenian-Rad, J. Huang, Vehicle-to-aggregator interaction game. IEEE Trans. Smart Grid **3**(1), 434–442 (2012)
149. D. Infield, J. Short, C. Home, L. Freris, Potential for domestic dynamic demand-side management in the UK, in *IEEE Power Engineering Society General Meeting, 2007* (2007), pp. 1–6
150. S. Jang, Optimal decision on contract size for V2G aggregator regarding frequency regulation, in *Optimization of Electrical and Electronic Equipment (OPTIM)* (2010), pp. 54–62
151. H. Sekyung, H. Soohee, K. Sezaki, Development of an optimal vehicle-to-grid aggregator for frequency regulation. IEEE Trans. Smart Grid **1**(1), 65–72 (2010)
152. C. Hutson, G. Venayagamoorthy, K. Corzine, Intelligent scheduling of hybrid and electric vehicle storage capacity in a parking lot for profit maximization in grid power transactions, in *Energy 2030 Conference*, 17–18 November 2008, pp. 1–8
153. M. Baran, F. Wu, Optimal sizing of capacitors placed on a radial distribution system. IEEE Trans. Power Deliv. **4**(1), 735–743 (1989)
154. R. Gallego, A. Monticelli, R. Romero, Optimal capacitor placement in radial distribution networks. IEEE Trans. Power Syst. **16**(4), 630–637 (2001)
155. S. Sundhararajan, A. Pahwa, Optimal selection of capacitors for radial distribution systems using a genetic algorithm. IEEE Trans. Power Syst. **9**(3), 1499–1507 (1994)
156. Y.C. Huang, H.T. Yang, C.L. Huang, Solving the capacitor placement problem in a radial distribution system using tabu search approach. IEEE Trans. Power Syst. **11**(4), 1868–1873 (1996)
157. C. Wu, H. Mohsenian-Rad, J. Huang, J. Jatskevich, PEV-based combined frequency and voltage regulation for smart grid, in *Innovative Smart Grid Technologies (ISGT)* (2012), pp. 1–6
158. A.A. Lazar, N. Semret, The progressive second price auction mechanism for network resource sharing, in *8th International Symposium on Dynamic Games* (1998)
159. A.A. Lazar, N. Semret, Design and analysis of the progressive second price auction for network bandwidth sharing. Telecommun. Syst. (1999) (Special issue on Network Economics)
160. W. Vickrey, Counterspeculation, auctions, and competitive sealed tenders. J. Financ. **16**(1), 8–37 (1961)
161. E.H. Clark, Multipart pricing of public goods. Public Choice **11**(1), 19–33 (1971)
162. T. Groves, Incentives in teams. Econometrica **41**(4), 617–631 (1973)
163. L. Ausubel, P. Cramton, Demand reduction and inefficiency in multi-unit auctions. Working papers, University of Maryland, 2002
164. G. Federico, D. Rahman, Bidding in an electricity pay-as-bid auction. J. Regul. Econ. **24**(2), 175–211 (2003)
165. F. Wen, A.K. David, Optimal bidding strategies and modeling of imperfect information among competitive generators. IEEE Trans. Power Syst. **16**(1), 15–21 (2001)
166. R. Jain, J. Walrand, An efficient nash-implementation mechanism for network resource allocation. Automatica **46**, 1276–1283 (2010)
167. P. Jia, P. Caines, Analysis of decentralized decision processes in competitive markets: quantized single and double-sided auctions, in *49th IEEE Conference on Decision and Control* (2010), pp. 237–243
168. P. Jia, P.E. Caines, Analysis of decentralized quantized auctions on cooperative networks. IEEE Trans. Autom. Control. **58**(2), 529–534 (2013)

169. P. Jia, P. Caines, Analysis of quantized double auctions with application to competitive electricity markets. INFOR: Inf. Syst. Oper. Res. **48**(4), 239–250 (2010)
170. J. Forman, J. Stein, H. Fathy, Optimization of dynamic battery paramter characterization experiments via differential evolution, in *American Control Conference (ACC)*, Washington, DC, USA, 17–19 June 2013, pp. 867–874
171. J. Forman, S. Moura, J. Stein, H. Fathy, Optimal experimental design for modeling battery degradation, in *Proceedings of Dynamic Systems and Control Conference* (2012), pp. 309–318
172. S. Moura, J. Forman, S. Bashash, J. Stein, H. Fathy, Optimal control of film growth in lithium-ion battery packs via relay switches. IEEE Trans. Ind. Electron. **58**(8), 3555–3566 (2011)

Chapter 2
Charging Coordination via Non-cooperative Games

Abstract This chapter develops a strategy to coordinate the charging of autonomous plug-in electric vehicles (PEVs) using concepts from non-cooperative games. The foundation of the chapter is a model that assumes PEVs are cost-minimizing and weakly coupled via a common electricity price. At a Nash equilibrium, each PEV reacts optimally with respect to a commonly observed charging trajectory that is the average of all PEV strategies. This average is given by the solution of a fixed point problem in the limit of infinite population size. The ideal solution minimizes electricity generation costs by scheduling PEV demand to fill the overnight non-PEV demand "valley". The chapter's central theoretical result is a proof of the existence of a unique Nash equilibrium that almost satisfies that ideal. This result is accompanied by a decentralized computational algorithm and a proof that the algorithm converges to the Nash equilibrium in the infinite system limit. Several numerical examples are used to illustrate the performance of the solution strategy for finite populations. The examples demonstrate that convergence to the Nash equilibrium occurs very quickly over a broad range of parameters, and suggest this method could be useful in situations where frequent communication with PEVs is not possible. The method is useful in applications where fully centralized coordination is not possible, but where optimal or near-optimal charging patterns are essential to system operation.

2.1 Introduction

A number of recent studies have explored the potential impacts of high penetrations of PEVs on the power grid [1–4]. In general, these studies assume that PEV charging patterns "fill the valley" of night-time demand. For example, the overnight dip in Fig. 2.1 would be replaced by a total demand that remained relatively constant during the charging period. However, these studies do not address the issue of how to coordinate PEV charging patterns. Possible coordination strategies can be divided into two categories:

In centralized strategies, a central operator dictates precisely when and at what rate every individual PEV will charge. Decisions could be made on the basis of system-level considerations only, or they could factor vehicle-level preferences, for

Z. Ma, *Decentralized Charging Coordination of Large-scale Plug-in Electric Vehicles in Power Systems*, https://doi.org/10.1007/978-981-13-7652-8_2

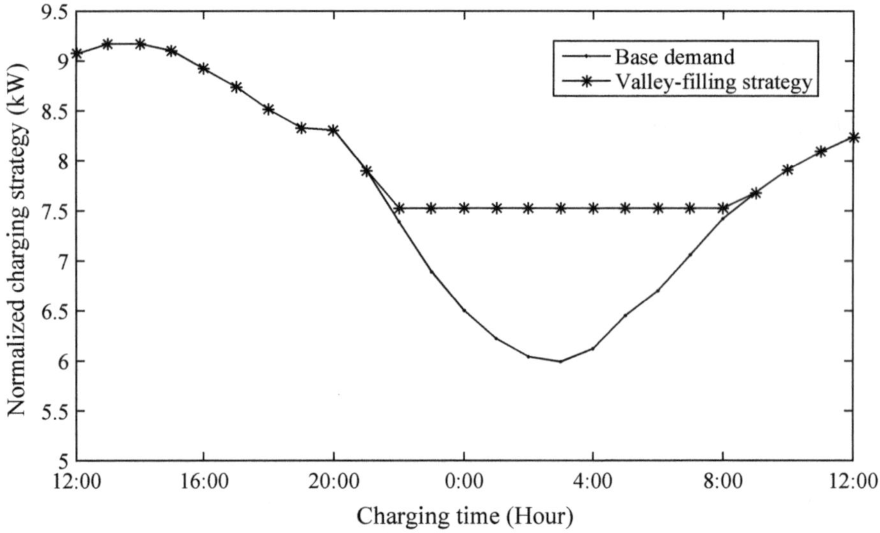

Fig. 2.1 Illustration of a typical base demand in summer for the region managed by the MISO, and a valley-filling profile

example desired charging interval, final state of charge, and budget. These strategies could be further distinguished by whether they attempt to identify charge patterns that are in some way optimal, or instead follow rules-of-thumb that seek to achieve aggregate charging patterns that are reasonably close to optimal.

Decentralized strategies allow individual PEVs to determine their own charging pattern. Vehicle charging decisions could, for example, be made on the basis of time-of-day or electricity price. The outcome of a decentralized approach may or may not be optimal, depending on the information and methods used to determine local charging patterns. Care must be taken to ensure charging strategies cannot inadvertently synchronize the responses of large numbers of PEVs, as the resulting abrupt changes in aggregate demand could potentially destabilize grid operations [5].

Time-based and price-based strategies that follow fixed schedules would, at best, not reflect day-to-day variation in system conditions and, at worst, produce abrupt changes in aggregate demand if all vehicles respond to the same time or price signal [5]. On the other hand, charging decisions could be made on the basis of variables that vary with system conditions, such as real-time price.

Though the valley-filling charging pattern is conceptually straightforward, the process of achieving it in a practical manner with a large population of independent PEVs presents a number of challenges. For example, a centralized approach may not be palatable to consumers, who are accustomed to having complete decision-making authority over their electricity consumption patterns. On the other hand, a decentralized strategy may preserve individual authority. However, as it will be discussed below, the challenge of coordinating many autonomous agents to achieve an optimal or near-optimal outcome is non-trivial.

This chapter explores the potential of decentralized strategies. Because time-based and fixed schedule price-based strategies have difficulty effectively filling the night-time valley [5], it instead focuses on methods that utilize real-time marginal electricity price information. It allows each PEV to choose and implement its own local charging coordination strategy, with the aim of minimizing its individual charging cost.

The decentralized PEV charging coordination problem studied in this work is a form of non-cooperative game, where a large number of selfish PEVs share electricity resources on a finite collection of charging intervals. It is further supposed that PEVs are coupled through a common price signal which is determined by the average charging strategy of the PEV population. Therefore each PEV effectively interacts with the average charging strategy of the rest of the PEV population. As the population grows substantially, the influence of each individual PEV on that average charging strategy becomes negligible. Accordingly, in the infinite population limit, all PEVs will observe the same average strategy as they calculate their optimal local strategy. In this situation, a collection of local charging controls is a Nash equilibrium, in case: Each PEV's charging strategy is optimal with respect to a single commonly observed charging trajectory, and the average of all the local optimal charging strategies is equal to that common trajectory.

In this chapter, it presents a novel procedure that identifies a Nash equilibrium by simultaneously updating each PEV's best individual strategy with respect to the average charging strategy of the whole population. It is proposed that this procedure would be undertaken prior to the actual charging interval. The computation time of this algorithm is unrelated to the number of PEVs, since they simultaneously and independently update their charging strategy. Under certain mild conditions, the proposed decentralized charging coordination procedure drives the system asymptotically to a unique Nash equilibrium that is nearly globally optimal (valley filling). In the case of homogeneous populations, where all vehicles have identical parameters, this unique Nash equilibrium becomes a perfect valley-filling charging strategy.

However due to the heterogeneity among individual PEVs, by adapting proper deviation cost and following the simultaneous best strategy update mechanism, the game systems converge to an equilibrium which is not a valley-filling in general.

In order to improve the performance, in this work, it presents an improved charging behavior update mechanism. More specifically, all individuals in a homogeneous subpopulation simultaneously update their best strategies considering the tradeoff between the low charging price and a low deviation cost penalizing against the deviation of the individual strategy from the average behavior in this subpopulation. This procedure continues until no individuals update their best strategies any longer. The proposed strategy update procedure is applied to all sub-populations successively.

By adapting the proposed update mechanism, at the equilibrium, the deviation cost vanishes, that is to say, the deviation cost is only introduced to enforce individuals to update their behavior towards the average behavior and vanishes at the equilibrium since all individuals in a subpopulation possess an identical strategy. Hence in case the update procedure converges, the system reaches the socially optimal valley-filling equilibrium. It also specifies technical sufficient condition under which the update procedure converges.

The rest of the chapter is organized as follows. Charging coordination problems for large populations of PEVs are formulated in Sect. 2.2. Section 2.3 develops a decentralized coordination strategy for optimally charging an infinite population of PEVs, and establishes existence and uniqueness properties of the resulting Nash equilibrium. Section 2.4 presents the coordination strategy as an algorithm that also applies to finite systems. The valley filling property of the Nash equilibrium is verified in Sect. 2.5. Finite population numerical examples are presented in Sect. 2.6. In Sect. 2.10, it lists the conclusions of this part.

2.2 Decentralized Charging Coordination of Large PEV Populations

In this section it introduces the basic PEV charging dynamics, and optimization models that are relevant for both the centralized and decentralized framework. In the decentralized case, it will also formally define the conditions for a Nash equilibrium.

2.2.1 Model and Notation

It studies the charging coordination of a population of PEVs $\mathcal{N}$ over a multi-time interval $\mathcal{T} \triangleq \{0, \ldots, T-1\}$.

Consider $N \equiv |\mathcal{N}|$ and $T \equiv |\mathcal{T}|$, say N and T represent the cardinality of $\mathcal{N}$ and $\mathcal{T}$ respectively.

Denote by soc_{nt} the state of charge (SOC) of the battery of PEV n at instant t, and it considers that the SOC shall be in the range of $[\mathrm{soc}_{n,\min}, \mathrm{soc}_{n,\max}]$ with $0 \leq \mathrm{soc}_{n,\min} < \mathrm{soc}_{n,\max} \leq 1$.

Denote by $\boldsymbol{u}_n \equiv (u_{nt}; t \in \mathcal{T})$, with units of kW, a charging strategy of PEV n, with $n \in \mathcal{N}$.

Remark: If the tth time period has length ΔT, then the energy delivered over that period is $u_{nt}\Delta T$. And unless it is specified, ΔT is set as 1 hour. Hence without loss of generality, it considers that $\Delta T = 1$.

The SOC of PEV n subject to an admissible strategy $\boldsymbol{u}_n$, evolves as below:

$$\mathrm{soc}_{n,t+1} = \mathrm{soc}_{nt} + \frac{\varsigma_n}{\Gamma_n} u_{nt}, \text{ for all } t \in \mathcal{T}, \tag{2.1}$$

with initial SOC value soc_{n0}, charger efficiency $\varsigma_n \in (0, 1]$, and energy capacity of battery Γ_n which is determined by external factors such as driving style and vehicle type [6].

Call $\boldsymbol{u}_n$ is admissible, if

$$u_{nt} \in [-\Upsilon_n^-, \Upsilon_n^+], \tag{2.2a}$$

$$\text{soc}_{nt} \in [\text{soc}_{n,\min}, \text{soc}_{n,\max}], \tag{2.2b}$$

$$\text{soc}_{nT} = \text{soc}_{n,\max}, \tag{2.2c}$$

for all $t \in \mathscr{T}$, where Υ_n^- and Υ_n^+ represent respectively uniform maximum discharging and charging rates of PEV n over $\mathscr{T}$.

Note: Except of Chap. 5, in this book, it considers that $\Upsilon_n^- = 0$, that is to say, each of individual PEVs isn't allowed to be discharged during the period of time except of Chap. 5.

Remark: By (2.4), the PEV n is coordinated over the period of $\mathscr{T}$ such that at each instant the SOC value has to be in a feasible range and at the terminal instant it has to be fully charged.

Moreover, by (2.1), it is straightforward to verify that, in case $\Upsilon_n^- = 0$, (2.2) is equivalent to the following

$$u_{nt} \in [0, \Upsilon_n^+], \text{ for all } t \in \mathscr{T}, \tag{2.3a}$$

$$\Sigma(\mathbf{u}_n) \equiv \sum_{t \in \mathscr{T}} u_{nt} = \frac{\Gamma_n}{\varsigma_n}(\text{soc}_{n,\max} - soc_{n0}). \tag{2.3b}$$

The set of admissible charging strategies is denoted by $\mathscr{U}_n$; then by the above, $\mathscr{U}_n$ is specified as below:

$$\mathscr{U}_n \triangleq \left\{ \mathbf{u}_n \in \mathbb{R}^T; \text{ s.t. (2.3a) and (2.3b) are satisfied} \right\}. \tag{2.4}$$

Subject to an admissible charging strategy $\mathbf{u} \in \mathscr{U}$, the cost associated with delivering the total system demand is given by,

$$J(\mathbf{u}) = \sum_{t \in \mathscr{T}} p(\mathfrak{r}_t^N(\mathbf{u}_t)) \left(d_t^N + \sum_{n \in \mathscr{N}} u_{nt} \right), \tag{2.5}$$

where $\mathbf{u}_t \equiv (u_{nt}; n \in \mathscr{N})$ denotes the collection of PEV charging rates at time t, $p(\mathfrak{r}_t^N(\mathbf{u}_t))$ is the electricity charging price at instant t, and d_t^N is the total inelastic non-PEV demand at instant t. It assumes the electricity charging price $p(\mathfrak{r}_t^N(\mathbf{u}_t))$ is determined by the ratio between the total demand and the total generation capacity, so

$$\mathfrak{r}_t^N(\mathbf{u}_t) \triangleq \frac{1}{\Lambda^N} \left(d_t^N + \sum_{n \in \mathscr{N}} u_{nt} \right), \tag{2.6}$$

where Λ^N denotes the total generation capacity. The importance of the dependence of $\bar{\Lambda}$ and d_t on N is discussed later in this section.

Note that this definition of price differs from typical retail electricity tariffs, which are either constant or change according to a fixed (demand-independent) schedule in time. Instead, price varies in proportion to total demand; this is how wholesale electricity market prices (and therefore the real-time marginal price of electricity) vary. In this formulation, it is assumed for simplicity that electricity price is a function only of instantaneous demand. It is aware that in practice, price in any given hour will be influenced by demand in all hours. This is because the mix of units available in a given hour is determined by the unit commitment process [7], which is a function of demand over the entire commitment interval. We note that this price definition does not consider transmission congestion or losses, under the assumption that they are negligible at night.

This chapter studies systems where the number of PEVs is sufficiently large that the action of each individual PEV on the system is negligible, but the action of the aggregation of PEVs may be significant. For this reason it will examine asymptotic properties of the system in the large N limit. To ensure that key properties are preserved at that limit, it is assumed that non-PEV demand and total generation capacity vary with the number of PEVs and make the following asymptotic assumptions as PEV population size approaches infinity,

$$\lim_{N\to\infty}\frac{d_t^N}{N}=\bar{d}_t,\qquad \lim_{N\to\infty}\frac{\Lambda^N}{N}=\bar{\Lambda}. \tag{2.7}$$

The implication inherent in (2.7) is that larger power systems, with greater capacity and base demand, are required to support large numbers of PEVs. Direct substitution into (2.6) gives,

$$\lim_{N\to\infty}\mathfrak{r}_t^N(\mathbf{u}_t)=\frac{1}{\bar{\Lambda}}(\bar{d}_t+\bar{\mathbf{u}}_t)\triangleq\mathfrak{r}_t(\bar{\mathbf{u}}_t) \tag{2.8}$$

where

$$\bar{\mathbf{u}}_t\triangleq\lim_{N\to\infty}\frac{1}{N}\sum_{n\in\mathcal{N}}u_{nt}. \tag{2.9}$$

2.2.2 *Definition of Valley-Filling*

Valley-filling charging is defined as follows:

$$\bar{\mathbf{u}}_t^{\mathrm{vf}}=\max\{0,\vartheta-\bar{d}_t\}, \tag{2.10}$$

for some constant ϑ. In words, in hours when $\vartheta>\bar{d}_t$ this strategy chooses total PEV demand such that system-wide demand is equal to ϑ; otherwise PEV demand is zero. The value of ϑ uniquely determines the total energy supplied for charging. This is illustrated in Fig. 2.1, where ϑ corresponds to the level of the horizontal line.

The following lemma establishes sufficient conditions for optimal centrally-determined charging strategies to be valley-filling.

Lemma 2.1 *If $p(\mathfrak{r}_t(\overline{\mathbf{u}}_t))$ is convex and increasing on $\overline{\mathbf{u}}_t$, then given non-PEV demand $\bar{d}_t$, the optimal charging strategy $(\overline{\mathbf{u}}_t^*;\, t \in \mathscr{T})$ is valley-filling.*

Proof Define a Lagrangian

$$\mathfrak{L}(\mathbf{u}, \lambda, \nu) = \mathbb{J}(\mathbf{u}) - \sum_{t\in\mathscr{T}} \lambda_t \overline{\mathbf{u}}_t + \nu\left(\frac{1}{N}\sum_n \Gamma_n - \sum_t \overline{\mathbf{u}}_t\right)$$

where $\lambda_t \geq 0$ and ν are Lagrange multipliers. Since this is a convex optimization problem the Karush–Kuhn–Tucker conditions ensure the following optimality conditions:

$$p'\left(\frac{1}{\Lambda}(\bar{d}_t + \overline{\mathbf{u}}_t^*)\right)\frac{1}{\Lambda}(\bar{d}_t + \overline{\mathbf{u}}_t^*) + p\left(\frac{1}{\Lambda}(\bar{d}_t + \overline{\mathbf{u}}_t^*)\right) - \lambda_t^* = \nu^* \qquad \forall t \in \mathscr{T} \tag{2.11}$$

$$\overline{\mathbf{u}}_t^* \geq 0 \qquad \forall t \in \mathscr{T} \tag{2.12}$$

$$\lambda_t^* \overline{\mathbf{u}}_t^* = 0 \qquad \forall t \in \mathscr{T} \tag{2.13}$$

$$\frac{1}{N}\sum_n \Gamma_n^* = \sum_t \overline{\mathbf{u}}_t^* \tag{2.14}$$

where (2.13) holds with complementary slackness. The right hand side of (2.11) is independent of t, and complementary slackness requires that $\lambda_t^* = 0$ when $\overline{\mathbf{u}}_t^* > 0$. Therefore, because $p(\cdot)$ is convex and increasing, $\bar{d}_t + \overline{\mathbf{u}}_t^*$ must be constant for all t when $\overline{\mathbf{u}}_t^* > 0$. It follows that (2.10) must hold. ■

It will be shown in Sect. 2.5 that in the case of a homogeneous population of PEVs, the proposed decentralized coordination process achieves this same minimum-cost strategy.

2.2.3 Decentralized Charging Coordination Method

In this part, it designs decentralized strategies by applying the method which utilizes real-time marginal electricity price information. It allows each PEV to choose and implement its own local charging strategy, with the aim of minimizing its individual charging cost. The underlying decentralized charging coordination problem formulated here is a form of non-cooperative game, where a large number of selfish PEVs share electricity resources on a finite collection of charging intervals.

Actually non-cooperative game theory has been widely used to study the supply side of electricity markets, especially in the context of imperfect competition, see [8–11]. Some game theoretic work has been done to understand demand-side behavior in the face of dynamic pricing tariffs, primarily in the context of demand aggregators [12, 13]. Though this work also applies game-theoretic principles to understand outcomes in electricity markets, it considers multi-period demand-side behavior with a local energy constraint applied to the total energy consumed across all periods, and analyzes the problem from a decentralized perspective.

PEVs are supposed to be coupled with each other through a common price signal which is determined by the average charging strategy of the PEV population. Therefore each PEV effectively interacts with the average charging strategy of the rest of the PEV population. As the population grows substantially, the influence of each individual PEV on that average charging strategy becomes negligible. Accordingly, in the infinite population limit, all PEVs will observe the same average strategy as they calculate their optimal local strategy. In this situation, a collection of local charging strategies is a Nash equilibrium, in case: Each PEV's charging strategy is optimal with respect to a single commonly observed charging trajectory, and the average of all the local optimal charging strategies is equal to that common trajectory.

In the rest of the chapter, it will study a decentralized game-based charging strategy for large PEV populations. This section develops the mathematical framework for this analysis, and establishes sufficient conditions for the decentralized problem to achieve a Nash equilibrium.

Consider the local cost function $\mathscr{J}_n$ for an individual PEV n subject to a collection of charging strategies $\mathbf{u}$,

$$\mathscr{J}_n(\mathbf{u}) \triangleq \sum_{t\in\mathscr{T}} p\big(\mathfrak{r}_t(\overline{\mathbf{u}}_t)\big)u_{nt}. \tag{2.15}$$

The locally optimal charging coordination problem with respect to a fixed collection of strategies $\mathbf{u}_{-n}$ is given by the minimization,

$$\mathbf{u}_n^*(\mathbf{u}_{-n}) \triangleq \operatorname*{argmin}_{\mathbf{u}_n\in\mathscr{U}_n} \mathscr{J}_n(\mathbf{u}_n;\mathbf{u}_{-n}), \tag{2.16}$$

where $\mathbf{u}_{-n} \triangleq \{\mathbf{u}_m; m \in \mathscr{N}, m \neq n\}$, in other words $\mathbf{u}_{-n}$ denotes the collection of strategies of all PEVs except PEV n. If a minimizing function exists, it will be referred to as an optimal strategy for the local charging coordination problem.

It is now in a position to formalize the definition of a Nash equilibrium in the context of PEV charging strategies.

Definition 2.1 A collection of PEV strategies $\{\mathbf{u}_n^*; n \in \mathscr{N}\}$ is a *Nash equilibrium* if each PEV n cannot benefit by unilaterally deviating from its individual strategy $\mathbf{u}_n^*$, i.e.,

$$\mathscr{J}_n(\mathbf{u}_n^*;\mathbf{u}_{-n}^*) \leq \mathscr{J}_n(\mathbf{u}_n;\mathbf{u}_{-n}^*), \quad \text{for all } \mathbf{u}_n \in \mathscr{U}_n, \text{ and all } n \in \mathscr{N}.$$

■

In Sect. 2.4 it proposes an iterative algorithm for obtaining the Nash equilibrium. At each iteration, every PEV optimizes its strategy relative to $\mathbf{u}$ determined in the previous iteration. As it will be shown in Sect. 2.6, specifically Fig. 2.4, charging intervals with high PEV demand at one iteration tend to induce low PEV demand at the following iteration, and vice versa. This occurs because PEVs move their charging requirements from expensive to inexpensive intervals; the resulting changes in demand reduce the marginal electricity price in the previously expensive intervals and raise the price in previously inexpensive intervals. This establishes an oscillatory pattern from one iteration to the next, preventing convergence to the Nash equilibrium.

To mitigate these oscillations, the local cost function (2.15) is modified to include a quadratic term that penalizes the deviation of an individual strategy from the population average:

$$\mathscr{J}_n(\mathbf{u}) \triangleq \sum_{t \in \mathscr{T}} \left(p \left(\frac{1}{\Lambda} (\bar{d}_t + \overline{\mathbf{u}}_t) \right) u_{nt} + \delta (u_{nt} - \overline{\mathbf{u}}_t)^2 \right), \tag{2.17}$$

where δ determines the magnitude of the penalty for deviating from the mass average. It will be shown that the presence of the squared deviation term ensures convergence to a unique collection of locally optimal charging strategies that is a Nash equilibrium. Unfortunately, because of the penalty term, this Nash equilibrium only coincides with the globally optimal strategy (2.10) when all PEVs are identical (homogeneous). Nevertheless, it will be seen that the cost added due to this term can be quite small compared with the electricity price p.

Given the formal definition of a Nash equilibrium provided by Definition 2.1, and the local cost function (2.17) for each PEV, it can now establish the conditions governing a Nash equilibrium for an infinite population of PEVs.

Theorem 2.1 *A collection of charging strategies $\mathbf{u} \in \mathscr{U}$ for an infinite population of PEVs is a Nash equilibrium, if*

(i) for all $n \in \mathscr{N}$, $\mathbf{u}_n$ minimizes the cost function,

$$J_n(\mathbf{u}_n; \mathbf{z}) = \sum_{t \in \mathscr{T}} \left(p \left(\frac{1}{\Lambda} (\bar{d}_t + z_t) \right) u_{nt} + \delta (u_{nt} - z_t)^2 \right) \tag{2.18}$$

with respect to a fixed $\mathbf{z}$, and

(ii) $z_t = \overline{\mathbf{u}}_t$, for all $t \in \mathscr{T}$, i.e., $\mathbf{z}$ can be reproduced by averaging the individual optimal charging trajectories of all PEVs.

Proof Consider the collection of PEV charging strategies $\{\mathbf{u}_n^*; n \in \mathscr{N}\}$ where each $\mathbf{u}_n^* \in \mathscr{U}_n$ minimizes its corresponding cost function (2.18) with respect to the common, fixed trajectory $\mathbf{z}^*$, and $z_t^* = \overline{\mathbf{u}}_t^*$ for all $t \in \mathscr{T}$.

As the population size N approaches infinity, each individual PEV's charging strategy u_{nt} has negligible influence on the population average $\overline{\mathbf{u}}_t$. Therefore, for every $n \in \mathscr{N}$,

$$\lim_{N\to\infty}\frac{1}{N}\left(u_{nt}+\sum_{m\in\mathscr{N}\backslash n}u^*_{mt}\right)=\lim_{N\to\infty}\frac{1}{N}\sum_{m\in\mathscr{N}}u^*_{mt}=\overline{\mathbf{u}}^*_t=z^*_t. \tag{2.19}$$

Using this relationship in (2.17) gives for every $n\in\mathscr{N}$,

$$\begin{aligned}\mathscr{J}_n(\mathbf{u}_n;\mathbf{u}^*_{-n})&=\sum_{t\in\mathscr{T}}\left(p\left(\frac{1}{\Lambda}(\bar{d}_t+\overline{\mathbf{u}}^*_t)\right)u_{nt}+\delta(u_{nt}-\overline{\mathbf{u}}^*_t)^2\right)\\&=J_n(\mathbf{u}_n;\mathbf{z}^*).\end{aligned} \tag{2.20}$$

Each $\mathbf{u}^*_n$ minimizes $J_n(\mathbf{u}_n;\mathbf{z}^*)$, and so by (2.20) also minimizes $\mathscr{J}_n(\mathbf{u}_n;\mathbf{u}^*_{-n})$. Hence, by Definition 2.1, $\{\mathbf{u}^*_n;\, n\in\mathscr{N}\}$ is a Nash equilibrium. ■

As mentioned earlier, PEV charging games are consistent with the Nash certainty equivalence principle (also known as mean-field games). The key similarity is that individual agents do not consider the behavior of other individuals. Instead individuals are influenced by the so-called "mass effect", i.e., the overall effect of the population on a given agent. In the case of PEV charging, the effect felt by all individuals is the electricity price, which is specified as a function of the mass average charging trajectory $\overline{\mathbf{u}}_t$.

2.3 Implementation for Infinite PEV Population Systems

In this section it studies the existence and uniqueness of the Nash equilibrium for the decentralized charging optimization defined by (2.17). Section 2.3.1 derives the local optimum with respect to an arbitrary mass average. Proofs of existence and uniqueness are presented in Sect. 2.3.2.

To obtain these results it will assume that the PEV population size is infinite. Obviously, any implementation of the strategy must work for finite PEV populations. This issue will be further discussed in Sects. 2.4 and 2.6, and will be shown that the infinite population results apply to finite systems.

2.3.1 Locally Optimal Charging Strategy

Lemma 2.2 determines the optimal charging trajectory $\mathbf{u}^*_n(\mathbf{z})$ for an individual PEV n when it is subjected to a fixed trajectory $\mathbf{z}$.

Lemma 2.2 *Consider a fixed trajectory* $\mathbf{z}$*, and the charging trajectory* $\mathbf{u}_n(\mathbf{z}, A)\in\mathscr{U}_n$ *defined by,*

$$u_{nt}(\mathbf{z}, A) = \frac{1}{2\delta} \max\left\{0, A - p\left(\frac{1}{\bar{A}}(\bar{d}_t + z_t)\right) + 2\delta z_t)\right\}, \quad \textit{for all } t \in \mathcal{T}. \tag{2.21}$$

For a particular value of A, uniquely dependent upon $\mathbf{z}$ *and denoted* $A^*(\mathbf{z})$*, the trajectory (2.21) provides the unique optimal strategy minimizing the cost function* $J_n(\mathbf{u}_n; \mathbf{z})$ *given in (2.18), subject to the admissible charging requirement* $\mathbf{u}_n \in \mathcal{U}_n$ *defined in (2.4).*

Proof It defines a Lagrangian as below:

$$L_n(\mathbf{u}_n; \mathbf{z}) \triangleq J_n(\mathbf{u}_n; \mathbf{z}) + A\big(\Gamma_n - \Sigma(\mathbf{u}_n)\big)$$

where A is the Lagrange multiplier. Since $J_n(\mathbf{u}_n; \mathbf{z})$ is convex with respect to $\mathbf{u}_n$, the local charging strategy that minimizes $J_n(\mathbf{u}_n; \mathbf{z})$, subject to $\mathbf{u}_n \in \mathcal{U}_n$, must satisfy,

(i) $\frac{\partial L_n}{\partial A} = 0$, and
(ii) $\frac{\partial L_n}{\partial u_{nt}} \leq 0$, $u_{nt} \geq 0$, with complementary slackness.

Condition (i) recovers the constraint $\Sigma(\mathbf{u}_n) = \Gamma_n$. It can be derived from condition (ii),

$$p\left(\frac{1}{\bar{A}}(\bar{d}_t + z_t)\right) + 2\delta(u_{nt} - z_t) - A \begin{cases} = 0, & \text{when } u_{nt} > 0 \\ < 0, & \text{otherwise} \end{cases} \tag{2.22}$$

which is equivalent to (2.21).

The form of the dependence of $u_{nt}(\mathbf{z}, A)$ on A expressed in (2.21) ensures that, for any fixed $\mathbf{z}$,

- There exists an A_- such that for $A \leq A_-$, $\Sigma(\mathbf{u}_n(\mathbf{z}, A)) = 0$.
- For $A > A_-$, $\Sigma(\mathbf{u}_n(\mathbf{z}, A))$ is strictly increasing with A, with the relationship continuous, though not smooth. Hence, $\Sigma(\mathbf{u}_n(\mathbf{z}, A))$ is invertible over this domain.

Therefore a constraint $\Sigma(\mathbf{u}_n(\mathbf{z}, A)) = K > 0$ defines a unique $A > A_-$ for each fixed $\mathbf{z}$, which may be written $A(\mathbf{z})$. The particular value of A that ensures satisfaction of the constraint $\Sigma(\mathbf{u}_n) = \Gamma_n$ shall be denoted $A^*(\mathbf{z})$. The resulting charging trajectory can be written $\mathbf{u}_n(\mathbf{z}, A^*(\mathbf{z})) = \mathbf{u}_n^*(\mathbf{z})$.

Since $J_n(\mathbf{u}_n; \mathbf{z})$ is convex with respect to $\mathbf{u}_n$, the minimizing strategy defined by (2.21) must be unique for a given $\mathcal{U}_n$. ■

It now turns to identifying conditions that guarantee existence and uniqueness of the Nash equilibrium identified in Sect. 2.2.3. Before doing so it will introduce additional notation. Denote by

$$\mathbf{u}_n^*(\mathbf{z}) \triangleq \underset{\mathbf{u}_n \in \mathcal{U}_n}{\operatorname{argmin}}\, J_n(\mathbf{u}_n; \mathbf{z})$$

the charging strategy that minimizes the local cost function (2.18) with respect to a fixed $\mathbf{z}$. By Lemma 2.2, it obtains that $\mathbf{u}_n^*(\mathbf{z}) = \mathbf{u}_n(\mathbf{z}, A^*(\mathbf{z}))$. Define another local coordination strategy $\mathbf{v}_n(\mathbf{z}, \widehat{\mathbf{z}}) \equiv \mathbf{u}_n(\widehat{\mathbf{z}}, A^*(\mathbf{z}))$ for PEV n. This $\mathbf{v}_n(\mathbf{z}, \widehat{\mathbf{z}})$ describes a local charging strategy satisfying (2.21) with respect to $\widehat{\mathbf{z}}$ and $A^*(\mathbf{z})$. There is no guarantee that $\Sigma(\mathbf{v}_n(\mathbf{z}, \widehat{\mathbf{z}})) = \Gamma_n$.

2.3.2 Existence and Uniqueness of the Nash Equilibrium

Lemma 2.4, presented in Appendix 2.11.1, establishes several important properties of the charging trajectories $\mathbf{u}_n^*(\mathbf{z})$, $\mathbf{u}_n^*(\widehat{\mathbf{z}})$ and $\mathbf{v}_n(\mathbf{z}, \widehat{\mathbf{z}})$. In this section we will apply this key technical lemma to show existence and uniqueness of the Nash equilibrium.

Theorem 2.2 *Assume $p(\mathfrak{r})$ is continuous on $\mathfrak{r}$. Then there exists a Nash equilibrium for the infinite population charging coordination system.*

Proof From Lemma 2.4, it obtains that

$$\begin{aligned} |\mathbf{u}_n^*(\mathbf{z}) - \mathbf{u}_n^*(\widehat{\mathbf{z}})|_1 &\leq 2|\mathbf{u}_n^*(\mathbf{z}) - \mathbf{v}_n(\mathbf{z}, \widehat{\mathbf{z}})|_1 \\ &\leq 2\sum_{t\in\mathscr{T}} |(z_t - \widehat{z}_t) - \frac{1}{2\delta}(p(\mathfrak{r}_t) - p(\widehat{\mathfrak{r}}_t))|, \end{aligned}$$

where $\mathfrak{r}_t = \frac{1}{\Lambda}(\bar{d}_t + z_t)$ and $\widehat{\mathfrak{r}}_t = \frac{1}{\Lambda}(\bar{d}_t + \widehat{z}_t)$. This implies that $\mathbf{u}_n^*(\mathbf{z})$ is continuous in $\mathbf{z}$ if $p(\mathfrak{r})$ is continuous in r. It follows that $\overline{\mathbf{u}}^*(\mathbf{z})$ is continuous in $\mathbf{z}$, since the average of a group of continuous functions is also continuous.

Define a convex compact set below:

$$\mathscr{Y} \triangleq \left\{\mathbf{y} \equiv (y_t; t \in \mathscr{T});\ \text{such that } 0 \leq y_t \leq \max_{n\in\mathscr{N}}\{\widehat{\Gamma}_n\}\right\},$$

with $\widehat{\Gamma}_n = \frac{\Gamma_n}{\varsigma_n}(\mathrm{soc}_{n,\max} - soc_{n0})$.

By the specifications of admissible strategies given in (2.4), it obtains that $\mathscr{U}_n \subset \mathscr{Y}$ and so by extension $\overline{\mathbf{u}}^*(\mathbf{z}) \in \mathscr{Y}$. Therefore, for any $\mathbf{z} \in \mathscr{Y}$, we have $\overline{\mathbf{u}}^*(\mathbf{z}) \in \mathscr{Y}$, so $\overline{\mathbf{u}}^*(\cdot)$ maps a convex compact set to itself. Consequently, by the Brouwer fixed point theorem [14], there must be a fixed point $\mathbf{z} \in \mathscr{Y}$ such that $\overline{\mathbf{u}}^*(\mathbf{z}) = \mathbf{z}$. Since $\{\mathbf{u}_n^*(\mathbf{z}); n \in \mathscr{N}\}$ is the set of locally optimal charging strategies, by Theorem 2.1 the fixed point $\overline{\mathbf{u}}^*(\mathbf{z}) = \mathbf{z}$ is a Nash equilibrium. ■

The uniqueness of Nash equilibrium for the underlying charging game systems will be discussed in Theorem 2.3 below. Before that, it gives a couple of assumptions.

It considers the following assumptions:

Assumption 2.1 $p(\mathfrak{r})$ is continuously differentiable and strictly increasing on $\mathfrak{r}$. ■

Remark on $p(\mathfrak{r})$ and its Assumption 2.1. Consider that the real-time price $p(\cdot)$ represents the generation marginal cost, and hence is the derivative of the generation cost on the supply. Suppose the electricity generation cost, denoted by $c(\cdot)$, in a quadratic form on the supply, which has be widely applied in the literature, e.g., [15–17] and references therein, say

$$c(\mathfrak{r}) = \frac{1}{2}\mathsf{a}\mathfrak{r}^2 + \mathsf{b}\mathfrak{r} + \mathsf{c}, \tag{2.23}$$

with properly valued parameters a, b and c; then in this quadratic form of the generation cost, the generation marginal cost at instant t, which may be represented by ρ_t for notational simplicity, evolves linearly with respect to the total demand at that time, say

$$\rho_t \equiv p(\mathfrak{r}_t) = \mathsf{a}\mathfrak{r}_t + \mathsf{b}, \tag{2.24}$$

with respect to the total demand $\mathfrak{r}_t$ at instant t, which satisfies Assumption 2.1.

Assumption 2.2 The parameter δ satisfies the following:

$$\frac{1}{2\Lambda}\sup\nolimits_{\mathfrak{r}\in[\mathfrak{r}_{min},\mathfrak{r}_{max}]}\left\{\frac{dp(\mathfrak{r})}{d\mathfrak{r}}\right\} \le \delta \le \frac{a}{\Lambda}\inf\nolimits_{\mathfrak{r}\in[\mathfrak{r}_{min},\mathfrak{r}_{max}]}\left\{\frac{dp(\mathfrak{r})}{d\mathfrak{r}}\right\}, \tag{2.25}$$

for some a in the range $\frac{1}{2} < a < 1$, where $\mathfrak{r}_{min}$ and $\mathfrak{r}_{max}$ denote, respectively, the minimum and maximum possible $\mathfrak{r}$ over the charging interval $\mathscr{T}$, subject to the admissible charging set $\mathscr{U}_n$. ■

Remarks on Assumptions 2.1 and 2.2. Given the above specification of price curve $p(\cdot)$, Assumption 2.2 degenerates to the following:

$$\delta \in \left[\frac{\mathsf{a}}{2}, a\mathsf{a}\right], \quad \text{for some } a \text{ with } \frac{1}{2} < a < 1.$$

Theorem 2.3 *Under Assumptions 2.1 and 2.2, the infinite population charging system possesses a unique Nash equilibrium.*

The proof is provided in Appendix 2.11.2.

Theorem 2.3 establishes a sufficient condition for a range of values of δ for which the system will converge to a unique Nash equilibrium. It may be difficult to satisfy this condition over a large demand range $[\mathfrak{r}_{min}, \mathfrak{r}_{max}]$, especially if the higher demand value approaches the capacity limits of the system (the supply curve is usually very steep there). However, for overnight charging this is less likely to be a binding factor. Moreover, as we will show using a numerical example in Sect. 2.6, convergence is still possible even when condition (2.25) is slightly violated.

2.4 Decentralized Computational Algorithm

Assuming the technical conditions underpinning Theorem 2.3 are satisfied, $\overline{\mathbf{u}}^*(\mathbf{z})$ is a contraction mapping with respect to $\mathbf{z}$. This result motivates an iterative algorithm for computing the unique Nash equilibrium associated with the decentralized charging coordination system:

(S1). The utility broadcasts the prediction of non-PEV base demand $(\bar{d}_t; t \in \mathscr{T})$ to all the PEVs.

(S2). Each of the PEVs proposes an optimal charging strategy minimizing its charging cost with respect to a common aggregate PEV demand broadcast by the utility.

(S3). The utility collects all the optimal charging strategies proposed in (S2), and updates the aggregate PEV demand. This updated aggregate PEV demand is rebroadcast to all PEVs.

(S4). Repeat (S2) and (S3) until the optimal strategies proposed by all PEVs no longer change.

A more formal expression of this procedure is given by Algorithm 2.1. At convergence, the collection of optimal charging strategies is a Nash equilibrium. Some time later, when the actual charging period occurs, each PEV implements its optimal strategy.

Algorithm 2.1 (*Implementation of decentralized charging strategy*)

Require:

Initialize a positive ε, and define a tolerance $\varepsilon_{\text{stop}}$ required to terminate iterations. Provide an initial average charging strategy forecast $\mathbf{z}^{(1)}$, and set $i = 1$.

Ensure:

Charging strategies of PEVs;

1: **while** $\varepsilon > \varepsilon_{\text{stop}}$ **do**
2: Obtain optimal charging strategy $\mathbf{u}_n^{*(i)}$, w.r.t. $\mathbf{z}^{(i)}$, for all $n \in \mathscr{N}$;
Set $\mathbf{z}^{(i+1)}$ equal to $\overline{\mathbf{u}}^{*(i)}$, where $\mathbf{u}^{*(i)} \equiv (\mathbf{u}_n^{*(i)}; n \in \mathscr{N})$;
Update $\varepsilon = |\mathbf{z}^{(i+1)} - \mathbf{z}^{(i)}|_1$;
$i = i + 1$;
3: **end while**

■

During iterations, the optimal charging trajectories proposed by PEVs may result in large average demand $\overline{\mathbf{u}}_t$ at some charging instants. Consequently, the demand $\bar{d}_t + \overline{\mathbf{u}}_t$ at those instants may exceed generation capacity $\bar{\Lambda}$, giving $\mathfrak{r}_t > 1$. This would, however, only occur as the population iterated towards the Nash equilibrium. It will be shown in Sect. 2.5 that the Nash equilibrium corresponds to a charging strategy that is almost valley-filling, implying that large excursions in demand are not likely at the equilibrium.

Implementation of the charging strategy must, of course, work for finite groups of PEVs. To understand the consequences of a finite population N, it refers to

Theorem 2.1. The infinite population limit is required in (2.19) to establish equality in (2.20) between $\mathscr{J}_n(\mathbf{u}_n; \mathbf{u}^*_{-n})$, which quantifies the Nash equilibrium concept in Definition 2.1, and $J_n(\mathbf{u}_n; \mathbf{z}^*)$, which underpins Algorithm 2.1. For finite N, the equality (2.19) reverts to the approximation,

$$\frac{1}{N}\mathbf{u}_n + \frac{N-1}{N}\overline{\mathbf{u}}^*_{-n} \approx \overline{\mathbf{u}}^*.$$

Nevertheless, for large N, this approximation is sufficiently accurate.

The proposed algorithm converges to an ε-Nash equilibrium for a PEV charging system with a finite population, and ε tends to zero as the population size approaches infinity. In this limiting case, the Nash equilibrium corresponds to a Wardrop equilibrium [18]. This method also has connections with the Nash certainty equivalence principle (or mean-field games), proposed by Huang et al. [19, 20] in the context of large-scale games for sets of weakly coupled linear stochastic control systems.

2.5 Valley-Filling Property of the Nash Equilibrium

Having proven existence, uniqueness and convergence for the Nash equilibrium obtained from the PEV charging process of Algorithm 2.1, this section establishes that the Nash equilibrium is *valley filling*. In its simplest form, the valley filling property appears as in Fig. 2.1. However there are several special cases to consider. The following key points capture the various cases that are formalized in Theorem 2.4:

(i) For any pair of charging instants, the one with the smaller non-PEV base demand is assigned a larger charging rate (for individual PEVs as well as for the average over all PEVs), and possesses an equal or lower total aggregate demand.

(ii) The total demand, consisting of aggregate PEV charging load together with non-PEV demand, is constant during charging subintervals when all PEV charging rates are strictly positive. This is also true of the demand obtained by summing non-PEV demand with the charging load of any individual PEV.

For a homogeneous population of PEVs, this second outcome guarantees perfect valley filling. That is not the case, however, for heterogeneous populations because there may be charging subintervals when the charging rate for some PEVs is zero. The examples of Sect. 2.6 illustrate these outcomes.

Theorem 2.4 *Suppose that the collection of charging trajectories* $\mathbf{u}^* \equiv \{\mathbf{u}^*_n; n \in \mathscr{N}\}$ *is a Nash equilibrium, and consider Assumption 2.1. Then* $\mathbf{u}^*_n$ *and the average* $\mathbf{z} = \overline{\mathbf{u}}^*$ *satisfy the following valley filling properties for all* $\delta > 0$*:*

(i) $z_t \geq z_s, \quad \bar{d}_t + z_t \leq d_s + z_s, \quad u^*_{nt} \geq u^*_{ns},$ *when* $\bar{d}_t \leq d_s$, *with* $t, s \in \mathscr{T}$ (2.26a)

(ii) $\bar{d}_t + z_t = \theta, \quad \bar{d}_t + u^*_{nt} = \theta_n,$ *for some* $\theta, \theta_n > 0$, *with* $t \in \widehat{\mathscr{T}}$, (2.26b)

where $\widehat{\mathscr{T}} \equiv \{t \in \mathscr{T};\, u^*_{nt} > 0 \text{ for all } n \in \mathscr{N}\}$.

The proof is provided in Appendix 2.11.3.

In case of homogeneous PEV populations, each of the individual optimal strategies $\mathbf{u}^*_n$ is coincident with their average strategy $\overline{\mathbf{u}}^*$. It follows that the properties of the Nash equilibrium specified in (2.26) are equivalent to,

$$\overline{\mathbf{u}}^*_t = \max\{0, \theta - \bar{d}_t\}, \quad \text{for some } \theta > 0, \tag{2.27}$$

which is the normalized form of the optimal valley-filling strategy given in (2.10). In other words, in the case of a homogeneous PEV population, the Nash equilibrium coincides with the charging strategy given by centralized method, and is therefore globally optimal.

2.6 Numerical Examples

2.6.1 Background

A range of examples will be used in this section to illustrate the main results of the work. In particular, it will consider the conditions for convergence of Algorithm 2.1, and explore the nature of valley filling.

The examples use the non-PEV demand profile of Fig. 2.1, which shows the load of the Midwest ISO region for a typical summer day during 2007. It is assumed that the total generation capacity is 1.2×10^8 kW. Furthermore, the simulations are based on assuming $N = 10^7$, which corresponds to roughly 30% of vehicles in the MISO footprint. This gives $c^N \triangleq \frac{1}{N}\Lambda^N = 12$ kW, where the superscript N indicates finite population size. Also, define $d^N_t \triangleq \frac{1}{N}d^N_t$, with d^N_t displayed in Fig. 2.1.

The following PEV population parameters will be used for all the examples. All PEVs have an initial SOC of 15%, i.e., $soc_{n0} = 0.10$ for all n, $soc_{n,\max} = 90\%$, $\Upsilon^+_n = 5$ kW and 90% charging efficiency, i.e., $\varsigma_n = 0.90$ for all n. The charging interval $\mathscr{T}$ covers the 12-hour period from 8:00 pm on one day to 8:00 am on the next. The continuously differentiable and strictly increasing price function

$$p(\tau) = 0.15\tau^{1.5}/\text{kWh} \tag{2.28}$$

is used in all cases.

Other parameters, such as PEV battery size Γ_n and the tracking cost parameter δ, are specified within each of the examples.

2.6.2 Computation of Nash Equilibrium for Homogeneous PEVs

This section considers the computation of the Nash equilibrium for a homogeneous population of PEVs, each of which possesses an identical battery size of $\Gamma^n = 10\,\mathrm{kWh}$. Firstly, it may be verified from Fig. 2.1 that

$$\mathfrak{r}_{min} = \min_{t\in\mathscr{T}}\{d_t^N\}/\Lambda^N \approx 0.5.$$

To determine $\mathfrak{r}_{max}$, it assumes that the entire energy requirement Γ_n is delivered over a single time step, so

$$r_{max} = (\max_{t\in\mathscr{T}}\{d_t^N\} + \Gamma_n)/\Lambda^N \approx 1.5.$$

Referring to (2.28), this gives

$$\frac{1}{2c^N}\max_{[\mathfrak{r}_{min},\mathfrak{r}_{max}]}\frac{dp(\mathfrak{r})}{d\mathfrak{r}} = 0.012 \le \frac{a}{\Lambda^N}\min_{[\mathfrak{r}_{min},\mathfrak{r}_{max}]}\frac{dp(\mathfrak{r})}{d\mathfrak{r}} = 0.013a,$$

which can be satisfied for some a in the range $\frac{1}{2} < a < 1$. Therefore a tracking parameter δ exists such that condition (2.25) of Theorem 2.3 holds.

Figure 2.2 provides simulation results for the decentralized computation algorithm of Sect. 2.3, for the homogeneous PEV population of this section. The tracking

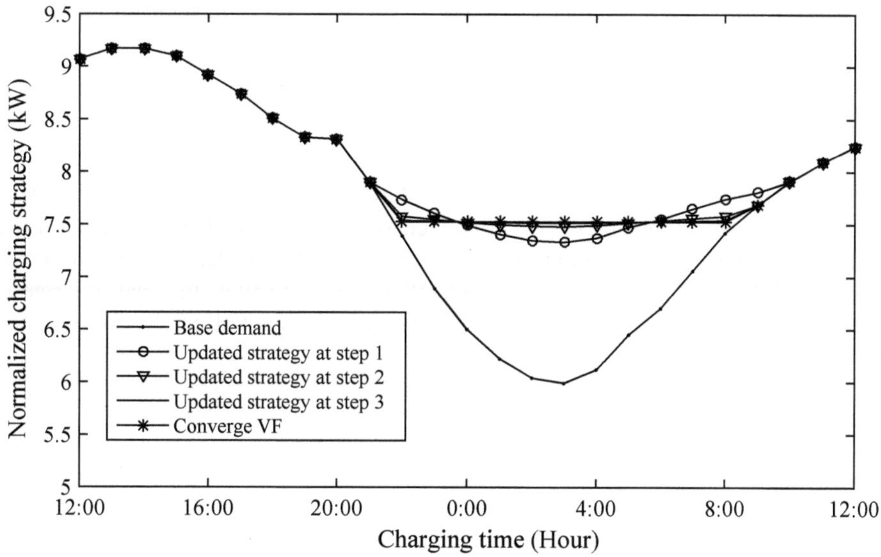

Fig. 2.2 Convergence of Algorithm 2.1 for a homogenous PEV population, with $\delta = 0.012$

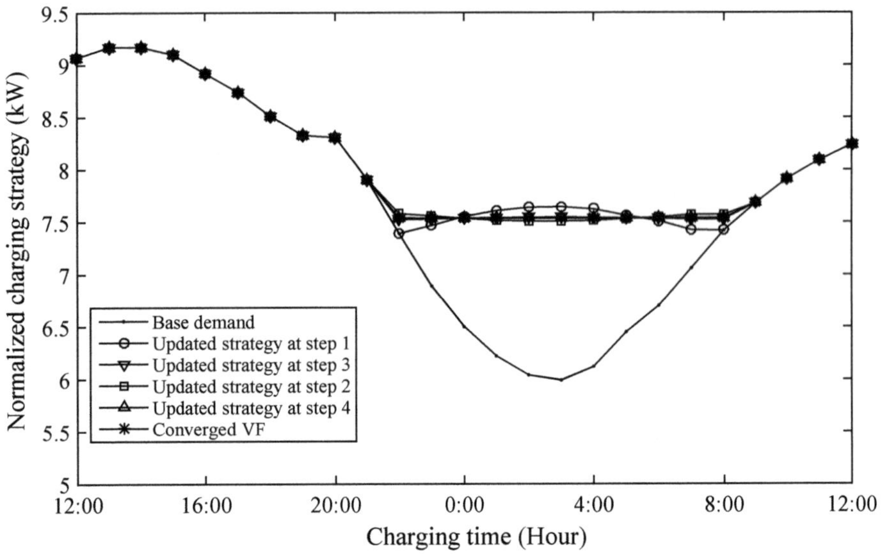

Fig. 2.3 Convergence of Algorithm 2.1 for a homogenous PEV population, with $\delta = 0.007$. This violates the condition given in Theorem 2.3

parameter $\delta = 0.012$ was used for this case. Each line in the figure corresponds to an iterate of the algorithm. It can be observed that convergence to the Nash equilibrium (shown by the solid flat black curve) is achieved in a few cycles. This Nash equilibrium is clearly the globally optimal valley-filling strategy.

The condition on δ established in Theorem 2.3 is sufficient, but not necessary. Figure 2.3 confirms this. Here,

$$\delta = 0.007 < \frac{1}{2\Lambda^N} \max_{[\mathfrak{r}_{min}, \mathfrak{r}_{max}]} \frac{dp(\mathfrak{r})}{d\mathfrak{r}} = 0.012,$$

yet the system still converges to the same valley-filling solution as in Fig. 2.2. As δ decreases, however, the process eventually ceases to converge. This can be observed in Fig. 2.4, which shows the iterations when $\delta = 0.003$. In order to avoid an unreasonably high charging rate during the non-PEV demand valley, the charging rate is constrained to a maximum of 3 kW. This constraint does not effect the convergence property of the algorithm.

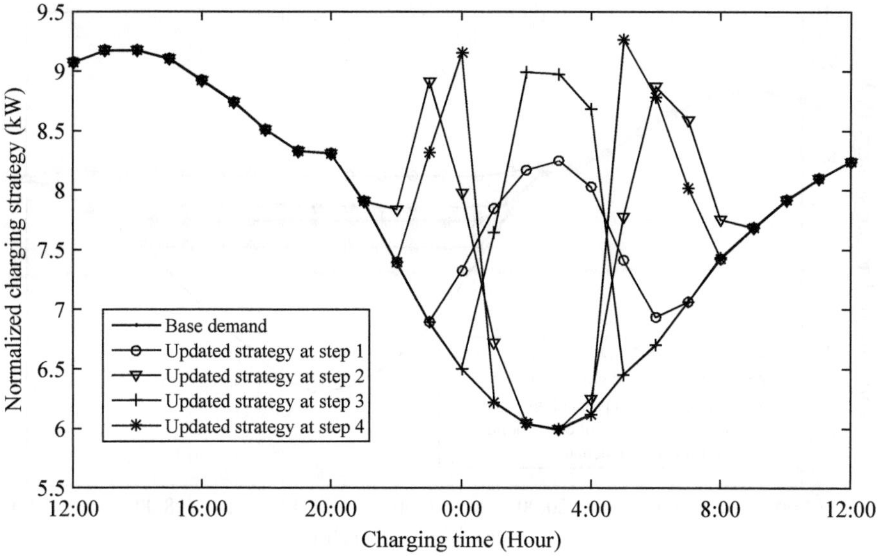

Fig. 2.4 Non-convergence of Algorithm 2.1 for a homogenous PEV population, with $\delta = 0.003$. This significantly violates the condition given in Theorem 2.3

2.6.3 Computation of Nash Equilibrium for Heterogeneous PEVs

PEV populations are heterogeneous if vehicles do not have identical charging requirements. To examine optimal charging outcomes for heterogeneous populations, a simplified case is constructed such that PEVs have one of three charging energy requirements: 10, 15 or 20 kWh. It is further supposed that the number of PEVs in each group accounted for about 50, 30 and 20% of the population respectively. It can be verified that

$$\mathfrak{r}_{min} = \min_{t \in \mathcal{T}}\{d_t^N\}/\Lambda^N \approx 0.5, \qquad \mathfrak{r}_{max} = (\max_{t \in \mathcal{T}}\{d_t^N\} + \overline{\Gamma})/\Lambda^N \approx 1.8,$$

where $\overline{\Gamma} = 13.5$ kWh denotes the energy delivery requirement of each PEV averaged across the entire population. It follows that

$$\frac{1}{2\Lambda^N} \max_{[\mathfrak{r}_{min}, \mathfrak{r}_{max}]} \frac{dp(\mathfrak{r})}{d\mathfrak{r}} = 0.0125 \le \frac{a}{\Lambda^N} \min_{[\mathfrak{r}_{min}, \mathfrak{r}_{max}]} \frac{dp(\mathfrak{r})}{d\mathfrak{r}} = 0.013a,$$

which can be satisfied for some a in the range $\frac{1}{2} < a < 1$. Figure 2.5 shows the results for this heterogeneous case, with tracking parameter $\delta = 0.0125$. In particular the dashed curves with marks show the optimal charging strategies for the 1st, 2nd and 3rd class of PEVs, and the solid curve provides the average demand value across the

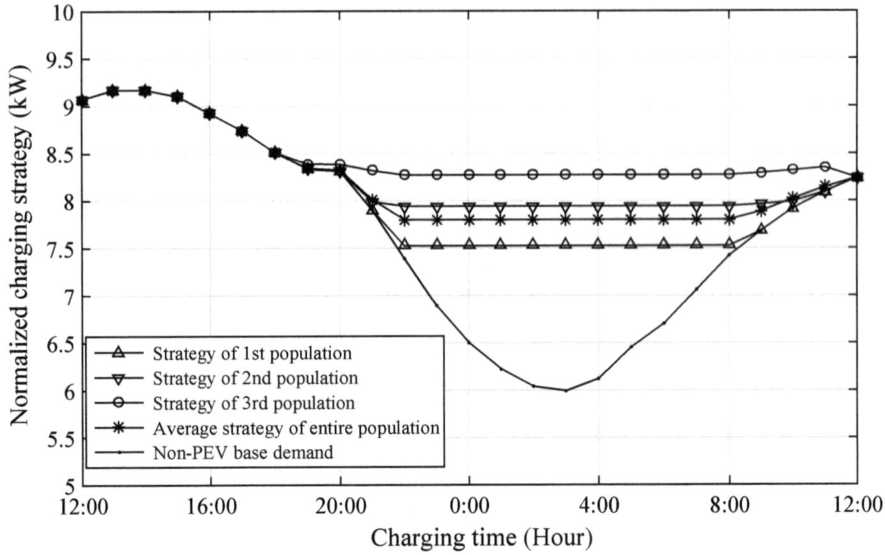

Fig. 2.5 Convergence of Algorithm 2.1 for a heterogeneous PEV population

entire population. Notice that this curve of average demand is flat between 10 pm and 8 am, where all PEVs are charging, in accordance with Theorem 2.4.

As observed in Fig. 2.5, for the heterogeneous PEV populations, the implemented charging strategy by applying Algorithm 2.1 developed in this part is not optimal. Alternatively, in Sect. 2.7, Algorithm 2.1 will be slightly updated such that the implemented strategy is optimal for heterogeneous classes of PEV populations.

2.7 Decentralized Valley-Filling Method for Heterogeneous PEVs

It first develops the mathematical framework for the decentralized method. Suppose that the whole population $\mathcal{N}$ is partitioned into L distinct classes of homogeneous subpopulations $\mathcal{N}_k$, with $l \in \mathcal{K}$, where each of the PEVs possesses identical Γ_k, i.e. $\Gamma_n = \Gamma_k$ for all $n \in \mathcal{N}_k$. Define $\mathfrak{a}_k^N$, such that $\mathfrak{a}_k^N = \frac{N_k}{N}$, where N_k denotes the size of subpopulation $\mathcal{N}_k$; then it obtains that $\mathcal{N} = \dot{\bigcup}_{k \in \mathcal{K}} \mathcal{N}_k$, where $\dot{\bigcup}$ represents a disjoint union operator, and $\sum_{k=1}^{K} \mathfrak{a}_k^N = 1$.

Suppose that all of the agents in a subpopulation simultaneously update their strategies; then as discussed in earlier parts, the cheap and expensive resources during the charging interval may become expensive and cheap ones respectively. This oscillating behavior implies that the system can't converge to any Nash equilibrium. To mitigate this oscillation behavior it introduces an individual deviation cost, denoted by $J_n^d(\boldsymbol{u})$, which measures the penalty of the individual strategy deviated from the average strategy in the subpopulation, such that $J_n^d(\boldsymbol{u}) \triangleq \delta||\boldsymbol{u}_n - \overline{\boldsymbol{u}}_k||^2$, with a positive valued δ, for all $n \in \mathscr{N}_k$, where $\overline{\boldsymbol{u}}_{kt}^N \triangleq \frac{1}{N_k} \sum_{m \in \mathscr{N}_k} u_{mt}$, i.e. $\overline{\boldsymbol{u}}_{kt}^N$ denotes the average value of charging rate among all PEVs in $\mathscr{N}_k$ at t. In the rest of this part, it considers the combined individual cost function:

$$J_n(\boldsymbol{u}) \triangleq \mathscr{J}_n(\boldsymbol{u}) + J_n^d(\boldsymbol{u}), \tag{2.29}$$

where $\mathscr{J}_n(\boldsymbol{u})$ is given in (2.15), and $\mathsf{r}_n(\boldsymbol{u}) \triangleq \dfrac{J_n^d(\boldsymbol{u})}{J_n(\boldsymbol{u})}$, i.e. $\mathsf{r}_n(\boldsymbol{u})$ represents the ratio between the deviation cost and the total cost of agent n subject to strategy $\boldsymbol{u}$.

$\{\boldsymbol{u}_n^0; n \in \mathscr{N}\}$ is a *Nash equilibrium* if each PEV n can't benefit by unilaterally deviating from its individual strategy $\boldsymbol{u}_n^0$, i.e.

$$J_n(\boldsymbol{u}_n^0; \boldsymbol{u}_{-n}^0) \leq J_n(\boldsymbol{u}_n; \boldsymbol{u}_{-n}^0), \quad \text{for all } \boldsymbol{u}_n \in \mathscr{U}_n;$$

then at any Nash equilibrium $\boldsymbol{u}^0$, with (2.29), the deviation cost $J_n^d(\boldsymbol{u})$ vanishes, since $\boldsymbol{u}_n^0 = \overline{\boldsymbol{u}}_k^0$, for all $n \in \mathscr{N}_k$.

Same as the part studied in Sect. 2.2, this part also studies systems where the size of the PEV population is sufficiently large that the effect of each of the PEVs on the system is negligible. More specifically, it examines the asymptotic properties of the system in the population size limit, such that the asymptotic property specified in (2.7) holds. Moreover, it is supposed that:

$$\mathsf{a}_k \triangleq \lim_{N \to \infty} \frac{N_k}{N}. \tag{2.30}$$

Define $\overline{\boldsymbol{u}}_k \triangleq \lim_{N \to \infty} \frac{1}{N_k} \sum_{n \in \mathscr{N}_k} \boldsymbol{u}_n$; then by (2.30), it implies that $\overline{\boldsymbol{u}}^N = \sum_{k \in \mathscr{K}} \mathsf{a}_k \overline{\boldsymbol{u}}_k^N$ and $\overline{\boldsymbol{u}} = \sum_{k \in \mathscr{K}} \mathsf{a}_k \overline{\boldsymbol{u}}_k$.

Following the formulation above, the charging coordination game will be setup.

The system collects the PEV population partition information $\mathscr{N}_k$ according to the submitted PEV charging characteristics, and informs each PEV, $n \in \mathscr{N}_k$, that the individual cost function (2.29) is composed of the charging price cost and a quadratic deviation cost, with parameter δ, from the average behavior over in $\mathscr{N}_k$. The system also designs a strategy update mechanism for PEVs, with updated price information and average behavior with respect to the submitted PEV strategies, as below:

Algorithm 2.2 Mixed update mechanism for the Nash equilibrium implementation:

(S0) Consider an initial collection of average strategies of subpopulations $(\overline{\boldsymbol{u}}_k^{(0)}; l \in \mathscr{K})$;

(S1) Each agent n in subpopulation $\mathscr{N}_k$ simultaneously implements his best strategy $\boldsymbol{u}_n^*$ which minimizes the cost function (2.31) with respect to $(\overline{\boldsymbol{u}}_k^{(0)}; k \in \mathscr{K})$,

(S2) Update $\overline{\boldsymbol{u}}_k$ with the average value of collection of best strategies $(\boldsymbol{u}_n^*; n \in \mathcal{N}_k)$ implemented in (S1);
(S3) Repeat (S1) and (S2) until $\overline{\boldsymbol{u}}_k$ does not update any longer;
(S4) Implement (S1)–(S3) for all of the subpopulations successively.

■

Note: Both the simultaneous and successive updates are involved in Algorithm 2.2. In this sense and to distinguish from the method proposed in Algorithm 2.1, it may call Algorithm 2.2 as a *mixed strategy update algorithm.*

By Theorem 2.5, the implemented strategy following Algorithm 2.2 is a Nash equilibrium.

Theorem 2.5 *$\boldsymbol{u}^* \in \mathscr{U}$ is a Nash equilibrium for infinite-population charging games, if*

(i) $\boldsymbol{u}_n^*$, for all $n \in \mathcal{N}_k$, minimizes the following

$$J_n(\boldsymbol{u}_n; z) \triangleq \sum_{t \in \mathscr{T}} \left(p(\mathfrak{r}_t)u_{nt} + \delta(u_{nt} - z_{kt})^2\right), \tag{2.31}$$

with $\mathfrak{r}_t = \frac{1}{\Lambda}(\bar{d}_t + \sum_{k \in \mathscr{K}} \mathfrak{a}_k z_{kt})$;
(ii) $\overline{\boldsymbol{u}}_k = z_k$, for all k, i.e. z_k can be reproduced by averaging the individual best strategies given in (i) of all PEVs in $\mathcal{N}_k$.

Proof Outline. *Each of individuals is only influenced by the average behavior of the subpopulations, and the behavior of a single agent is negligible. Hence the game system is at a Nash equilibrium, if (i) and (ii) hold.* ■

Suppose that $\boldsymbol{u}^*$ is a Nash equilibrium; then

$$J_n(\boldsymbol{u}_n^*; \overline{\boldsymbol{u}}_k^*, k \in \mathscr{K}) = \sum_{t \in \mathscr{T}} p(\mathfrak{r}_t)u_{nt}^*,$$

since the deviation cost $\delta(\boldsymbol{u}_n^* - \overline{\boldsymbol{u}}_k^*)^2$ vanishes at equilibrium due to $\boldsymbol{u}_n^* = \overline{\boldsymbol{u}}_k^*$, for all $n \in \mathcal{N}_k$.

Theorem 2.6 *Suppose that $\boldsymbol{u}^*$ is a Nash equilibrium, and consider Assumption 2.1; then $\boldsymbol{u}^*$ satisfies the following valley-filling property*

$$\overline{\boldsymbol{u}}_t^* = \max\left\{0, w - \bar{d}_t\right\}, \quad \text{with } \overline{\boldsymbol{u}}^* \triangleq \sum_{k \in \mathscr{K}} \mathfrak{a}_k \overline{\boldsymbol{u}}_k^*, \tag{2.32}$$

such that $\sum_{t \in \mathscr{T}} \overline{\boldsymbol{u}}_t^* = \sum_{k \in \mathscr{K}} \mathfrak{a}_k \Gamma_k$.

Proof By Theorem 2.6, $(\boldsymbol{u}_n^*; n \in \mathcal{N}_k)$, for any $k \in \mathscr{K}$, can be considered as a Nash equilibrium of the charging coordination game of the subpopulation $\mathcal{N}_k$, with the cost function given in (2.31) where $\mathfrak{r}_t = \frac{1}{\Lambda}(\widehat{\boldsymbol{d}}_t + \mathfrak{a}_k z_{kt})$ with $\widehat{\boldsymbol{d}} = \boldsymbol{d} + \sum_{m \in \mathscr{K}_{-k}} \mathfrak{a}_m \overline{\boldsymbol{u}}_m^*$;

then by Theorem 2.4, it obtains that the $\boldsymbol{u}_k$ satisfies the following valley-filling property

$$\overline{u}_{kt}^{*} = \frac{1}{\mathfrak{a}_k} \max\{0, w - \widehat{d}_t\}, \text{ with a constant } w, \tag{2.33}$$

such that $\sum_{t\in\mathcal{T}} \overline{u}_{kt}^{*} = \Gamma_k$, i.e. $\overline{\boldsymbol{u}}_k^{*}$ is a valley-filling strategy which fills the valley of the demand $\boldsymbol{d}$.

Hence the Nash equilibrium for the whole PEV population $\mathcal{N}$ satisfies the valley-filling property specified in (2.32). ■

2.8 Implementation of Nash Equilibrium

In Theorem 2.7 below, it will be shown that by adopting the Algorithm 2.2 the infinite-population game system converges to a unique valley-filling Nash equilibrium under Assumptions 2.1 and 2.2. Before Theorem 2.7, it firstly gives a lemma below.

Lemma 2.3 *Considering Assumptions 2.1 and 2.2, the asymptotic property of (2.7), and by adapting the strategy update mechanism (S1)–(S3) for subpopulation $\mathcal{N}_k$, all of the agents in $\mathcal{N}_k$ converge to a unique charging strategy $\boldsymbol{u}_k^{*}$, such that*

$$\overline{u}_{kt}^{*} = \frac{1}{\mathfrak{a}_k} \max\left\{0, w - \sum_{m\in\mathcal{K}_{-k}} \mathfrak{a}_m \overline{u}_{mt} - \bar{d}_t\right\}, \tag{2.34}$$

with a constant w and $\sum_{t\in\mathcal{T}} \overline{\boldsymbol{u}}_{kt}^{*} = \Gamma_k$, i.e $\overline{\boldsymbol{u}}_k^{*}$ is a valley-filling strategy with respect to collection of average behavior $\{\overline{\boldsymbol{u}}_m; m \neq k, m \in \mathcal{K}\}$.

Proof The statement of Lemma 2.3 is essentially equivalent to the problems below:

To study the computation of Nash equilibrium for following iterative computation algorithm (S1)–(S3), of infinite-population of homogeneous agents $\mathcal{N}_k$ with the individual cost function given below

$$J_n(\boldsymbol{u}_n; (\overline{\boldsymbol{u}}_k; k \in \mathcal{K})) = \sum_{t\in\mathcal{T}} (p(\mathfrak{r}_t) u_{nt} + \delta(u_{nt} - \overline{\boldsymbol{u}}_{kt})^2),$$

with $\mathfrak{r}_t = \frac{1}{\Lambda}(\bar{d}_t + \overline{u}_t)$,

Fix the charging strategies of all of the PEVs except those from subpopulation $\mathcal{N}_k$; then the best strategy update procedure (S1)–(S3) proposed in Algorithm 2.2 is essentially the implementation of the Nash equilibrium among PEV agents in $\mathcal{N}_k$ with respect to the base demand $\sum_{m\in\mathcal{K}_{-k}} \mathfrak{a}_m \overline{u}_{mt} - \bar{d}_t$. Hence by applying Theorem 2.3, it can obtain that under Assumptions 2.1 and 2.2, the system converges to a unique equilibrium $\boldsymbol{u}^{*}$ which is a valley-filling strategy such that (2.34) holds, by Theorem 2.6. ■

Theorem 2.7 *Considering Assumptions 2.1 and 2.2, the asymptotic property of (2.7), the game system (2.31) reaches a unique valley-filling Nash equilibrium by applying Algorithm 2.2.*

Proof Consider an initial average strategy $\overline{\boldsymbol{u}}_m^{(0)}$, for each subpopulation $\mathscr{N}_m$. By applying Algorithm 2.2, subpopulations successively update their best strategies. Denote $\boldsymbol{u}_n^*$, with $n \in \mathscr{N}_m$, and $\overline{\boldsymbol{u}}_m^*$ as the updated best strategy of PEV n in subpopulation $\mathscr{N}_m$ and the average behavior of $\{\boldsymbol{u}_n^*; n \in \mathscr{N}_m\}$ respectively.

Suppose that all of subpopulations $\mathscr{N}_m$, with $m \leq k-1$, already implemented their strategies following (S1)–(S3) in Algorithm 2.2; then similarly next, all of the agents in subpopulation $\mathscr{N}_k$ shall update their best strategies following (S1)–(S3) with respect to collection of mass strategies of all subpopulations, $(\overline{\boldsymbol{u}}_m^*; m \leq k-1)$ and $(\overline{\boldsymbol{u}}_j^{(0)}; j \geq k+1)$. Define $\widehat{\boldsymbol{d}}_{-k}$ as the total demand trajectory subject to collection of mass charging behaviors $\{\overline{\boldsymbol{u}}_m^*; m \leq k-1\}$ and $\{\overline{\boldsymbol{u}}_j^{(0)}; j \geq k+1\}$, i.e.

$$\widehat{\boldsymbol{d}}_{-k} \triangleq \boldsymbol{d} + \sum_{m=1}^{k-1} \mathfrak{a}_m \overline{\boldsymbol{u}}_m^* + \sum_{j=k+1}^{K} \mathfrak{a}_j \overline{\boldsymbol{u}}_j^{(0)}.$$

By Lemma 2.3, the average best strategy of subpopulation $\mathscr{N}_k$ converges to a valley-filling strategy $\overline{\boldsymbol{u}}_k^*$ with respect to $\widehat{\boldsymbol{d}}_{-k}$, such that

$$\overline{u}_{kt}^* = \frac{1}{\mathfrak{a}_k} \max\{0, w_k - \widehat{d}_{-k,t}\}, \tag{2.35}$$

where w_k is a constant value such that $\sum_{t\in\mathscr{T}} \overline{u}_{kt}^* = \Gamma_k$, i.e. following (S1)–(S3), the initial average strategy $\overline{\boldsymbol{u}}_k^{(0)}$ of $\mathscr{N}_k$, is updated by $\overline{\boldsymbol{u}}_k^*$ which is a valley-filling strategy with respect to $\widehat{\boldsymbol{d}}_{-k}$. As a result, after the subpopulation $\mathscr{N}_k$ proposed their best strategy, the average behaviors of PEV populations are updated with $(\overline{\boldsymbol{u}}_m^*, \overline{\boldsymbol{u}}_j^{(0)}; m \leq k, j \geq k+1)$.

Define a set of instants $\mathscr{T}_k$ such that

$$\mathscr{T}_k \triangleq \{t \in \mathscr{T}; \text{ s.t. } \widehat{d}_{-k,t} \leq w_k\}, \tag{2.36}$$

and show, in (i, ii) below, the following holds

$$\mathscr{T}_{k+1} \supseteq \mathscr{T}_k, \text{ and } w_{k+1} \geq w_k, \qquad \text{for all } k \leq K-1. \tag{2.37}$$

(i). Suppose that $\overline{u}_{k+1,t}^{(0)} = 0$ for all t, $t \in \mathscr{T}$, at which $\widehat{d}_{-k,t} + \mathfrak{a}_k \overline{u}_{kt}^* > w_k$; then the initial strategy $\overline{\boldsymbol{u}}_{k+1}^{(0)}$ is a valley-filling strategy with respect to the demand $\widehat{\boldsymbol{d}}_{-(k+1)} \equiv \boldsymbol{d} + \sum_{m=1}^{k} \mathfrak{a}_m \overline{\boldsymbol{u}}_m^* + \sum_{j=k+2}^{K} \mathfrak{a}_j \overline{\boldsymbol{u}}_j^{(0)}$; then by Lemma 2.3, $\overline{\boldsymbol{u}}_{k+1}^* = \overline{\boldsymbol{u}}_{k+1}^{(0)}$. It implies that $\mathscr{T}_{k+1} = \mathscr{T}_k$ and $w_{k+1} = w_k$.

(ii). Otherwise, suppose that there exists at least an instant s, $s \in \mathscr{T}$, at which $\widehat{d}_{-k,s} + \mathfrak{a}_k \overline{u}^*_{ks} > w_k$, and $\overline{u}^{(0)}_{k+1,s} > 0$. That is to say, the initial $\overline{\boldsymbol{u}}^{(0)}_{k+1}$ is not a valley-filling strategy with respect to $\widehat{\boldsymbol{d}}_{-(k+1)}$ specified above.

$w_{k+1} > w_k$ is verified using the proof by contradiction. Suppose that $w_{k+1} \leq w_k$. By the definition of w_{k+1} it obtains that $\overline{u}^*_{k+1,t} = 0 < \overline{u}^{(0)}_{k+1,t}$ in case $t = s$, and $\overline{u}^*_{k+1,t} \leq \overline{u}^{(0)}_{k+1,t}$ otherwise. It implies that $\sum_{t\in\mathscr{T}} \overline{u}^*_{k+1,t} < \sum_{t\in\mathscr{T}} \overline{u}^{(0)}_{k+1,t} = \Gamma_{k+1}$ which is contradictive with the full charging constraint of $\sum_{t\in\mathscr{T}} \overline{u}^*_{k+1,t} = \Gamma_{k+1}$. Also by (2.36) and $\overline{\boldsymbol{u}}^*_l$ specified in (2.35), it gets that $\widehat{d}_{-k,t} + \mathfrak{a}_k \overline{u}^*_{kt} = w_k$ at each of instants $t \in \mathscr{T}_k$; then the following holds

$$\widehat{d}_{-(k+1),t} = \widehat{d}_{-k,t} + \mathfrak{a}_k \overline{u}^*_{kt} - \mathfrak{a}_{k+1} \overline{u}^{(0)}_{k+1,t} < w_k < w_{k+1},$$

for all $t \in \mathscr{T}_k$, which implies that $\mathscr{T}_{k+1} \supseteq \mathscr{T}_k$.

By (2.37), after each of the subpopulations updated their best strategy following (S1)–(S3) successively, it obtains that

$$\mathscr{T}_K \supseteq \mathscr{T}_k, \text{ and } w_K \geq w_k, \qquad \text{for each } k \leq K. \tag{2.38}$$

As a result, by applying Algorithm 2.2, the system reaches the valley-filling strategy. ■

Remark: The decentralized method, proposed in Algorithm 2.2, is implemented ahead of the charging interval based upon the predicted non-PEV base demand. As a consequence, the valley-filling performance under the proposed method, as shown in Theorem 2.7, is largely based upon the accuracy of the prediction on the base demand.

2.9 Numerical Examples

The examples use the load of the Midwest ISO (MISO) region for a typical summer/spring day as the non-PEV demand profile and consider the total population size of $N = 10^7$ which corresponds to roughly 30% of vehicles in the MISO region. See dot marked curves in Figs. 2.6 and 2.11 for an illustration of the normalized base demands, $\bar{d}_t \triangleq \frac{1}{N} d^N_t$, in a summer and spring day respectively.

For the purpose of demonstration, it considers simplified parameters for PEV populations. The battery capacity of an individual PEV is either 5 or 20 kWh, and assume that the number of PEVs in each group accounted for about 50% and 50% of the population respectively. All of the PEVs have a common initial SOC of 15% and an identical charging efficiency of 85%. The charging interval $\mathscr{T}$ covers a 14-hour period from 8:00 PM on one day to 10:00 AM on the next day. Consider a linear price curve, such that $p(\mathfrak{r}) = 0.15\mathfrak{r}$\$/kWh, and a deviation parameter $\delta = 0.015$. An initial charging strategy is displayed in Fig. 2.6.

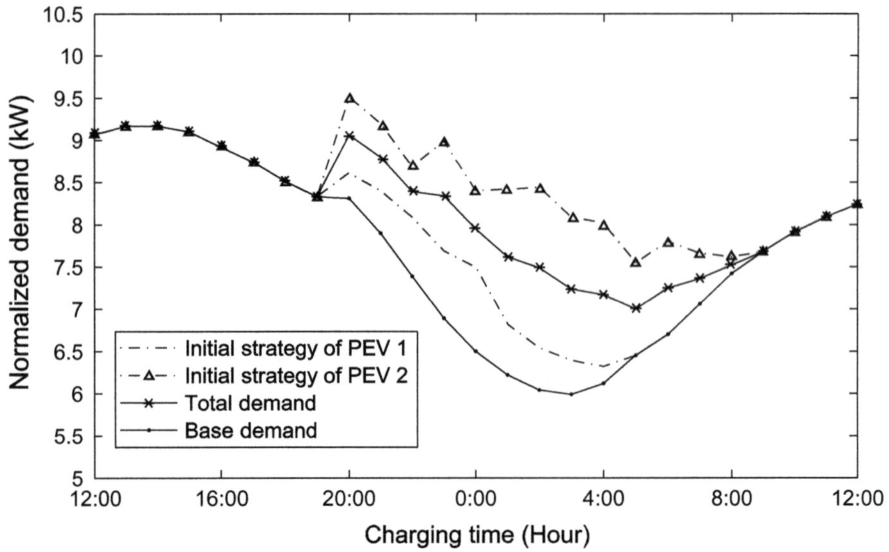

Fig. 2.6 An initial normalized average behavior of subpopulations

2.9.1 Implementation of Nash Equilibrium Under Algorithm 2.2

Firstly following the best strategy update procedure (S1)–(S3) of Algorithm 2.2, the 1st PEV subpopulation with respect to $\widehat{\boldsymbol{d}}_{-1} = \boldsymbol{d} + a_2\overline{\boldsymbol{u}}_2^0$. The solid blue curves in Fig. 2.7 illustrate the normalized best strategy update procedure of PEV 1 w.r.t. the total demand $\widehat{\boldsymbol{d}}_{-1}$. As demonstrated in Fig. 2.7, the aggregate behavior converges to a valley-filling strategy w.r.t. $\widehat{\boldsymbol{d}}_{-1}$, denoted by $\overline{\boldsymbol{u}}_1^*$, in about 20 iteration steps. This convergence procedure is consistent with the statement of Lemma 2.3.

Similarly, Fig. 2.8 illustrates the converged updates of normalized best strategy of the 2nd subpopulation.

In summary, after all of the subpopulations updated their best strategies, the system reaches the valley-filling Nash equilibrium.

2.9.2 Performance Comparisons

2.9.2.1 Example I

Figure 2.9 illustrates the implemented equilibrium for the charging problems specified in last section with the method proposed in Algorithm 2.1. It is obvious that

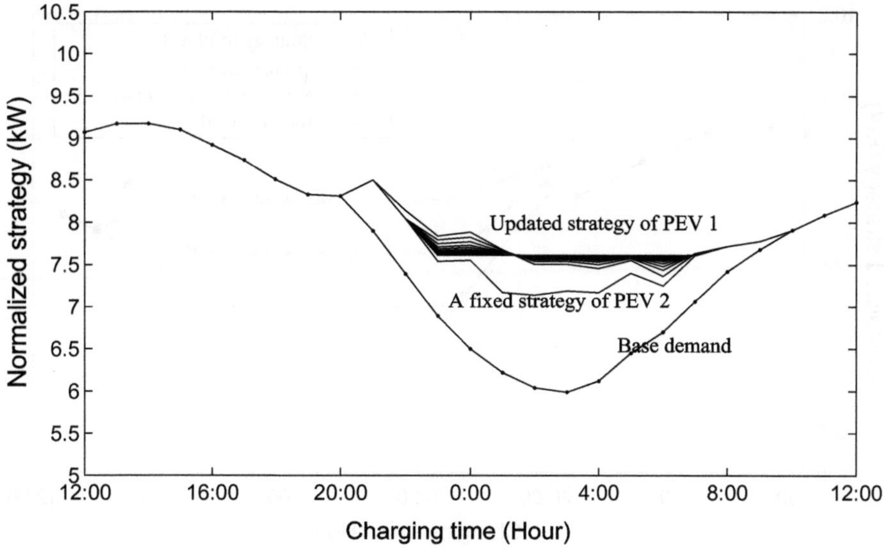

Fig. 2.7 Updates of best normalized strategy of PEV 1

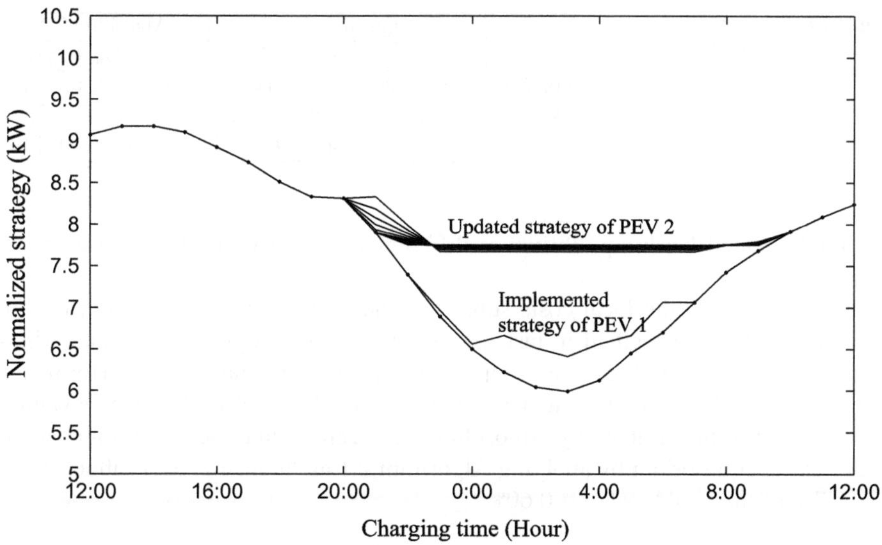

Fig. 2.8 Updates of best normalized strategy of PEV 2, $\alpha_2\overline{\boldsymbol{u}}_2^*$, w.r.t. total demand $\widehat{\boldsymbol{d}}_{-2}$, with $\widehat{\boldsymbol{d}}_{-2} = \boldsymbol{d} + \alpha_1\overline{\boldsymbol{u}}_1^*$

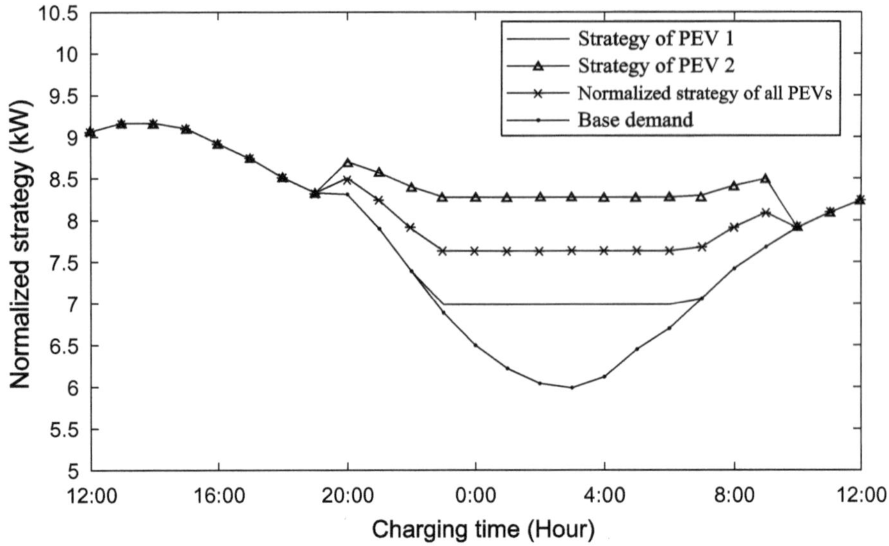

Fig. 2.9 Implemented Nash equilibrium by applying Algorithm 2.1

Fig. 2.10 Comparison between Nash equilibria

	Algorithm 2.1				Algorithm 2.2			
	J_n^p	J_n^d	$J_n^p + J_n^d$	Γ_n^*	J_n^p	J_n^d	$J_n^p + J_n^d$	Γ_n^*
PEV 1	0.54	0.068	0.608	10.85%	0.595	0	0.595	0%
PEV 2	2.25	0.068	2.318	2.91%	2.38	0	2.38	0%
Overall	1.40	0.068	1.468	4.64%	1.49	0	1.49	0%

the implemented Nash equilibrium is different from the valley-filling strategy as displayed in Fig. 2.8.

Figure 2.10 lists the local costs subject to the implemented equilibrium for the charging problems specified in last section by applying Algorithm 2.2 and Algorithm 2.1 respectively. It can be observed as expected that the overall performance with Algorithm 2.2 is better than that with Algorithm 2.1. Each of the agent deviation costs related to the strategy by Algorithm 2.2 is zero, while the deviation costs of PEVs at the equilibrium by applying Algorithm 2.1 are high, e.g. J_n^p is about 0.07$ or 11.7% of the total local cost 0.60$.

2.9.2.2 Example II

Figure 2.11 illustrates the implemented equilibria under Algorithm 2.1 and Algorithm 2.2 on a typical spring day respectively.

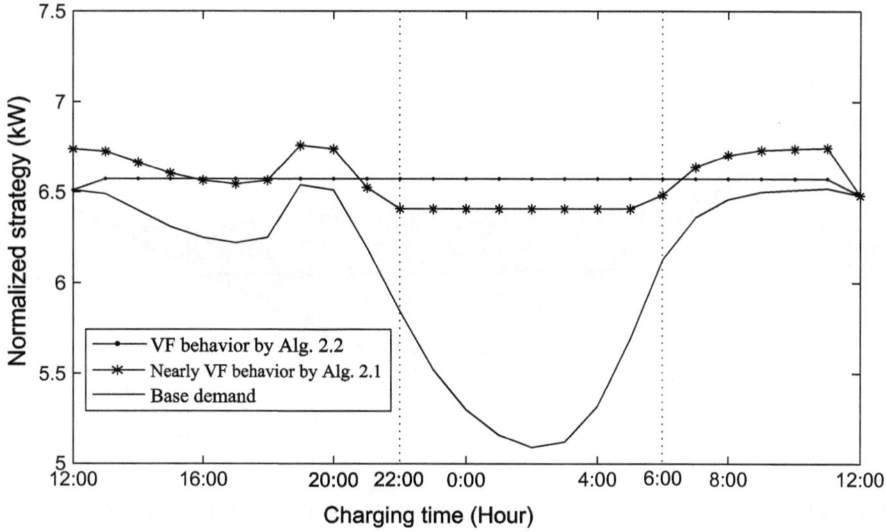

Fig. 2.11 Implemented equilibria of the game systems on a spring day by adapting different methods

2.9.2.3 Example III

Denote by J^*_{VF} the overall cost corresponding to the valley-filling equilibrium by adapting Algorithm 2.2; while J^*_{sub} represents the cost subject to the near valley-filling equilibrium by applying Algorithm 2.1. Figure 2.12 illustrates the values of $\frac{J^*_{\text{sub}}}{J^*_{\text{VF}}}$ on a pair of parameters (Γ_1, Γ_2) in the range of $[5, 40] \times [5, 40]$. As observed, the value of $\frac{J^*_{\text{sub}}}{J^*_{\text{VF}}}$ equals one in case Γ_1 and Γ_2 are identical with each other and increases as the difference between Γ_1 and Γ_2 increases.

2.10 Conclusions

This chapter introduces a class of decentralized charging coordination problems for large populations of PEVs. These problems are formulated as large-population games on a finite charging interval. It studies the existence, uniqueness and optimality of the Nash equilibrium of the charging problems. In particular, following the decentralized computational mechanism established in the chapter, it shows that, under certain mild conditions, the large-population charging games will converge to a unique Nash equilibrium which is either globally optimal (for homogeneous populations) or nearly globally optimal (for the heterogeneous case). These results are demonstrated with illustrative examples.

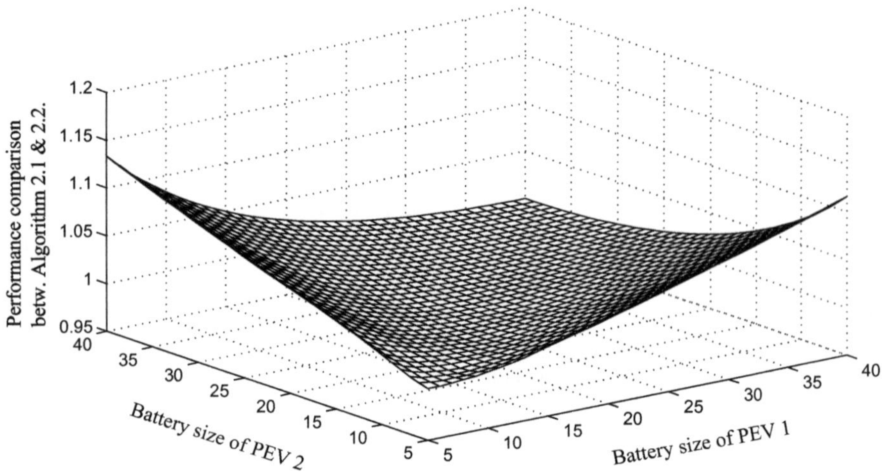

Fig. 2.12 Ratio between the system cost by adopting Algorithm 2.1 and that under Algorithm 2.2

These examples demonstrate that convergence to the Nash equilibrium occurs very quickly over a broad range of parameters. Therefore, the method may be particularly useful in applications where fully centralized coordination method is not possible, yet optimal or near-optimal charging patterns are essential to system operation. The results in this chapter will be important when PEV market penetration becomes sufficiently large that electricity demand patterns change significantly with PEV charging. This is because the algorithm allows users to choose their own locally optimal charging pattern while still achieving near-optimal global conditions. The strategy may improve PEV market penetration, especially relative to centralized strategies that could deter consumers who wish to independently determine their charging strategy.

To improve the performance, a revised method is proposed, such that a deviation cost from the average value of the homogeneous subpopulation is used, and each of homogeneous populations updates their strategies successively. By adopting the revised update mechanism, the system converges to the exact valley-filling equilibrium and the artificially introduced deviation cost vanishes at the equilibrium.

2.11 Appendices

2.11.1 Statement and Proof of Lemma 2.4

Lemma 2.4 *Control trajectories* $\mathbf{u}_n^*(\mathbf{z})$, $\mathbf{u}_n^*(\widehat{\mathbf{z}})$ *and* $\mathbf{v}_n(\mathbf{z}, \widehat{\mathbf{z}})$ *satisfy the following inequalities for all* $\delta > 0$*:*

$$|u_{nt}^*(\mathbf{z}) - v_{nt}(\mathbf{z}, \widehat{\mathbf{z}})| \le \left|(z_t - \widehat{z}_t) - \frac{1}{2\delta}\big(p(\mathfrak{r}_t) - p(\widehat{\mathfrak{r}}_t)\big)\right|, \quad \textit{for all } t \in \mathscr{T}, \tag{2.39}$$

$$|\mathbf{u}_n^*(\mathbf{z}) - \mathbf{u}_n^*(\widehat{\mathbf{z}})|_1 \le 2|\mathbf{u}_n^*(\mathbf{z}) - \mathbf{v}_n(\mathbf{z}, \widehat{\mathbf{z}})|_1, \tag{2.40}$$

where $\mathfrak{r}_t = \frac{1}{\Lambda}(\bar{d}_t + z_t)$, $\widehat{\mathfrak{r}}_t = \frac{1}{\Lambda}(\bar{d}_t + \widehat{z}_t)$, *and* $|\cdot|_1$ *denotes the* l_1 *norm of the associated vector.*

Proof For notational simplicity, $\mathbf{v_n} \equiv \mathbf{v_n}(\mathbf{z}, \widehat{\mathbf{z}})$ will be used throughout the proof.

Equation (2.39) *proof:* There are four cases to consider:

(i) $v_{nt} = u_{nt}^*(\mathbf{z}) = 0$. It follows immediately that $v_{nt} - u_{nt}^*(\mathbf{z}) = 0$.

(ii) $v_{nt} > 0$ and $u_{nt}^*(\mathbf{z}) = 0$. By (2.21), $v_{nt} > 0$ implies $v_{nt} = \frac{1}{2\delta}(A^*(\mathbf{z}) - p(\widehat{\mathfrak{r}}_t) + 2\delta\widehat{z}_t)$, and $u_{nt}^*(\mathbf{z}) = 0$ implies $A^*(\mathbf{z}) - p(\mathfrak{r}_t) + 2\delta z_t \le 0$. Together these give

$$0 < v_{nt} - u_{nt}^*(\mathbf{z}) \le \frac{1}{2\delta}(A^*(\mathbf{z}) - p(\widehat{\mathfrak{r}}_t) + 2\delta\widehat{z}_t) - \frac{1}{2\delta}(A^*(\mathbf{z}) - p(\mathfrak{r}_t) + 2\delta z_t)$$

with the last term equal to $(\widehat{z}_t - z_t) - \frac{1}{2\delta}(p(\widehat{\mathfrak{r}}_t) - p(\mathfrak{r}_t))$.

(iii) $v_{nt} = 0$ and $u_{nt}^*(\mathbf{z}) > 0$. Similarly to (ii), it can derive,

$$0 < u_{nt}^*(\mathbf{z}) - v_{nt} \le (z_t - \widehat{z}_t) - \frac{1}{2\delta}(p(\mathfrak{r}_t) - p(\widehat{\mathfrak{r}}_t)).$$

(iv) $v_{nt} > 0$ and $u_{nt}^*(\mathbf{z}) > 0$. By (2.21), it is straightforward to have

$$v_{nt} - u_{nt}^*(\mathbf{z}) = (\widehat{z}_t - z_t) - \frac{1}{2\delta}(p(\widehat{\mathfrak{r}}_t) - p(\mathfrak{r}_t)).$$

Equation (2.40) *proof:* There are three cases to consider:

(i) $\Sigma(\mathbf{v}_n) = \Sigma(\mathbf{u}_n^*(\mathbf{z}))$. This equality ensures $\mathbf{v}_n \in \mathscr{U}_n$. Also, charging strategy $\mathbf{v}_n$ has the form (2.21) with $A = A^*(\mathbf{z})$. Therefore, by Lemma 2.2, $\mathbf{v}_n$ is the local optimal strategy with respect to $\widehat{\mathbf{z}}$, and hence $\mathbf{u}_n^*(\widehat{\mathbf{z}}) = \mathbf{v}_n$. It follows that

$$|\mathbf{u}_n^*(\widehat{\mathbf{z}}) - \mathbf{u}_n^*(\mathbf{z})|_1 = |\mathbf{v}_n(\mathbf{z}, \widehat{\mathbf{z}}) - \mathbf{u}_n^*(\mathbf{z})|_1 \le 2|\mathbf{v}_n(\mathbf{z}, \widehat{\mathbf{z}}) - \mathbf{u}_n^*(\mathbf{z})|_1.$$

(ii) $\Sigma(\mathbf{v}_n) > \Sigma(\mathbf{u}_n^*(\mathbf{z}))$. By (2.4) it gets that $\Sigma(\mathbf{u}_n^*(\widehat{\mathbf{z}})) = \Sigma(\mathbf{u}_n^*(\mathbf{z})) = \Gamma_n$. Therefore $\Sigma(\mathbf{u}_n^*(\widehat{\mathbf{z}})) < \Sigma(\mathbf{v}_n)$ which, together with (2.21) and the definitions of $\mathbf{u}_n^*(\widehat{\mathbf{z}})$ and $\mathbf{v}_n$, implies,

$$A^*(\widehat{\mathbf{z}}) < A^*(\mathbf{z}), \text{ and } u_{nt}^*(\widehat{\mathbf{z}}) \le v_{nt}, \text{ for all } t.$$

Hence

$$\begin{aligned} 0 \le |\mathbf{v}_n - \mathbf{u}_n^*(\widehat{\mathbf{z}})|_1 &= \Sigma(\mathbf{v}_n) - \Sigma(\mathbf{u}_n^*(\widehat{\mathbf{z}})) \\ &= \Sigma(\mathbf{v}_n) - \Sigma(\mathbf{u}_n^*(\mathbf{z})) \le |\mathbf{v}_n - \mathbf{u}_n^*(\mathbf{z})|_1 \end{aligned}$$

with the last line a consequence of the triangle inequality for norms, taking into account that $\Sigma(\cdot) = |\cdot|_1$ for all valid charging trajectories. Then

$$|\mathbf{u}_n^*(\widehat{\mathbf{z}}) - \mathbf{u}_n^*(\mathbf{z})|_1 \leq |\mathbf{v}_n - \mathbf{u}_n^*(\mathbf{z})|_1 + |\mathbf{v}_n - \mathbf{u}_n^*(\widehat{\mathbf{z}})|_1 \leq 2|\mathbf{v}_n - \mathbf{u}_n^*(\mathbf{z})|_1. \quad (2.41)$$

(iii) $\Sigma(\mathbf{v}_n) < \Sigma(\mathbf{u}_n^*(\mathbf{z}))$. A similar argument to (ii) can be used to show that (2.41) holds in this case. ■

2.11.2 *Proof of Theorem 2.3*

Proof of Theorem 2.3. First notice that $\mathfrak{r}_t - \widehat{\mathfrak{r}}_t = \frac{1}{\Lambda}(z_t - \widehat{z}_t)$. Therefore,

$$\begin{aligned}|p(\mathfrak{r}_t) - p(\widehat{\mathfrak{r}}_t)| &\leq \max_{\mathfrak{r}\in[\mathfrak{r}_{min},\mathfrak{r}_{max}]} \frac{dp(\mathfrak{r})}{d\mathfrak{r}} \times |\mathfrak{r}_t - \widehat{\mathfrak{r}}_t| = \max_{\mathfrak{r}\in[\mathfrak{r}_{min},\mathfrak{r}_{max}]} \frac{dp(\mathfrak{r})}{d\mathfrak{r}} \times \frac{1}{\bar{\Lambda}}|z_t - \widehat{z}_t| \\ &\leq 2\delta|z_t - \widehat{z}_t|\end{aligned}$$

where the final inequality follows directly from (2.25). This result, together with a similar argument in terms of $\min_{\mathfrak{r}\in[\mathfrak{r}_{min},\mathfrak{r}_{max}]} \frac{dp(\mathfrak{r})}{d\mathfrak{r}}$ gives,

$$\frac{1}{2a}|z_t - \widehat{z}_t| \leq \frac{1}{2\delta}\big|p(\mathfrak{r}_t) - p(\widehat{\mathfrak{r}}_t)\big| \leq |z_t - \widehat{z}_t|. \quad (2.42)$$

Manipulation of (2.42) results in,

$$\left(1 - \frac{1}{2a}\right)|z_t - \widehat{z}_t| \geq |z_t - \widehat{z}_t| - \frac{1}{2\delta}\big|p(\mathfrak{r}_t) - p(\widehat{\mathfrak{r}}_t)\big| \geq 0. \quad (2.43)$$

Because $p(\mathfrak{r}_t)$ is strictly increasing with $\mathfrak{r}_t = \frac{1}{\Lambda}(\bar{d}_t + z_t)$, (2.43) can be rewritten

$$\left(1 - \frac{1}{2a}\right)|z_t - \widehat{z}_t| \geq \left|(z_t - \widehat{z}_t) - \frac{1}{2\delta}\big(p(\mathfrak{r}_t) - p(\widehat{\mathfrak{r}}_t)\big)\right| \geq 0.$$

This inequality, in conjunction with (2.39) and (2.40) of Lemma 2.4, gives

$$|\mathbf{u}_n^*(\mathbf{z}) - \mathbf{u}_n^*(\widehat{\mathbf{z}})|_1 \leq (2 - \frac{1}{a})|\mathbf{z} - \widehat{\mathbf{z}}|_1$$

and hence

$$|\overline{\mathbf{u}}^*(\mathbf{z}) - \overline{\mathbf{u}}^*(\widehat{\mathbf{z}})|_1 \leq (2 - \frac{1}{a})|\mathbf{z} - \widehat{\mathbf{z}}|_1.$$

Since $\frac{1}{2} < a < 1$, it follows that $\overline{\mathbf{u}}^*(\mathbf{z})$ is a contraction mapping with respect to $\mathbf{z}$. It may be concluded from the contraction mapping theorem [21] that the infinite population of PEVs possesses a unique fixed point which is the unique Nash equilibrium for the infinite population charging coordination system. ■

2.11.3 Proof of Theorem 2.4

Proof of Theorem 2.4. Consider any pair of time instants $t, s \in \mathscr{T}$, and denote by $\mathscr{U}_n(\{t,s\})$ the set of charging strategies u_{nt} and u_{ns} that satisfy $u_{nt}, u_{ns} \geq 0$ and $u_{nt} + u_{ns} \leq \Gamma_n$. Let

$$a = \frac{u_{ns} - u_{nt}}{2}, \qquad b = \frac{u_{ns} + u_{nt}}{2} \tag{2.44}$$

so that $u_{nt} = b - a$, and $u_{ns} = b + a$. It follows that $\mathscr{U}_n(\{t,s\})$ is equivalent to $\mathscr{S} \triangleq \{(a,b);\ \text{s.t. } a \in [-b,b],\ b \in [0, \Gamma_n/2]\}$.

It is proceeded by writing the minimum of the cost function (2.18) as a Bellman equation [22]. To do so, define

$$\begin{aligned} V_n(\Gamma', \mathscr{T}') = \min_{u_{n\hat{t}};\hat{t}\in\mathscr{T}'} & \left\{ \sum_{\hat{t}\in\mathscr{T}'} \left(p(\mathfrak{r}_{\hat{t}}) u_{n\hat{t}} + \delta (u_{n\hat{t}} - z_{\hat{t}})^2 \right) \right\} \\ \text{s.t.} \quad & u_{n\hat{t}} \leq 0 \quad \text{for all } \hat{t} \in \mathscr{T}' \\ & \sum_{\hat{t}\in\mathscr{T}'} u_{n\hat{t}} = \Gamma'. \end{aligned}$$

The minimum over the entire charging period $\mathscr{T}$ can then be written,

$$\begin{aligned} V_n(\Gamma_n, \mathscr{T}) = \min_{u_{nt}, u_{ns} \in \mathscr{U}_n(\{t,s\})} & \left\{ \sum_{\hat{t}\in\{t,s\}} \left(p(\mathfrak{r}_{\hat{t}}) u_{n\hat{t}} + \delta (u_{n\hat{t}} - z_{\hat{t}})^2 \right) \right. \\ & \left. + V_n\big(\Gamma_n - (u_{nt} + u_{ns}), \mathscr{T} \setminus \{t,s\}\big) \right\}. \end{aligned} \tag{2.45}$$

In terms of a and b defined at (2.44), this becomes

$$V_n(\Gamma_n, \mathscr{T}) = \min_{(a,b)\in\mathscr{S}} \left\{ 2\delta \left(a - \frac{1}{2}(z_s - z_t) + \frac{1}{4\delta}(p(\mathfrak{r}_s) - p(\mathfrak{r}_t)) \right)^2 + g(b) \right\} \tag{2.46}$$

where $g(b)$ is an expression in b that is unrelated to a. Let a_n^* and b_n^* denote the values of a and b associated with the optimal strategies u_{nt}^* and u_{ns}^*. Then by (2.46), a_n^* is a function of b_n^* that satisfies,

$$a_n^*(b_n^*) = \underset{a \in [-b_n^*, b_n^*]}{\operatorname{argmin}} \left\{(a - \zeta)^2\right\}, \tag{2.47}$$

with

$$\zeta \equiv \frac{1}{2}(z_s - z_t) - \frac{1}{4\delta}(p(\mathfrak{r}_s) - p(\mathfrak{r}_t)). \tag{2.48}$$

It follows from (2.47) that

$$0 < a_n^* \leq \zeta \quad \text{if } \zeta > 0 \tag{2.49a}$$
$$a_n^* = 0 \quad \text{if } \zeta = 0 \tag{2.49b}$$
$$\zeta \leq a_n^* < 0 \quad \text{if } \zeta < 0. \tag{2.49c}$$

(i.1), first part of (2.26a). This result can be shown by establishing a contradiction. Suppose there exist two time instants t and s, such that $\bar{d}_t \leq d_s$ and $z_t < z_s$, which implies $\mathfrak{r}_t < \mathfrak{r}_s$. Since $p(\mathfrak{r})$ is strictly increasing on $\mathfrak{r}$, $p(\mathfrak{r}_t) < p(\mathfrak{r}_s)$, and so from (2.48), $\zeta < \frac{1}{2}(z_s - z_t)$. It follows from (2.49) that $a_n^* < \frac{1}{2}(z_s - z_t)$. Hence, $u_{ns}^*(\mathbf{z}) - u_{nt}^*(\mathbf{z}) = 2a_n^* < z_s - z_t$, for all $n \in \mathscr{N}$, which implies $\overline{\mathbf{u}}_s^*(\mathbf{z}) - \overline{\mathbf{u}}_t^*(\mathbf{z}) < z_s - z_t$. However $\mathbf{u}^*$ is a Nash equilibrium, so $\overline{\mathbf{u}}^*(\mathbf{z}) = \mathbf{z}$, hence a contradiction.

(i.2), second part of (2.26a). Proof by contradiction is again used. Suppose there exist two time instants t and s, such that $\bar{d}_t + z_t > d_s + z_s$ when $d_t \leq d_s$. It follows that $p(\mathfrak{r}_t) > p(\mathfrak{r}_s)$, and so from (2.48), $\zeta > \frac{1}{2}(z_s - z_t)$. But from (i.1), $z_s - z_t \leq 0$, so it follows from (2.49) that $u_{ns}^*(\mathbf{z}) - u_{nt}^*(\mathbf{z}) = 2a_n^* > z_s - z_t$, for all $n \in \mathscr{N}$, which implies $\overline{\mathbf{u}}_s^*(\mathbf{z}) - \overline{\mathbf{u}}_t^*(\mathbf{z}) > z_s - z_t$. However $\mathbf{u}^*$ is a Nash equilibrium, so $\overline{\mathbf{u}}^*(\mathbf{z}) = \mathbf{z}$, hence a contradiction.

(i.3), third part of (2.26a). Again consider two time instants t and s, where $\bar{d}_t \leq d_s$. From (i.1) and (i.2), it obtains that $z_t \geq z_s$ and $p(\mathfrak{r}_t) \leq p(\mathfrak{r}_s)$ respectively. Therefore (2.48) implies $\zeta \leq 0$, so it may be concluded from (2.49) that $u_{ns}^* - u_{nt}^* = 2a_n^* \leq 0$. Hence $u_{nt}^* \geq u_{ns}^*$ as desired.

(ii.1), first part of (2.26b). Proof by contradiction will again be used to establish this result. Suppose there exist two time instants $t, s \in \widehat{\mathscr{T}}$ such that $d_s + z_s \neq \bar{d}_t + z_t$. Without lose of generality, assume $d_s + z_s = \bar{d}_t + z_t + \eta$, for some $\eta > 0$. Then there exist n and $\widehat{\eta} \geq \eta$, such that $d_s + u_{ns}^* = \bar{d}_t + u_{nt}^* + \widehat{\eta}$. By the definition of $\widehat{\mathscr{T}}$, $u_{ns}^* > 0$ for $s \in \widehat{\mathscr{T}}$ and all $n \in \mathscr{N}$. Therefore there exists a sufficiently small $\varepsilon > 0$ such that $u_{ns}^* - \varepsilon > 0$.

Consider a revised charging strategy $\mathbf{u}_n^\varepsilon$, with

$$u_{nt}^\varepsilon = u_{nt}^* + \varepsilon$$
$$u_{ns}^\varepsilon = u_{ns}^* - \varepsilon$$
$$u_{n\widehat{t}}^\varepsilon = u_{n\widehat{t}}^* \quad \text{for } \widehat{t} \in \mathscr{T} \setminus \{t, s\}.$$

For the cost function $J_n(\mathbf{u}_n; \mathbf{z})$ defined at (2.18), it follows that

$$J_n(\mathbf{u}_n^\varepsilon; \mathbf{z}) - J_n(\mathbf{u}_n^*; \mathbf{z}) = \varepsilon(p(\mathfrak{r}_t) - p(\mathfrak{r}_s)) + 2\delta\varepsilon((u_{nt}^* - u_{ns}^*) - (z_t - z_s)) + 2\delta\varepsilon^2.$$

Notice that $(u^*_{nt} - u^*_{ns}) - (z_t - z_s) = \eta - \widehat{\eta} \le 0$. Also, because $\bar{d}_t + z_t < d_s + z_s$ and $p(\mathfrak{r})$ is strictly increasing, $p(\mathfrak{r}_t) - p(\mathfrak{r}_s) < 0$. Therefore $J_n(\mathbf{u}^\varepsilon_n; \mathbf{z}) < J_n(\mathbf{u}^*_n; \mathbf{z})$ for sufficiently small $\varepsilon > 0$. However, $\mathbf{u}^*$ is a Nash equilibrium, and therefore minimizes $J_n(\mathbf{u}_n; \mathbf{z})$. Hence a contradiction.

(ii.2), second part of (2.26b). The total energy delivered to PEV n over the period $\widehat{\mathscr{T}}$ by the optimal charging strategy $\mathbf{u}^*$ is given by $\Sigma_{\widehat{t} \in \widehat{\mathscr{T}}} u^*_{n\widehat{t}} = \omega^*_n > 0$, for every $n \in \mathscr{N}$. Provided fixed energy ω^*_n is delivered over $\widehat{\mathscr{T}}$, variation of the trajectory $\{u_{n\widehat{t}}; \widehat{t} \in \widehat{\mathscr{T}}\}$ has no influence on the cost over the balance of the charging period $\mathscr{T} \setminus \widehat{\mathscr{T}}$. The optimal choice for $\{u_{n\widehat{t}}; \widehat{t} \in \widehat{\mathscr{T}}\}$ is therefore given by,

$$\min_{u_{n\widehat{t}}, \widehat{t} \in \widehat{\mathscr{T}}} \sum_{\widehat{t} \in \widehat{\mathscr{T}}} \left(p(\mathfrak{r}_{\widehat{t}}) u_{n\widehat{t}} + \delta (u_{n\widehat{t}} - z_{\widehat{t}})^2 \right) \tag{2.50}$$

$$\text{s.t.} \quad \Sigma_{\widehat{t} \in \widehat{\mathscr{T}}} u_{n\widehat{t}} = \omega^*_n. \tag{2.51}$$

According to (ii.1), $\bar{d}_t + z_t = d_s + z_s$, for all $t, s \in \widehat{\mathscr{T}}$. Therefore the electricity charging price $p(\mathfrak{r}_{\widehat{t}})$, with $\mathfrak{r}_{\widehat{t}} = \frac{1}{\Lambda}(\bar{d}_{\widehat{t}} + z_{\widehat{t}})$, is a constant p for all $\widehat{t} \in \widehat{\mathscr{T}}$. This allows the cost function (2.50) to be rewritten as $p\omega + \delta \sum_{\widehat{t} \in \widehat{\mathscr{T}}} (u_{n\widehat{t}} - z_{\widehat{t}})^2$, so the minimum cost can be found from,

$$\min_{u_{n\widehat{t}}, \widehat{t} \in \widehat{\mathscr{T}}} \sum_{\widehat{t} \in \widehat{\mathscr{T}}} (u_{n\widehat{t}} - z_{\widehat{t}})^2$$

subject to (2.51). Using Lagrange multipliers, optimality is achieved when all $u^*_{n\widehat{t}} - z_{\widehat{t}}, \widehat{t} \in \widehat{\mathscr{T}}$ are equal. In conjunction with (ii.1), this gives $\bar{d}_{\widehat{t}} + u^*_{n\widehat{t}} = \theta_n$ for all $\widehat{t} \in \widehat{\mathscr{T}}$. ■

References

1. P. Denholm, W. Short, An evaluation of utility system impacts and benefits of optimally dispatched plug-in hybrid electric vehicles. Technical Report NREL/TP-620-40293 (National Renewable Energy Laboratory, 2006)
2. S. Rahman, G.B. Shrestha, An investigation into the impact of electric vehicle load on the electric utility distribution system. IEEE Trans. Power Deliv. **8**(2), 591–597 (1993)
3. F. Koyanagi, Y. Uriu, Modeling power consumption by electric vehicles and its impact on power demand. Electr. Eng. Jpn. **120**(4), 40–47 (1997)
4. F. Koyanagi, T. Inuzuka, Y. Uriu, R. Yokoyama, Monte Carlo simulation on the demand impact by quick chargers for electric vehicles, in *Proceedings of the IEEE Power Engineering Society Summer Meeting*, vol. 2 (1999), pp. 1031–1036
5. D.S. Callaway, I.A. Hiskens, Achieving controllability of electric loads. Proc. IEEE **99**(1), 184–199 (2011)
6. T. Lee, Z. Bareket, T. Gordon, Stochastic modeling for studies of real-world PHEV usage: driving schedule and daily temporal distributions. IEEE Trans. Veh. Technol. **61**(4), 1493–1502 (2012)
7. N.P. Padhy, Unit commitment-a bibliographical survey. IEEE Trans. Power Syst. **19**(2), 1196–1205 (2004)

8. S. Borenstein, J.B. Bushnell, F.A. Wolak, Measuring market inefficiencies in California's restructured wholesale electricity market. Am. Econ. Rev. **92**(5), 1376–1405 (2002)
9. J.B. Bushnell, E.T. Mansur, C. Saravia, Vertical arrangements, market structure, and competition: an analysis of restructured US electricity markets. Am. Econ. Rev. **98**(1), 237–266 (2008)
10. B.E. Hobbs, Linear complementarity models of Nash-Cournot competition in bilateral and POOLCO power markets. IEEE Trans. Power Syst. **16**(2), 194–202 (2002)
11. L.B. Cunningham, R. Baldick, M.L. Baughman, An empirical study of applied game theory: transmission constrained Cournot behavior. IEEE Trans. Power Syst. **17**(1), 166–172 (2002)
12. E. Pettersen, *Managing end-user flexibility in electricity markets* (Fakultet for samfunnsvitenskap og teknologiledelse, Institutt for industriell økonomi og teknologiledelse, NTNU, 2004)
13. A.B. Philpott, E. Pettersen, Optimizing demand-side bids in day-ahead electricity markets. IEEE Trans. Power Syst. **21**(2), 488–498 (2006)
14. K.C. Border, *Fixed Point Theorems with Applications to Economics and Game Theory* (Cambridge University Press, Cambridge, UK, 1985)
15. Ettore Bompard, Yuchao Ma, Roberto Napoli, Graziano Abrate, The demand elasticity impacts on the strategic bidding behavior of the electricity producers. IEEE Trans. Power Syst. **22**(1), 188–197 (2007)
16. V.P. Gountis, A.G. Bakirtzis, Bidding strategies for electricity producers in a competitive electricity marketplace. IEEE Trans. Power Syst. **19**(1), 356–365 (2004)
17. F.S. Wen, A.K. David, Strategic bidding for electricity supply in a day-ahead energy market. Electr. Power Syst. Res. **59**, 197–206 (2001)
18. J.G. Wardrop, Some theoretical aspects of road traffic research, in *Proceedings of the Institution of Civil Engineers, Part 2*, vol. 1 (1952), pp. 325–378
19. M. Huang, P.E. Caines, R.P. Malhamé, Individual and mass behaviour in large population stochastic wireless power control problems: centralized and Nash equilibrium solutions, in *Proceedings of the 42th IEEE International Conference on Decision and Control* (Maui, Hawaii, 2003), pp. 98–103
20. M. Huang, P.E. Caines, R.P. Malhamé, Large-population cost-coupled LQG problems with non-uniform agents: individual-mass behaviour and decentralized epsilon-Nash equilibria. IEEE Trans. Autom. Control **52**(9), 1560–1571 (2007)
21. D.R. Smart, *Fixed Point Theorems* (Cambridge University Press, London, UK, 1974)
22. D.P. Bertsekas, *Dynamic Programming and Optimal Control*, vol. I (Athena Scientific, Nashua, 1995)

Chapter 3
Decentralized Charging Coordination with Battery Degradation Cost

Abstract In Chap. 2, the decentralized charging method has been designed to effectively coordinate the charging behaviors of large-scale PEVs, like the valley-filling strategy, to minimize their impacts on the power grid. However high charging rates under the valley-filling strategy may result in high battery degradation cost. Consequently in this chapter, it formulates a class of PEV charging coordination problems which deals with the tradeoff between total generation cost and the accumulated battery degradation cost of PEV populations. Due to the autonomy of individual PEVs and the computational complexity of the system with large-scale PEVs, it is impractical to implement the solution in a centralized way. Alternatively in this part a decentralized method is proposed such that all of the individual PEVs simultaneously update their own best charging behaviors with respect to a common electricity price curve, which is updated as the generation marginal cost with respect to the aggregated charging behaviors of the PEV populations implemented at last step. The iteration procedure terminates in case the price curve does not update any longer. It has been shown that, by applying the proposed decentralized method and under certain mild conditions, the system can converge to a unique charging strategy which is nearly socially optimal or efficient. Simulation examples are studied to illustrate the results developed in this chapter.

3.1 Introduction

It has been widely studied, e.g., [1, 2], that some key characteristics of chemical batteries, like the state of health, the growth of resistance, the cycle life, etc., are effected by the charging behaviors. As a consequence, this chapter addresses the need for a charging coordination scheme which considers the tradeoffs between system-wide economic efficiency, distribution-level limitations and battery degradation concerns.

Charging behavior affects key battery characteristics, including the state of health, the resistance impedance growth and the cycle life, which are all strongly related to the energy capacity of a battery [1, 2]. Intermittent charging may also shorten the battery lifespan [3]. Optimal charging strategies that take into account both the total energy cost and the battery state of health have been studied for single PEVs [1, 4].

Z. Ma, *Decentralized Charging Coordination of Large-scale Plug-in Electric Vehicles in Power Systems*, https://doi.org/10.1007/978-981-13-7652-8_3

These ideas form the basis for the extension, undertaken in this chapter, to large-scale coordination. It analyzes the battery degradation cost of the $LiFePO_4$ battery with respect to the charging power.

In this part, it analyzes the battery degradation cost of the $LiFePO_4$ battery with respect to the charging power. The system considers the tradeoff between the total generation cost and the local costs associated with overloading and battery degradation. Furthermore, consider the charging flexibility of PEVs. PEVs can independently determine their total charging capacity according to their own characteristics and the environment in which they are charged. The optimal charging strategies are analyzed as the parameter varies. Some research work, e.g., [1, 4], studied the optimal charging behavior for a single PEV by taking into account both total the energy cost and the state of health of batteries.

Decentralized charging coordination methods are proposed for large-scale PEVs dealing with the total generation cost and the accumulated battery degradation cost over multi-time intervals. As the $LiFePO_4$ battery, a lithium-ion type battery, has been widely applied in the PEV market, e.g., GM Chevrolet Spark, Nissan Leaf, BYD (e6, F3DM, F6DM), Renault Clio, etc., the health model for $LiFePO_4$ cell units was specified in [5] based upon the evolution of battery cell characteristics developed in [6, 7]. By adopting the analysis given in [5], it analyzes the battery degradation cost for the $LiFePO_4$ battery with respect to the charging power.

In general it is challenging to achieve an optimal or near-optimal outcome by coordinating large-scale PEVs in a decentralized way. As discussed in [8], it is difficult to effectively fill the nigh-time valley based upon the time-based or fixed price; then in this chapter it will adopt a real-time price model which has been widely applied in the literature, e.g., [9, 10] for demand response management, and [11–14] for electric vehicle charging coordination. More specifically, the electricity price at an instant is determined by the total demand at that instant and represents the generation marginal cost.

In the decentralized approach to charging coordination proposed in this part, participating PEVs simultaneously determine their optimal charging strategy with respect to an energy price forecast. These proposed charging strategies are used to estimate the total demand over the charging horizon. An updated price forecast is obtained as a weighted average of the previous price forecast and the generation marginal cost evaluated at this latest demand forecast. The revised price is (re)broadcast to the PEVs, and the process repeats. This scheme is formalized in Sects. 3.3 and 3.4, for different energy demand requirements, where it is shown that convergence is guaranteed under mild conditions. Upon convergence, the price profile is coincident with the generation marginal cost over the charging horizon. As a consequence, the resulting collection of PEV charging strategies is efficient (socially optimal). Moreover, convergence is obtained without the need for artificial deviation costs to damp oscillations, as in [15, 16]. Cost terms introduced to mitigate the effects of local demand peaks and battery degradation play the same role as congestion pricing used for traffic control in communication networks [17], which has been adopted in [13] to schedule PEV charging.

It is worth to note that, in the decentralized framework proposed in this part, each of the individual PEVs deals with the tradeoff between the battery degradation cost and the electricity cost; then the system can improve the longevity of the batteries and may mitigate the oscillation caused by the greedy behavior for the cheap electricity resource by individual PEVs. As a result, the deviation cost artificially introduced to mitigate the oscillation in [15, 16] is no longer required. The battery degradation cost plays the same role as the congestion pricing which has been widely used in Internet traffic control, see [17], and has been adopted in [13] to schedule PEV charging strategies in the power grid.

The chapter is organized as follows. In Sect. 3.2, it formulates a class of charging coordination problems which deals with the tradeoff between total electricity cost and the accumulated battery degradation cost of all of the PEVs over the charging intervals. A decentralized charging coordination algorithm is presented in Sect. 3.3 and the convergence of the proposed method is analyzed. This method is updated and extended in Sect. 3.4 to the charging coordination problems with flexible energy demands. Simulations, given in Sect. 3.5, illustrate various characteristics of the algorithm. Section 3.6 concludes this chapter and discusses ongoing research.

3.2 Formulation of Charging Coordination with Battery Degradation Cost

3.2.1 Admissible Charging Strategies

In this section, it considers the charging coordination of a large population of PEVs with most of their specifications given in Sect. 2.2.1, except that, instead of the common charging horizon $\mathcal{T}$, the charging horizon of PEV n is $\mathcal{T}_n$ such that $\mathcal{T}_n \subset \mathcal{T}$.

Remark: The parameter $\mathcal{T}_n$ is determined by external factors such as driving style and vehicle type [18].

Hence, by (2.3), a charging strategy $\boldsymbol{u}_n \equiv (u_{nt}; t \in \mathcal{T})$ is *admissible* if,

$$u_{nt} \begin{cases} \in [0, \Upsilon_n^+], & t \in \mathcal{T}_n \\ = 0, & t \in \mathcal{T} \setminus \mathcal{T}_n \end{cases}, \tag{3.1a}$$

$$\|\boldsymbol{u}_n\|_1 \equiv \sum_{t \in \mathcal{T}} u_{nt} = \frac{\Gamma_n}{\varsigma_n}(\mathrm{soc}_{n,\max} - soc_{n0}), \tag{3.1b}$$

with Γ_n, ς_n and $\mathrm{soc}_{n,\max}$ defined in Sect. 2.2.1.

Same as last chapter, the set of admissible charging strategies for PEV n, as specified in (3.1) above, is denoted by $\mathcal{U}_n$ as well.

Coordination of PEV charging across a large population has generally sought to minimize total generation cost over the charging horizon, see for example [15, 16, 19]. In contrast, the coordination strategies developed in this chapter seek to manage the tradeoff between total generation cost and local ones arising from high distribution-level demand and PEV battery degradation. These latter costs will now be discussed.

3.2.2 Analysis on Battery Degradation Cost

The $LiFePO_4$ battery, a lithium-ion type of battery, has been widely applied in PEVs, e.g., GM Chevrolet Spark, Nissan Leaf, BYD (e6, F3DM, F6DM), Renault Clio etc. Hence it will analyzes the battery degradation cost for the $LiFePO_4$ batteries with respect to the charging rates, based on the health model of $LiFePO_4$ cell units given in [5].

3.2.2.1 Evolution of Open Circuit Voltage of a Lithium-Ion Battery Cell w.r.t. Its SOC Value

It firstly reviews the relation between the open circuit voltage of a lithium-ion cell unit, denoted by V_{cell}, and its SOC value. As studied in the literature, e.g., [20, 21] and references therein, V_{cell} evolves with respect to its SOC value, as displayed in Fig. 3.1.

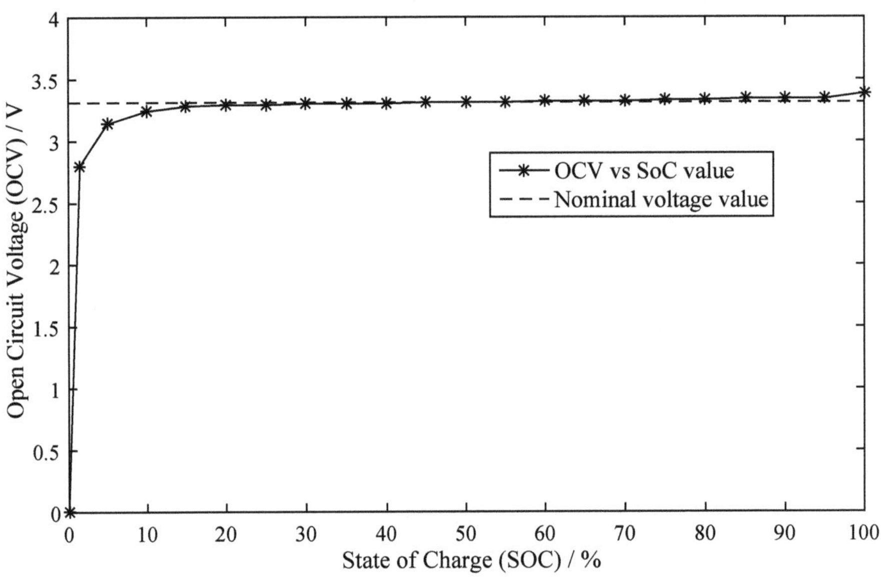

Fig. 3.1 Evolution of the open circuit voltage of $LiFePO_4$ battery cell with respect to the SOC value

More specifically, as studied in [4, 22], as the SOC value of the cell unit varies from 0% to a very small value, denoted by soc_{min} with $soc_{min} > 0$, V_{cell} increases rapidly from zero to its nominal value, denoted by V_{norm}, and stays near to V_{norm} until the SOC of the cell unit varies beyond a high level of SOC value, denoted by soc_{max}.

As a summary, for the lithium-ion type of cell units, V_{cell} approximately satisfies the following:

$$V_{cell} = V_{norm}, \quad \text{in case } SOC \in [soc_{min}, soc_{max}]. \tag{3.2}$$

Hence following the above analysis and as considered in the literature, e.g., [15, 16, 23], in this chapter, it is supposed that $SOC \in [soc_{min}, soc_{max}]$, i.e., the battery is neither allowed to be over discharged to any SOC value below soc_{min} nor allowed to be overcharged to any SOC value above soc_{max}.

3.2.2.2 Degradation Cost Modeling of $LiFePO_4$ Cell Units w.r.t. Charging Current and Voltage

In this work, it adopts the degradation modeling of the $LiFePO_4$ cell units, develops in [5], denoted by $h_{cell}(I, V)$, which measures the energy capacity loss per second (in $Amp \times Hour \times Sec^{-1}$) of a $LiFePO_4$ cell unit due to the resistance growth, see [1, 5], such that:

$$h_{cell} = a_1 + a_2 I + a_3 V + a_4 I^2 + a_5 V^2 + a_6 IV + a_7 V^3, \tag{3.3}$$

with parameters a_i, with $i = 1, \ldots, 7$, specified as -1.148×10^{-7}, 3.9984×10^{-8}, 1.3158×10^{-7}, 5.5487×10^{-10}, -4.968×10^{-8}, -1.1166×10^{-8}, 6.1675×10^{-9} respectively, see [5].

$V = V_{cell}$, i.e., the charging voltage V is nearly equal to the voltage of cell units; then by this together with (3.2), it can approximately set the charging voltage $V = V_{norm}$ in case $SOC \in [soc_{min}, soc_{max}]$. Moreover, by $u_{cell} = VI$ where u_{cell} represents the charging power on a cell unit, the charging current I satisfies the following:

$$I(u_{cell}) = u_{cell}/V_{norm}, \quad \text{in case } SOC \in [soc_{min}, soc_{max}].$$

Hence, based upon the above analysis, together with the battery degradation model specified in (3.3), it can obtain that, in case $SOC \in [soc_{min}, soc_{max}]$, the energy capacity degradation of a cell unit per second with respect to its charging power u_{cell}, denoted by $h_{cell}(u_{cell})$, is specified as:

$$h_{cell}(u_{cell}) = \mathfrak{a} u_{cell}^2 + \mathfrak{b} u_{cell} + \mathfrak{c}, \tag{3.4}$$

where $\mathfrak{a} = a_4/V_{norm}^2$, $\mathfrak{b} = a_2/V_{norm} + a_6$ and $\mathfrak{c} = a_1 + a_3 V_{norm} + a_5 V_{norm}^2 + a_7 V_{norm}^3$.

Denote by f_{cell} the *degradation cost* of a cell unit under a fixed charging power $u_{t,cell}$ over the charging interval t; then

$$f_{cell}(u_{t,cell}) = P_{cell} \cdot V_{norm} \cdot h_{cell}(u_{t,cell}) \cdot \Delta T, \tag{3.5}$$

where ΔT denotes the length of charging interval t (in seconds), and P_{cell} represents the price of a single energy unit of a battery cell.

Note: By (3.4) and (3.5), $f_{cell}(u_{cell})$ represents the monetary loss of a cell under a constant charging power u_{cell} over t.

Suppose that the battery stored in each of the PEVs is composed of a collection of identical cell units, and for the analytical simplicity, it is also supposed that all of the cell units share the same SOC values and are charged with a common charging power. Denote by Φ_n and Φ_{cell} the aggregated battery size of PEV n and the energy size of the common cell units in PEV n respectively; then the number of cell units in PEV n, denoted by M_n, is valued as $M_n = \Phi_n/\Phi_{cell}$. Under the charging power u_{nt} on the PEV n during interval t, the charging power on each of the cell units denoted by $u_{n,cell}$ is given as $u_{nt,cell} = 10^3 u_{nt}/M_n$ since u_{nt} represents the power rate in kW.

Denote by $f_n(u_{nt})$ the battery degradation cost of PEV n during interval t under a charging power u_{nt}; then the following holds

$$f_n(u_{nt}) = \mathfrak{a}_n u_{nt}^2 + \mathfrak{b}_n u_{nt} + \mathfrak{c}_n \tag{3.6}$$

with $\mathfrak{a}_n = \frac{10^6}{M_n} \cdot V_{norm} \cdot \Delta T \cdot P_{cell} \cdot \mathfrak{a}$, $\mathfrak{b}_n = 10^3 \cdot V_{norm} \cdot \Delta T \cdot P_{cell} \cdot \mathfrak{b}$, and $\mathfrak{c}_n = M_n \cdot V_{norm} \cdot \Delta T \cdot P_{cell} \cdot \mathfrak{c}$.

In [24], a quadratic form of $f_n(\cdot)$ is adopted as well.

Note: By (3.4) and (3.5), the degradation cost of the battery of PEV n, $f_n(u_{nt})$, as specified in (3.6), represents the total monetary losses of the battery package over interval t under the charging power u_{nt}.

For the purpose of demonstration, it is supposed that the battery package in the PEVs is composed of a collection of identical lithium battery cells, say ANR26650M1-B, from the A123 system, which has been widely applied in PEVs and grid stabilization energy storage systems. The nominal voltage and energy capacity of this type of cell units is 3.3 volts and 2.5 Ah (Amp × Hour) respectively, and the price of a single cell unit is about 15$.

As an example, it considers that the battery capacity in an individual PEV is 30 kWh; then it is straightforward to obtain that the battery is composed of about 3600 ANR26650M1-B units; then by (3.6), the degradation cost function, for the battery package specified above, is approximately given as below:

$$f_n(u_{nt}) = 0.004u_{nt}^2 + 0.075u_{nt} + 0.003. \tag{3.7}$$

Note: As discussed in Theorem 3.3, the best responses of individual PEVs are effected by the derivative of the degradation cost f_n, instead of f_n itself, such that $f_n'(u_{nt}) = 2\mathfrak{a}_n u_{nt} + \mathfrak{b}_n$ where the zero-order term $\mathfrak{c}_n$ is vanished. It implies that the small valued $\mathfrak{c}_n$ in (3.7) does not effect the results developed in this chapter.

3.3 Decentralized PEV Charging Coordination Method with Fixed Energy Demands

3.3.1 Charging Problems with Battery Degradation Costs

In this chapter, it is supposed that the system considers the tradeoff between the total electricity cost and the accumulated battery degradation cost formulated in Sect. 3.2.2 over the charging interval subject to the set of admissible charging strategies $\mathscr{U}$ of the PEV populations.

Denote, by $J(\boldsymbol{u})$, the system cost under $\boldsymbol{u}$ over the charging interval $\mathscr{T}$, such that

$$J(\boldsymbol{u}) \triangleq \sum_{t\in\mathscr{T}} \left\{ \sum_{n\in\mathscr{N}} f_n(u_{nt}) + \rho_t \cdot \left(d_t + \sum_{n\in\mathscr{N}} u_{nt} \right) \right\}, \tag{3.8}$$

where ρ_t represents the electricity price during interval t, and d_t denotes the aggregated inelastic base demand during instant t in the power grid.

In this part, a real time price model is adopted, such that

$$\rho_t = p(D_t(\boldsymbol{u}_t)), \quad \text{with } D_t(\boldsymbol{u}_t) \triangleq d_t + \sum_{n\in\mathscr{N}} u_{nt}, \tag{3.9}$$

i.e., the electricity price at instant t is determined by the total demand at this instant.

The real-time price model has been widely applied in the literature, e.g., [9, 10] for demand response management, and [11–13, 15, 16] for PEV charging coordination problems.

Note: The real-time price represents the generation marginal cost, and hence is the derivative of the generation cost on the supply. Suppose the electricity generation cost, denoted by $c(\cdot)$, in a quadratic form on the supply, say $c(D_t) = \frac{1}{2}\mathsf{a}D_t^2 + \mathsf{b}D_t + \mathsf{c}$, with properly valued parameters a, b and c, that has been widely considered, e.g., [25–27] and references therein; then the generation marginal cost evolves linearly with respect to the total demand, say $p_t(D_t) = \mathsf{a}D_t + \mathsf{b}$.

A class of centralized PEV charging coordination problems is formally formulated as follows:

Problem 3.1 $\min_{\boldsymbol{u}\in\mathscr{U}} \{J(\boldsymbol{u})\}$, that is to say, the objective of the PEV charging system is to implement a collection of socially optimal charging strategies, denoted by $\boldsymbol{u}^{**}$, by minimizing the system cost (3.8). ■

In this part, it considers the following assumption:

Assumption 3.1 $f_n(x)$, for all $n \in \mathscr{N}$, is monotonic increasing, strictly convex and differentiable on x. ■

Note: Assumption 3.1 is consistent with the battery degradation cost formulated in (3.7).

Lemma 3.1 *Under Assumption 3.1, the socially optimal strategy $\boldsymbol{u}^{**}$ satisfies the following properties:*

$$u_{nt} \geq u_{ns} \quad \text{and} \quad d_t + \sum_{m \in \mathscr{N}} u_{mt} \leq d_s + \sum_{m \in \mathscr{N}} u_{ms}, \tag{3.10}$$

in case $D_{-n,t} \leq D_{-n,s}$, for all $n \in \mathscr{N}$ and $t, s \in \mathscr{T}_n$, where $D_{-n,t} \equiv D_{-n,t}(\boldsymbol{u}) \triangleq d_t + \sum_{m \in \mathscr{N}/\{n\}} u_{mt}$, i.e., $\boldsymbol{D}_{-n}$ represents the aggregated demand trajectory composed of the base demand and the charging behaviors from all the others.

Proof The socially optimal charging strategy is implemented by applying the method of Lagrange multiplier, see [28]. Denote, by $\boldsymbol{\nu^{op}} \equiv (\nu_{nt}^{op}; n \in \mathscr{N}, t \in \mathscr{T}_n)$, $\boldsymbol{\lambda^{op}} \equiv (\lambda_{nt}^{op}; n \in \mathscr{N}, t \in \mathscr{T}_n)$ and $\boldsymbol{A^{op}} \equiv (A_n^{op}; n \in \mathscr{N})$, the Lagrange multipliers for the inequality constraints $\{u_{nt} \geq 0; n \in \mathscr{N}, t \in \mathscr{T}_n\}$, the inequality constraints $\{u_{nt} \leq \Upsilon_n^+; n \in \mathscr{N}, t \in \mathscr{T}_n\}$, and the equality constraints $\{\sum_{t \in \mathscr{T}} u_{nt} = \Gamma_n; n \in \mathscr{N}\}$, respectively; then the KKT conditions are given as below:

$$\nu_{nt}^{op} \geq 0,\ \nu_{nt}^{op} u_{nt} = 0,\ \forall n \in \mathscr{N}, \forall t \in \mathscr{T}_n, \tag{3.11a}$$

$$\lambda_{nt}^{op} \geq 0,\ \lambda_{nt}^{op}(u_{nt} - \Upsilon_n^+) = 0,\ \forall n \in \mathscr{N}, \forall t \in \mathscr{T}_n, \tag{3.11b}$$

$$[\nabla J(\boldsymbol{u})]_{nt} - \nu_{nt}^{op} + \lambda_{nt}^{op} - A_n^{op} = 0,\ \forall n \in \mathscr{N}, t \in \mathscr{T}_n, \tag{3.11c}$$

together with the inequality and equality constraints defined in (3.1), where $\nabla J(\boldsymbol{u})$ represents the gradient of $J(\boldsymbol{u})$ and $[\nabla J(\boldsymbol{u})]_{nt}$ denotes a component of $\nabla J(\boldsymbol{u})$, such that

$$[\nabla J(\boldsymbol{u})]_{nt} = f_n'(u_{nt}) + p'(D_t) \cdot D_t + p(D_t),$$

with $D_t \equiv d_t + \sum_{n \in \mathscr{N}} u_{nt}$. Equations (3.11a) and (3.11c) are equivalent with the following:

$$\eta(\boldsymbol{u}) \geq 0, \eta(\boldsymbol{u}) u_{nt} = 0,\ \forall n \in \mathscr{N}, \forall t \in \mathscr{T}_n, \tag{3.12}$$

with $\eta(\boldsymbol{u}) \equiv f_n'(u_{nt}) + \mathsf{a} D_t + p(D_t) + \lambda_{nt}^{op} - A_n^{op}$; then by (3.12) and (3.11b), it can derive that

$$\eta(\boldsymbol{u})\begin{cases}=0, & \text{in case } u_{nt}>0\\ \geq 0, & \text{in case } u_{nt}=0\end{cases}, \quad (3.13a)$$

$$u_{nt}\begin{cases}=\Upsilon_n^+, & \text{in case } \lambda_{nt}^{op}>0\\ \leq \Upsilon_n^+, & \text{in case } \lambda_{nt}^{op}=0\end{cases}. \quad (3.13b)$$

In the following the first part of (3.10) will be firstly verified using proof by contradiction.

Suppose that $u_{nt} < u_{ns}$ in case $D_{-n,t} \leq D_{-n,s}$, with $t, s \in \mathcal{T}_n$; then $D_{-n,t} + u_{nt} < D_{-n,s} + u_{ns}$. It implies that $p\left(D_{-n,t} + u_{nt}\right) < p\left(D_{-n,s} + u_{ns}^{**}\right)$. By (3.13), $u_{nt} \geq u_{ns}$, which is contradicted with $u_{nt} < u_{ns}$.

The second part of (3.10) can be shown following the same technique applied above. ■

Remark: By (3.10), it can be observed that the socially optimal charging coordination is different from the valley-filling strategy, denoted by $\boldsymbol{u}^{\text{vf}}$, defined below:

$$\sum_{n\in\mathcal{N}} u_{nt}^{\text{vf}}\begin{cases}= Np^{-1}(\nu) - d_t, \text{ in case } \sum\limits_{n\in\mathcal{N}} u_{nt}^{\text{vf}} < \sum\limits_{n\in\mathcal{N}} \Upsilon_n^+\\ = \sum_{n\in\mathcal{N}} \Upsilon_n^+, \quad \text{in case } p\left(D_t\right) \leq \nu\end{cases} \quad (3.14)$$

for all $t \in \mathcal{T}$, with $D_t = \sum_{n\in\mathcal{N}} u_{nt}^{\text{vf}} + d_t$.

In the following, as illustrated, with a numerical simulation, the difference between the optimal solution $\boldsymbol{u}^{**}$ for the underlying coordination problem and the valley-filling strategy $\boldsymbol{u}^{\text{vf}}$ since $\boldsymbol{u}^{**}$ deals with the tradeoff between the electricity cost and the aggregated battery degradation cost.

It specifies the charging coordination of a collection of PEVs with a population size of $N = 10^4$ over a charging interval from 12:00 AM on one day to 12:00 AM on the next day, and considers that the length of each of the charging intervals, denoted by ΔT, is 1 h. Figure 3.2 illustrates a typical non-PEV base demand in a summer season.

Consider a specific real-time electricity price, $p_t = \mathsf{a}D_t + \mathsf{b}$ (in \$/kWh), with $\mathsf{a} = 3.8 \times 10^{-7}$ and $\mathsf{b} = 0.06$, which corresponds to the marginal cost of a quadratic generation cost.

For the purpose of demonstration, here it is simply supposed that $\mathcal{T}_n = \mathcal{T}$ for all n, and all of the PEVs share an identical battery capacity size, common minimum and maximum SOCs which are equal to 27 kWh, 15% and 90% respectively, and it also considers that all of the PEVs share a common initial SOC value of 15%, and a common charger efficiency 90%; then the total required charging energy over $\mathcal{T}$ is specified as $\Gamma_n = \frac{\Phi_n}{\varsigma_n}(soc_{max} - soc_{n0})$, which is valued with 22.5 kWh, for all $n \in \mathcal{N}$.

Also suppose that all of the PEVs share a common degradation cost function specified in (3.7).

As illustrated in Fig. 3.2, the socially optimal charging strategy, which deals with the tradeoff between the electricity cost and the aggregated battery degradation cost, is

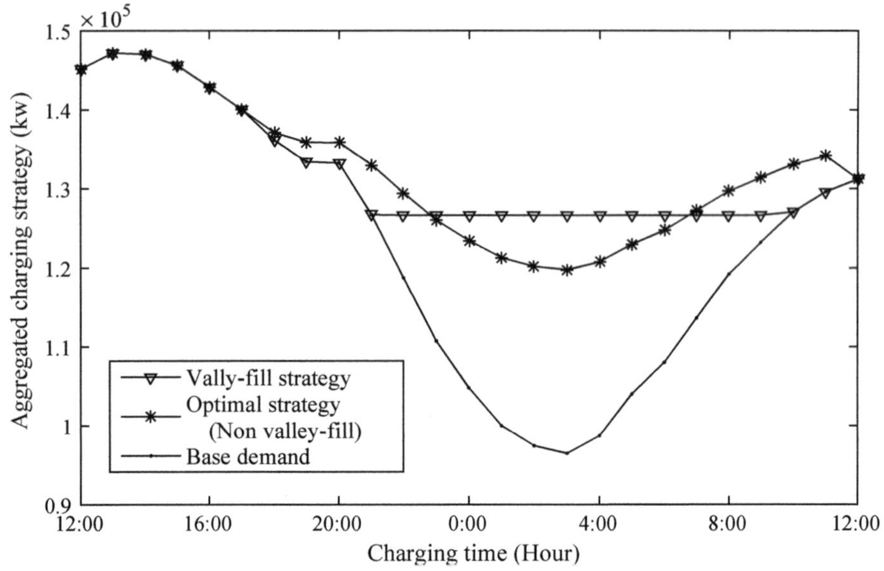

Fig. 3.2 Socially optimal strategy versus valley-filling strategy

different with the valley-filling strategy, since the valley-filling one may be penalized with high battery degradation cost.

The socially optimal charging strategy can be effectively implemented in the case that the system has complete information and has a permission to directly coordinate the behaviors of all of the individual PEVs. However, in practice the PEVs may not be willing to share their private information with others, and the transmission of complete information may create heavy communication signals. Hence the centralized coordination method is usually computationally infeasible.

3.3.2 Decentralized Charging Coordination Algorithm

It proposes a decentralized charging coordination method for the PEV populations such that each of the PEVs autonomously implements its own best charging strategy which deals with the tradeoff between the charging cost and the battery degradation cost over the charging intervals. More specifically, it firstly analyzes the best charging behaviors of PEVs with respect to a given fixed price curve in Sect. 3.3.2.1; then in Sect. 3.3.2.2, a decentralized best response update mechanism is proposed. In Sects. 3.3.2.3 and 3.3.2.4, it studies the convergence and performance of the implemented equilibrium under the proposed method.

3.3.2.1 Best Response of Individual PEVs w.r.t. Fixed Price Curve

Define an individual cost of PEV n, denoted by $J_n(\boldsymbol{u}_n; \boldsymbol{\rho})$, under a charging strategy $\boldsymbol{u}_n$ with a given fixed price curve $\boldsymbol{\rho} \equiv (\rho_t; t \in \mathscr{T})$, such that

$$J_n(\boldsymbol{u}_n; \boldsymbol{\rho}) \triangleq \sum_{t \in \mathscr{T}} \left\{ f_n(u_{nt}) + \rho_t u_{nt} \right\}, \tag{3.15}$$

i.e., PEV n implements its best charging strategy with respect to a presumed price curve $\boldsymbol{\rho}$.

Denote by $\boldsymbol{u}_n^*(\boldsymbol{\rho})$ the best charging strategy of PEV n minimizing the cost function defined in (3.15) with a given $\boldsymbol{\rho}$, that is $\boldsymbol{u}_n^*(\boldsymbol{\rho}) \triangleq \underset{\boldsymbol{u}_n \in \mathscr{U}_n}{\operatorname{argmin}} \{J_n(\boldsymbol{u}_n; \boldsymbol{\rho})\}$, which is determined in Lemma 3.2 below.

Lemma 3.2 *Under Assumption 3.1, the individual best strategy with respect to a given price curve $\boldsymbol{\rho}$ is specified as $\boldsymbol{u}_n(\boldsymbol{\rho}, A) \in \mathscr{U}_n$ such that*

$$u_{nt}(\boldsymbol{\rho}, A) = \begin{cases} \max \left\{0, \min \left\{ \Upsilon_n^+, [f_n']^{-1}(A - \rho_t) \right\} \right\}, & \text{if } t \in \mathscr{T}_n \\ 0, & \text{otherwise} \end{cases}, \tag{3.16}$$

where $[f_n']^{-1}$ represents an inverse operator of $f_n'(\cdot)$.

Proof The individual best charging strategy can be implemented with the method of Lagrange multiplier [28]. Denote, by $\boldsymbol{\nu} \equiv (\nu_t, t \in \mathscr{T}_n)$, $\boldsymbol{\lambda} \equiv (\lambda_t, t \in \mathscr{T}_n)$ and A, the Lagrange multipliers for the collection of inequality constraints $\{u_{nt} \geq 0; t \in \mathscr{T}_n\}$, the inequality constraints $\{u_{nt} \leq \Upsilon_n^+; t \in \mathscr{T}_n\}$, and the equality constraint $\sum_{t \in \mathscr{T}} u_{nt} = \Gamma_n$ respectively. The KKT conditions for the optimization problem are given by:

$$\nu_t \geq 0, \ \nu_t u_{nt} = 0, \ \forall t \in \mathscr{T}_n, \tag{3.17a}$$

$$\lambda_t \geq 0, \ \lambda_t(u_{nt} - \Upsilon_n^+) = 0, \ \forall t \in \mathscr{T}_n, \tag{3.17b}$$

$$[\nabla J_n(\boldsymbol{u}_n; \rho)]_t - \nu_t + \lambda_t - A = 0, \ \forall t \in \mathscr{T}_n, \tag{3.17c}$$

together with the inequality and equality constraints given in (3.1), where $\nabla J_n(\boldsymbol{u}_n; \rho)$ represents the gradient of $J_n(\boldsymbol{u}_n; \rho)$, and $[\nabla J_n(\boldsymbol{u}_n; \rho)]_t = f_n'(u_{nt}) + \rho_t$ denotes the tth component of $\nabla J_n(\boldsymbol{u}_n; \rho)$.

Equations (3.17a) and (3.17c) are equivalent with the following:

$$f_n'(u_{nt}) + \rho_t + \lambda_t - A \geq 0, (f_n'(u_{nt}) + \rho_t + \lambda_t - A)u_{nt} = 0, \tag{3.18}$$

for all $t \in \mathscr{T}_n$; then by (3.18) and (3.17b), it can derive that

$$f_n'(u_{nt}) + \rho_t + \lambda_t - A \begin{cases} = 0, & \text{in case } u_{nt} > 0 \\ \geq 0, & \text{in case } u_{nt} = 0 \end{cases}, \tag{3.19a}$$

$$u_{nt} \begin{cases} = \Upsilon_n^+, & \text{in case } \lambda_t > 0 \\ \leq \Upsilon_n^+, & \text{in case } \lambda_t = 0 \end{cases}, \tag{3.19b}$$

which is equivalent to the conclusion of (3.16).

Under Assumption 3.1, $J_n(\boldsymbol{u}_n; \boldsymbol{\rho})$ is convex with respect to $\boldsymbol{u}_n$; then the best strategy defined in (3.16) is unique. ■

The form of the dependence of $u_{nt}(\boldsymbol{\rho}, A)$ on A expressed in (3.16) ensures that, for any fixed $\boldsymbol{\rho}$,

- There exists an A_- such that for $A \leq A_-$, $\sum_{t \in \mathscr{T}} u_{nt}(\boldsymbol{\rho}, A) = 0$.
- For $A > A_-$, $\sum_{t \in \mathscr{T}} u_{nt}(\boldsymbol{\rho}, A)$ is strictly increasing and continuous on A. Hence, $\sum_{t \in \mathscr{T}} u_{nt}(\boldsymbol{\rho}, A)$ is invertible.

Therefore a constraint $\sum_{t \in \mathscr{T}} \boldsymbol{u}_{nt}(\boldsymbol{\rho}, A) = K > 0$ defines a unique $A > A_-$ for any fixed $\boldsymbol{\rho}$, which may be written as $A(\boldsymbol{\rho})$. The particular value of A, which ensures the constraint of $\sum_{t \in \mathscr{T}} u_{nt} = \Gamma_n$, shall be denoted by $A^*(\boldsymbol{\rho})$. The resulting charging trajectory can be written as $\boldsymbol{u}_n(\boldsymbol{\rho}, A^*(\boldsymbol{\rho})) = \boldsymbol{u}_n^*(\boldsymbol{\rho})$.

3.3.2.2 Decentralized Charging Coordination Method

In Sect. 3.3.2.1, it studies the best response of PEV n, denoted by $\boldsymbol{u}_n^*(\boldsymbol{\rho})$, with respect to a given price curve $\boldsymbol{\rho}$. By (3.9), the real-time electricity price is determined by the total demand. It implies that the presumed price curve has to satisfy the following fixed point equation:

$$\rho_t = p\left(d_t + \sum_{n \in \mathscr{N}} u_{nt}^*(\boldsymbol{\rho})\right), \qquad \text{for all } t \in \mathscr{T}. \tag{3.20}$$

Since it can not presumably know the solution to (3.20), it will specify an off-line decentralized iterative price curve update procedure prior to the charging interval $\mathscr{T}$ in Algorithm 3.1 later. Essentially each of the PEVs updates its best strategy by minimizing the cost function (3.15) with respect to a common price curve $\boldsymbol{\rho}$ which will be updated with $\widehat{\boldsymbol{\rho}}$, such that $\widehat{\boldsymbol{\rho}}_t = p\big(d_t + \sum_{n \in \mathscr{N}} u_{nt}^*(\boldsymbol{\rho})\big)$, for all $t \in \mathscr{T}$.

A collection of decentralized charging strategies is implemented in case the price curve update procedure converges. Before it formally proposes the decentralized charging coordination method in Algorithm 3.1 below, it firstly gives the following discussions.

Suppose that each of the PEVs is greedy for the cheap electricity as considered in the literature, e.g., [9, 16, 29]; then charging intervals with high price at one iteration tend to induce low price at the following iteration, and vice versa. This occurs because

PEVs move their charging requirements from expensive to inexpensive intervals; the resulting changes in demand reduce the generation marginal cost in the previously expensive intervals and raise the price in the previously inexpensive intervals. This results in an oscillatory pattern from one iteration to the next, preventing convergence to any price curve ρ. The deviation cost are artificially introduced to mitigate the oscillation, see [15, 16].

In this part, as specified in Sect. 3.3.2.1, each of individual PEVs deals with the tradeoff between battery degradation cost and electricity cost. As a consequence, the system can improve the longevity of the batteries and may mitigate the oscillation behavior caused by the greedy behaviors for the cheap electricity by individual PEVs.

Algorithm 3.1 (*Decentralized charging coordination method*)

- Set a presumed price curve $\boldsymbol{\rho}^{(0)} \equiv (\rho_t^{(0)}; t \in \mathscr{T})$;
- Set $k := 0$ and $\epsilon := \epsilon_0$ for some $\epsilon_0 > 0$;
- While $\epsilon > 0$

Obtain individual best response $\boldsymbol{u}_n^{*,(k+1)}$ w.r.t. $\boldsymbol{\rho}^{(k)}$, for all n, such that

$$\boldsymbol{u}_n^{*,(k+1)} := \underset{\boldsymbol{u}_n \in \mathscr{U}_n}{\text{argmin}} \sum_{t \in \mathscr{T}} \left\{ f_n(u_{nt}) + \rho_t^{(k)} u_{nt} \right\};$$

Set $\rho_t^{(k+1)} := p\Big(d_t + \sum_{n \in \mathscr{N}} u_{nt}^{*,(k+1)}\Big), \forall t \in \mathscr{T}$;
Update $\epsilon := ||\boldsymbol{\rho}^{(k+1)} - \boldsymbol{\rho}^{(k)}||_1$;
Update $k := k + 1$; ∎

Note: $||.||_1$ represents the l_1-norm of a vector.

3.3.2.3 Convergence of the Proposed Method

The sufficient conditions will be specified in Theorem 3.1, such that the convergence of the decentralized algorithm is guaranteed. Before that it will first introduce additional notions which will be used in Lemma 3.3 later.

Recall that $\boldsymbol{u}_n^*(\boldsymbol{\rho}) \triangleq \underset{\boldsymbol{u}_n \in \mathscr{U}_n}{\text{argmin}} \{J_n(\boldsymbol{u}_n; \boldsymbol{\rho})\}$, i.e., $\boldsymbol{u}_n^*(\boldsymbol{\rho})$ represents the charging strategy that minimizes the individual cost function (3.15) with respect to a given $\boldsymbol{\rho}$. Also by Lemma 3.2, $\boldsymbol{u}_n^*(\boldsymbol{\rho}) = \boldsymbol{u}_n(\boldsymbol{\rho}, A^*(\boldsymbol{\rho}))$. Here it defines another individual charging strategy $\boldsymbol{v}_n(\boldsymbol{\rho}, \widehat{\boldsymbol{\rho}})$ such that

$$v_{nt}(\boldsymbol{\rho}, \widehat{\boldsymbol{\rho}}) = \begin{cases} u_{nt}(\widehat{\boldsymbol{\rho}}, A^*(\boldsymbol{\rho})), & \text{with } t \in \mathscr{T}_n \\ 0, & \text{otherwise} \end{cases}.$$

Remark: The $\boldsymbol{v}_n(\boldsymbol{\rho}, \widehat{\boldsymbol{\rho}})$, specified above, describes an individual charging strategy satisfying (3.16) with respect to $\widehat{\boldsymbol{\rho}}$ and $A^*(\boldsymbol{\rho})$. There is no guarantee that $\boldsymbol{v}_n(\boldsymbol{\rho}, \widehat{\boldsymbol{\rho}})$ is an admissible charging strategy or $\Sigma(\boldsymbol{v}_n(\boldsymbol{\rho}, \widehat{\boldsymbol{\rho}})) = \Gamma_n$.

Lemma 3.3 *$\boldsymbol{u}_n^*(\boldsymbol{\rho})$ and $\boldsymbol{u}_n^*(\widehat{\boldsymbol{\rho}})$ satisfy the following inequality:*

$$||\boldsymbol{u}_n^*(\boldsymbol{\rho}) - \boldsymbol{u}_n^*(\widehat{\boldsymbol{\rho}})||_1 \\ \leq 2\sum_{t\in\mathcal{T}} \left|[f_n']^{-1}(A^*(\boldsymbol{\rho}) - \rho_t) - [f_n']^{-1}(A^*(\boldsymbol{\rho}) - \widehat{\rho}_t)\right|. \tag{3.21}$$

Proof (3.21) will be verified in (I) and (II) below.

(I) First it will show that the following holds, for all $t \in \mathcal{T}$,

$$|u_{nt}^*(\boldsymbol{\rho}) - v_{nt}(\boldsymbol{\rho}, \widehat{\boldsymbol{\rho}})| \\ \leq \left|[f_n']^{-1}(A^*(\boldsymbol{\rho}) - \rho_t) - [f_n']^{-1}(A^*(\boldsymbol{\rho}) - \widehat{\rho}_t)\right|. \tag{3.22}$$

For notational simplicity, it considers $\boldsymbol{v_n} \equiv \boldsymbol{v_n}(\boldsymbol{\rho}, \widehat{\boldsymbol{\rho}})$.
It is obvious that (3.22) holds for all $t \notin \mathcal{T}_n$, since by the specification of $u_{nt}^*(\boldsymbol{\rho})$ and $v_{nt}(\boldsymbol{\rho}, \widehat{\boldsymbol{\rho}})$, at each of these instants, $u_{nt}^*(\boldsymbol{\rho}) = v_{nt}(\boldsymbol{\rho}, \widehat{\boldsymbol{\rho}}) = 0$.

Show that (3.22) holds for all $t \in \mathcal{T}_n$ in (I.a)–(I.d) below:

(I.a) $v_{nt} = u_{nt}^*(\boldsymbol{\rho}) = 0$. It follows immediately that $v_{nt} - u_{nt}^*(\boldsymbol{\rho}) = 0$.
(I.b) $v_{nt} > 0$ and $u_{nt}^*(\boldsymbol{\rho}) = 0$. By (3.16), $v_{nt} > 0$ implies $[f_n']^{-1}(A^*(\boldsymbol{\rho}) - \widehat{\rho}_t) > 0$, and $u_{nt}^*(\boldsymbol{\rho}) = 0$ implies $[f_n']^{-1}(A^*(\boldsymbol{\rho}) - \rho_t) \leq 0$; then

$$0 < v_{nt} - u_{nt}^*(\boldsymbol{\rho}) \leq [f_n']^{-1}(A^*(\boldsymbol{\rho}) - \widehat{\rho}_t) - [f_n']^{-1}(A^*(\boldsymbol{\rho}) - \rho_t).$$

(I.c) $v_{nt} = 0$ and $u_{nt}^*(\boldsymbol{\rho}) > 0$. Similar to (I.b), it can derive,

$$0 < u_{nt}^*(\boldsymbol{\rho}) - v_{nt} \leq [f_n']^{-1}(A^*(\boldsymbol{\rho}) - \rho_t) - [f_n']^{-1}(A^*(\boldsymbol{\rho}) - \widehat{\rho}_t).$$

(I.d) $v_{nt} > 0$ and $u_{nt}^*(\boldsymbol{\rho}) > 0$. By (3.16),

$$|v_{nt} - u_{nt}^*(\boldsymbol{\rho})| \leq \left|[f_n']^{-1}(A^*(\boldsymbol{\rho}) - \widehat{\rho}_t) - [f_n']^{-1}(A^*(\boldsymbol{\rho}) - \rho_t)\right|.$$

(II) Verify the following inequality in (II.a)–(II.c)

$$||\boldsymbol{u}_n^*(\boldsymbol{\rho}) - \boldsymbol{u}_n^*(\widehat{\boldsymbol{\rho}})||_1 \leq 2||\boldsymbol{u}_n^*(\boldsymbol{\rho}) - \boldsymbol{v}_n(\boldsymbol{\rho}, \widehat{\boldsymbol{\rho}})||_1.$$

For notational simplicity, in (II.a)–(II.c) below, it considers $\Sigma(\boldsymbol{u}_n) \equiv \sum_{t\in\mathcal{T}} u_{nt}$.

(II.a) $\Sigma(\boldsymbol{v}_n) = \Sigma(\boldsymbol{u}_n^*(\boldsymbol{\rho}))$. This equality ensures $\boldsymbol{v}_n \in \mathcal{U}_n(\Upsilon_n^+)$. Also, the strategy $\boldsymbol{v}_n$ has the form (3.16) with $A = A^*(\boldsymbol{\rho})$. Therefore, by Lemma 3.2, $\boldsymbol{v}_n$ is the individual best strategy with respect to $\widehat{\boldsymbol{\rho}}$; then $\boldsymbol{u}_n^*(\widehat{\boldsymbol{\rho}}) = \boldsymbol{v}_n$. Hence

$$||\boldsymbol{u}_n^*(\widehat{\boldsymbol{\rho}}) - \boldsymbol{u}_n^*(\boldsymbol{\rho})||_1 = ||\boldsymbol{v}_n(\boldsymbol{\rho}, \widehat{\boldsymbol{\rho}}) - \boldsymbol{u}_n^*(\boldsymbol{\rho})||_1 \\ \leq 2||\boldsymbol{v}_n(\boldsymbol{\rho}, \widehat{\boldsymbol{\rho}}) - \boldsymbol{u}_n^*(\boldsymbol{\rho})||_1.$$

(II.b) $\Sigma(\boldsymbol{v}_n) > \Sigma(\boldsymbol{u}_n^*(\boldsymbol{\rho}))$. By (3.1), $\Sigma(\boldsymbol{u}_n^*(\widehat{\boldsymbol{\rho}})) = \Sigma(\boldsymbol{u}_n^*(\boldsymbol{\rho})) = \Upsilon_n^+$. Therefore $\Sigma(\boldsymbol{u}_n^*(\widehat{\boldsymbol{\rho}})) < \Sigma(\boldsymbol{v}_n)$, together with (3.16) and the definitions of $\boldsymbol{u}_n^*(\widehat{\boldsymbol{\rho}})$ and $\boldsymbol{v}_n$, implies, $A^*(\widehat{\boldsymbol{\rho}}) < A^*(\boldsymbol{\rho})$, and $u_{nt}^*(\widehat{\boldsymbol{\rho}}) \leq v_{nt}$, for all t. Hence

$$\begin{aligned}0 \leq &||\boldsymbol{v}_n - \boldsymbol{u}_n^*(\widehat{\boldsymbol{\rho}})||_1 = \Sigma(\boldsymbol{v}_n) - \Sigma(\boldsymbol{u}_n^*(\widehat{\boldsymbol{\rho}}))\\ =&\Sigma(\boldsymbol{v}_n) - \Sigma(\boldsymbol{u}_n^*(\boldsymbol{\rho})) \leq ||\boldsymbol{v}_n - \boldsymbol{u}_n^*(\boldsymbol{\rho})||_1\end{aligned}$$

where the last inequality is a consequence of the triangle inequality for norms, taking into account that $\Sigma(\cdot) = ||\cdot||_1$ for all admissible non-negative charging trajectories; then

$$\begin{aligned}||\boldsymbol{u}_n^*(\widehat{\boldsymbol{\rho}}) - \boldsymbol{u}_n^*(\boldsymbol{\rho})||_1 \leq &||\boldsymbol{v}_n - \boldsymbol{u}_n^*(\boldsymbol{\rho})||_1 + ||\boldsymbol{v}_n - \boldsymbol{u}_n^*(\widehat{\boldsymbol{\rho}})||_1\\ \leq &2|\boldsymbol{v}_n - \boldsymbol{u}_n^*(\boldsymbol{\rho})|_1.\end{aligned} \tag{3.23}$$

(II.c) $\Sigma(\boldsymbol{v}_n) < \Sigma(\boldsymbol{u}_n^*(\boldsymbol{\rho}))$. A similar argument to (II.b) can be applied to show that (3.23) holds in this case.

The conclusion of (3.21) can be obtained following (I) and (II). ■

Based upon Lemma 3.3, it will show, in Theorem 3.1 below, that by adopting the method proposed in Algorithm 3.1, the system converges to a unique collection of strategies under certain mild conditions.

Theorem 3.1 (Convergence of Algorithm) *Suppose that the generation marginal cost $p(.)$ is strictly increasing and consider Assumption 3.1, such that*

$$|f_n'(x) - f_n'(\widehat{x})| \geq \xi|x - \widehat{x}|, \; |p(D) - p(\widehat{D})| \leq \eta|D - \widehat{D}|, \tag{3.24}$$

with $\xi > 2\eta N > 0$; then, by adopting Algorithm 3.1, the system converges to a unique collection of charging strategies.

Proof Under Assumption 3.1, $[f_n']^{-1}$, the inverse of f_n', exists; moreover by adopting $|f_n'(x) - f_n'(\widehat{x})| \geq \xi|x - \widehat{x}|$, the following holds $|[f_n']^{-1}(z) - [f_n']^{-1}(\widehat{z})| \leq \frac{1}{\xi}|z - \widehat{z}|$; then it can obtain

$$\begin{aligned}&\left|[f_n']^{-1}(A^*(\boldsymbol{\rho}) - \rho_t) - [f_n']^{-1}(A^*(\boldsymbol{\rho}) - \widehat{\rho}_t)\right|\\ \leq &|(A^*(\boldsymbol{\rho}) - \rho_t) - (A^*(\boldsymbol{\rho}) - \widehat{\rho}_t)|/\xi = |\rho_t - \widehat{\rho}_t|/\xi;\end{aligned} \tag{3.25}$$

by which together with (3.21),

$$||\boldsymbol{u}_n^*(\boldsymbol{\rho}) - \boldsymbol{u}_n^*(\widehat{\boldsymbol{\rho}})||_1 \leq 2||\boldsymbol{\rho} - \widehat{\boldsymbol{\rho}}||_1/\xi. \tag{3.26}$$

Also by (3.24), it obtains that

$$\begin{aligned}&\left|p\Big(d_t + \sum_{n\in\mathcal{N}} u^*_{nt}(\boldsymbol{\rho})\Big) - p\Big(d_t + \sum_{n\in\mathcal{N}} \boldsymbol{u}^*_{nt}(\widehat{\boldsymbol{\rho}})\Big)\right| \\ \le&\eta \sum_{n\in\mathcal{N}} \left|u^*_{nt}(\boldsymbol{\rho}) - u^*_{nt}(\widehat{\boldsymbol{\rho}})\right| \end{aligned} \tag{3.27}$$

Denote by $\boldsymbol{p}(\boldsymbol{\rho}) \equiv (p_t(\boldsymbol{\rho}); t \in \mathcal{T})$ with $p_t(\boldsymbol{\rho}) \equiv p\big(d_t + \sum_{n\in\mathcal{N}} u^*_{nt}(\boldsymbol{\rho})\big)$; then by (3.27),

$$||\boldsymbol{p}(\boldsymbol{\rho}) - \boldsymbol{p}(\widehat{\boldsymbol{\rho}})||_1 \le \eta \sum_{n\in\mathcal{N}} ||\boldsymbol{u}^*_n(\boldsymbol{\rho}) - \boldsymbol{u}^*_n(\widehat{\boldsymbol{\rho}})||_1. \tag{3.28}$$

Hence by (3.26) and (3.28),

$$||\boldsymbol{p}(\boldsymbol{\rho}) - \boldsymbol{p}(\widehat{\boldsymbol{\rho}})||_1 \le 2\eta N ||\boldsymbol{\rho} - \widehat{\boldsymbol{\rho}}||_1/\xi; \tag{3.29}$$

then by applying the contraction mapping theorem, [30], the system can converge to a unique price curve $\boldsymbol{\rho}^*$. In summary, by applying the proposed decentralized algorithm, the system converges to a unique collection of charging strategies $(\boldsymbol{u}^*_n(\boldsymbol{\rho}^*); n \in \mathcal{N})$ under the constraints of (3.24). ∎

3.3.2.4 Performance of the Proposed Method

Theorem 3.2 (Performance of Implemented Strategy) *Suppose that the generation marginal cost $p(\cdot)$ is strictly increasing, and $\boldsymbol{u}^* \equiv \{\boldsymbol{u}^*_n; n \in \mathcal{N}\}$ is the implemented strategy by applying Algorithm 3.1; then $\boldsymbol{u}^*$ satisfies (3.10) under Assumption 3.1.*

Remark. By Lemma 3.1, *Theorems* 3.1 *and* 3.2, *it can obtain that, by applying the proposed method, the system can converge to a collection of unique charging strategies which is nearly socially optimal.*

Proof of Theorem 3.2. Consider a collection of strategies $\boldsymbol{u}^\dagger \equiv (\boldsymbol{u}^\dagger_n; n \in \mathcal{N})$, and denote by $\boldsymbol{\rho}(\boldsymbol{u}^\dagger)$ the updated price curve under $\boldsymbol{u}^\dagger$; then $\rho_t = p\Big(d_t + \sum_{m\in\mathcal{N}} u^\dagger_{mt}\Big)$. Hence, at the convergence, the updated best strategy with respect to the price curve $\boldsymbol{\rho}(\boldsymbol{u}^\dagger)$, denoted by $\boldsymbol{u}^* \equiv \boldsymbol{u}^*(\boldsymbol{\rho}(\boldsymbol{u}^\dagger))$, satisfies the following:

$$\boldsymbol{u}^*_n = \operatorname*{argmin}_{u_n\in\mathcal{U}_n} \sum_{t\in\mathcal{T}} \Big\{ f_n(u_{nt}) + \rho_t \cdot u_{nt} \Big\}, \quad \text{and} \quad \boldsymbol{u}^*_n = \boldsymbol{u}^\dagger_n.$$

(i). Proof of the 1st part of (3.10): The proof by contradiction will be used. Suppose that $u^\dagger_{nt} < u^\dagger_{ns}$ in case $D^\dagger_{-n,t} \le D^\dagger_{-n,s}$, with $t, s \in \mathcal{T}_n$, where $D^\dagger_{-n,t} \equiv d_t + \sum_{m\in\mathcal{N}/\{n\}} u^\dagger_{mt}$; then $D^\dagger_{-n,t} + u^\dagger_{nt} < D^\dagger_{-n,s} + u^\dagger_{ns}$; hence $p(D^\dagger_{-n,t} + u^\dagger_{nt}) < p(D^\dagger_{-n,s} + u^\dagger_{ns})$. By (3.19) in Lemma 3.2, $u^*_{nt} \ge u^*_{ns}$ which, together with the assumed $u^\dagger_{nt} < u^\dagger_{ns}$, implies $\boldsymbol{u}^*_n \ne \boldsymbol{u}^\dagger_n$ which is contradicted with $\boldsymbol{u}^*_n = \boldsymbol{u}^\dagger_n$ since the $\boldsymbol{u}^*_n$ is assumed to be an implemented strategy of PEV n.

(ii). Proof of the 2nd part of (3.10): The proof by contradiction will be used again. Suppose that $d_t + \sum_{n\in\mathcal{N}} u^{\dagger}_{nt} > d_s + \sum_{n\in\mathcal{N}} u^{\dagger}_{ns}$ in case $D^{\dagger}_{-n,t} \leq D^{\dagger}_{-n,s}$, with $t, s \in \mathcal{T}_n$; then it implies that $p(d_t + \sum_{n\in\mathcal{N}} u^{\dagger}_{nt}) > p(d_s + \sum_{n\in\mathcal{N}} u^{\dagger}_{ns})$. By (3.19) in Lemma 3.2, $u^*_{nt} < u^*_{ns}$ which is contradicted with the 1st part of (3.10) verified in *(i)*. ■

In Corollary 3.1 below, it will study the performance of the implemented strategy $\boldsymbol{u}^*$ in case $\mathcal{T}_n = \mathcal{T}$ for all n.

Corollary 3.1 *Consider $\mathcal{T}_n = \mathcal{T}$ for all n; then*

$$u^*_{nt} \geq u^*_{ns}, \quad and \; d_t + \sum_{n\in\mathcal{N}} u^*_{nt} \leq d_s + \sum_{n\in\mathcal{N}} u^*_{ns}, \tag{3.30}$$

in case $d_t \leq d_s$, for all $n \in \mathcal{N}$ and for any pair of distinct instants $t, s \in \mathcal{T}$.

Proof by contradiction. Suppose that there exists a PEV n satisfies $u^*_{nt} < u^*_{ns}$; then by (3.19) in Lemma 3.2 and the convexity of f_n, $p(D_t) > p(D_s)$, by which together with the increasing property of $p(D)$, it obtains that $d_t + \sum_{n\in\mathcal{N}} u^*_{nt} > d_s + \sum_{n\in\mathcal{N}} u^*_{ns}$. Hence there exists an $m \in \mathcal{N}/\{n\}$, such that $u^*_{mt} > u^*_{ms}$. By (3.19) in Lemma 3.2, $p(D_t) < p(D_s)$ which is contradicted with $p(D_t) > p(D_s)$ in case $d_t \leq d_s$. By $u^*_{nt} \geq u^*_{ns}$, $\forall n \in \mathcal{N}$, and (3.19) stated in Lemma 3.2, $d_t + \sum_{n\in\mathcal{N}} u^*_{nt} \leq d_s + \sum_{n\in\mathcal{N}} u^*_{ns}$. ■

3.3.3 Numerical Examples

In this section, unless specified, it adopts the parameters considered in the example in Sect. 3.3.1.

Consider the price $p(\cdot)$ (in \$/kWh) specified in Sect. 3.3.1, $|p(D) - p(\widehat{D})| = \mathsf{a}|D - \widehat{D}|$, with $\mathsf{a} = 3.8 \times 10^{-7}$. Also by the battery degradation cost f_n given in (3.7) for the PEVs specified in Sect. 3.3.1, it obtains that $|f'_n(x_1) - f'_n(x_2)| = 2\alpha|x_1 - x_2|$, with $\alpha = 0.004$; then it is straightforward to verify that f_n and p_t satisfy (3.24).

As illustrated in Fig. 3.3, the system converges to a nearly socially optimal charging strategy by adopting the method proposed in Algorithm 3.1. This is consistent with Theorem 3.1 and Corollary 3.1. Moreover it can be observed that the performance under the decentralized method is better than that under the valley-filling strategy.

The iteration updates of the associated price curve are displayed in Fig. 3.4.

The above simulation verifies the results developed in this chapter. Nevertheless the system may still converge in case the sufficient condition (3.24) is not satisfied. For example, suppose that the battery degradation cost for some other PEVs is specified as $f_n = 0.0027u^2_{nt} + 0.05u_{nt}$; then though it can verify that (3.24) does not hold any longer, as illustrated in Fig. 3.5, the system still converges.

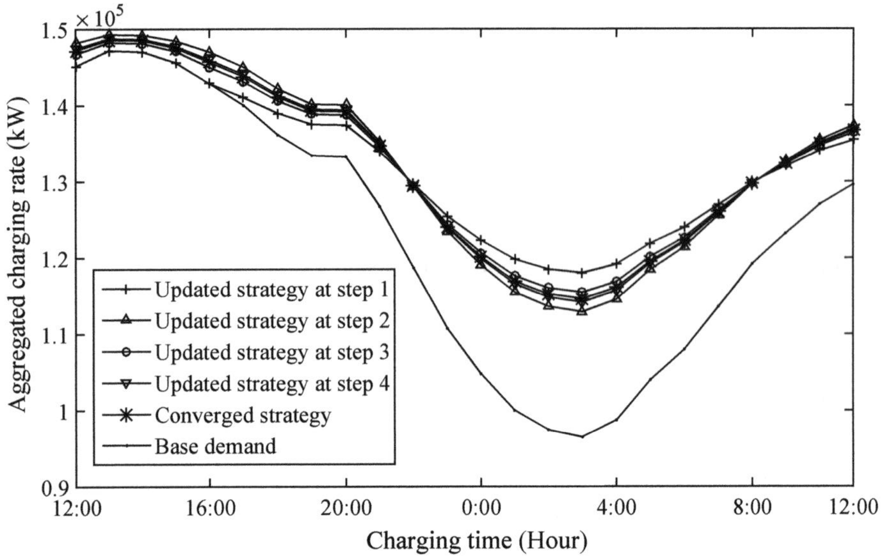

Fig. 3.3 A converged best response update procedure of PEVs in case f_n and p satisfy the condition of (3.24)

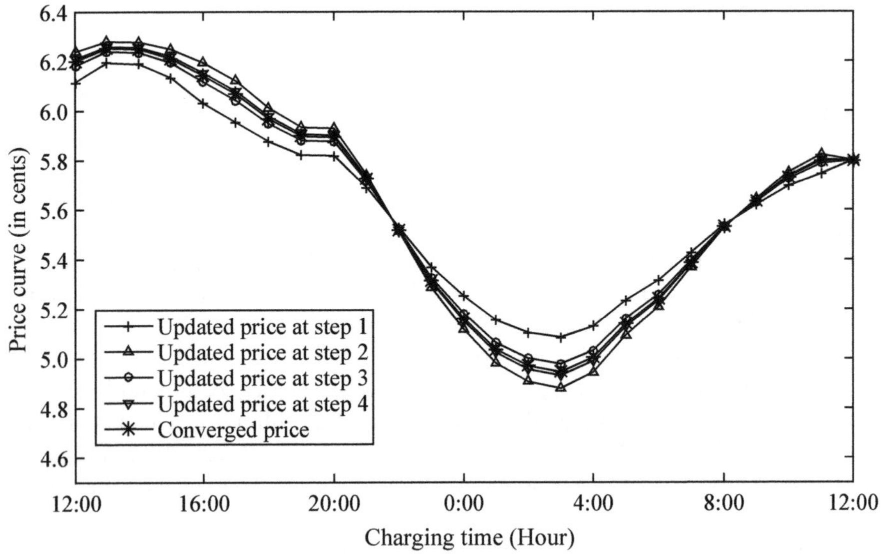

Fig. 3.4 A converged price curve update procedure in case f_n and p satisfy the condition of (3.24)

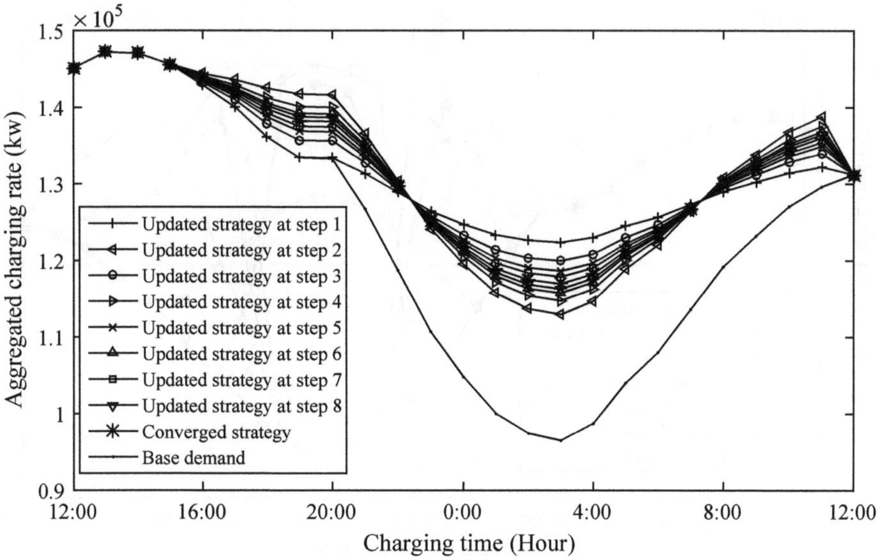

Fig. 3.5 A converged iteration update procedure in case the sufficient condition stated in Theorem 3.1 is unsatisfied

However if it is supposed that the value of $\mathfrak{a}$, the coefficient in battery degradation cost f_n, is much smaller, e.g., $\mathfrak{a} = 0.001$; then as illustrated in Fig. 3.6, the iterative update procedure does not converge any longer.

For approaching a realistic situation of PEV populations, suppose that the initial SOC values of PEV populations, $\{soc_{n0}; n \in \mathcal{N}\}$, approximately satisfy a Gaussian distribution, $\mathrm{N}(\hat{\mu}, \hat{\gamma})$, see [31, 32]. For the purpose of demonstration, it considers $\hat{\mu} = 0.5$ and $\hat{\gamma} = 0.1$. Figure 3.7 illustrates the iteration updates of the decentralized strategies following Algorithm 3.1 such that the system converges in a few of iteration steps.

3.4 Decentralized Methods with Flexible Energy Demands

In last section, it studies the charging coordination problems with degradation cost function $f_n(\cdot)$ and a fixed charging energy demand Γ_n.

Besides the individual degradation cost, due to the distribution-level impacts of PEV charging including line and transformer overloading, low voltages and increased losses, all these effects are a consequence of coincident high charger power demand u_{nt}. Therefore undesirable distribution-grid effects can be minimized by encouraging PEVs to charge at lower power levels. This can be achieved by introducing a demand charge,

$$Cost_{demand,nt} = f_{demand,nt}(u_{nt}) \tag{3.31}$$

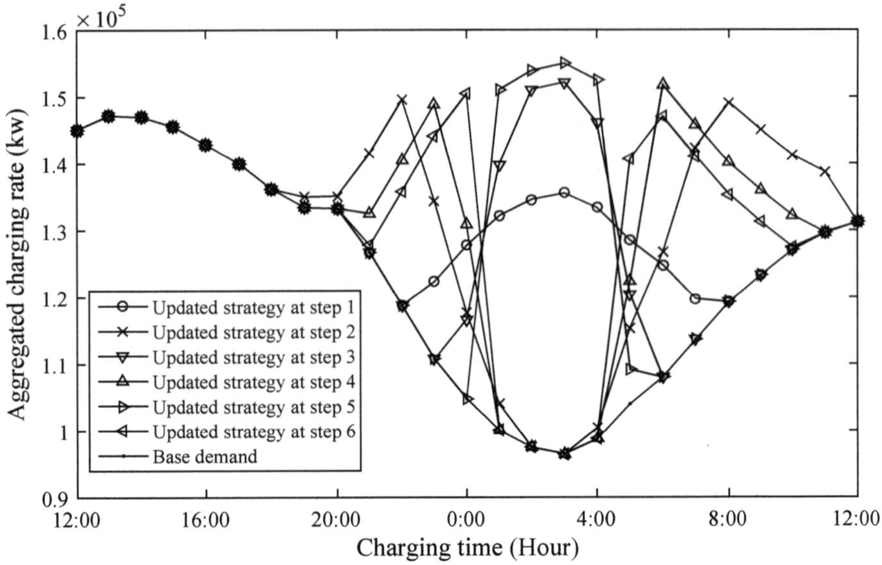

Fig. 3.6 An unconverged iteration update procedure in case the sufficient condition stated in Theorem 3.1 is unsatisfied

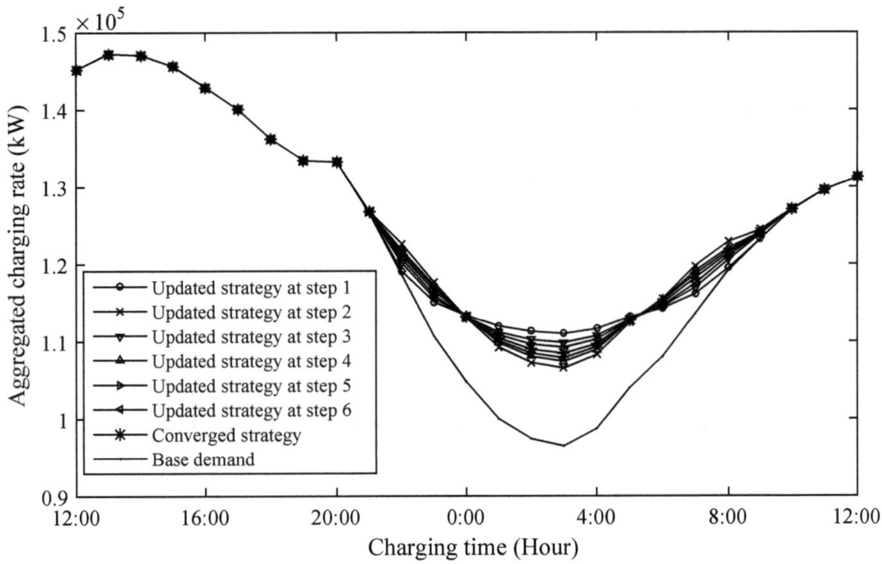

Fig. 3.7 An illustration of the equilibrium of the two-stage market

whereby PEVs incur a higher cost as their charging power increases, i.e. $f_{demand,nt}(\cdot)$ is a strictly increasing function. This charge is in addition to the cost of the energy delivered to the battery, and is consistent with existing tariff structures for larger consumers [33].

Moreover, it may allow the individual PEVs to consider the tradeoff between the charging cost and the total charged energy during the charing interval.

Consequently, in this section, it is worth to design the decentralized method to solve the PEV charging problems with flexible charging energy demands, and involving another individual cost for each PEV which is specified below.

3.4.1 Centralized PEV Charging Coordination

The coordination problem of interest considers the tradeoff between the total cost of supplying energy to the PEV population and the benefit derived from doing so. The total cost is composed of the generation cost, the demand charge discussed earlier in Sect. 3.4, and the PEV battery degradation cost formulated in Sect. 3.2.2. Coordination must ensure that all charging strategies are admissible, $\boldsymbol{u}_n \in \mathscr{U}_n$ for all $n \in \mathscr{N}$.

Given a collection of admissible charging strategies $\boldsymbol{u} \in \mathscr{U}$, the system cost function can be expressed as,

$$J(\boldsymbol{u}) \triangleq \sum_{t\in\mathscr{T}} \left\{ c\left(d_t + \sum_{n\in\mathscr{N}} u_{nt}\right) + \sum_{n\in\mathscr{N}} g_{nt}(u_{nt}) \right\} - \sum_{n\in\mathscr{N}} \left\{ h_n\left(\|\boldsymbol{u}_n\|_1\right) \right\}, \quad (3.32)$$

where:

- $c(\cdot)$ gives the generation cost with respect to the total demand $d_t + \sum_{n\in\mathscr{N}} u_{nt}$, and d_t denotes the aggregate inelastic base demand at time t;
- $g_{nt}(u_{nt}) = f_{demand,nt}(u_{nt}) + f_n(u_{nt})$ captures the demand charge (3.31) and battery degradation cost (3.6) of PEV n; and,
- $h_n\left(\|\boldsymbol{u}_n\|_1\right)$ denotes the benefit function of PEV n with respect to the total energy delivered over the charging horizon. In [34], this function has the quadratic form,

$$h_n\left(\|\boldsymbol{u}_n\|_1\right) = -\delta_n(\|\boldsymbol{u}_n\|_1 - \Gamma_n)^2, \quad (3.33)$$

with the factor δ_n reflecting the relative importance of delivering the full charge to the PEV over the charging horizon.

The utility function of PEV n, for a charging strategy $\boldsymbol{u}_n \in \mathscr{U}_n$, can be written,

$$v_n(\boldsymbol{u}_n) \triangleq h_n(\|\boldsymbol{u}_n\|_1) - \sum_{t\in\mathscr{T}} g_{nt}(u_{nt}). \quad (3.34)$$

The system cost $J(\boldsymbol{u})$ given by (3.32) can then be rewritten:

$$J(\boldsymbol{u}) = \sum_{t \in \mathscr{T}} c\left(d_t + \sum_{n \in \mathscr{N}} u_{nt}\right) - \sum_{n \in \mathscr{N}} v_n(\boldsymbol{u}_n). \tag{3.35}$$

The individual utility function (3.34) is similar to that specified in [13], where a decentralized PEV charging algorithm ia developed based on congestion pricing concepts from internet traffic control [17].

Remarks on generation cost $c(\cdot)$: $c(\cdot)$ is briefly discussed on the paragraph following Assumption 2.1. In Assumption 3.2 below, it gives a formal consideration on $c(\cdot)$.

Assumption 3.2 $c(\cdot)$ is monotonically increasing, strictly convex and differentiable. ■

Actually, as stated earlier in Chap. 2, it is commonly assumed, see [25–27] and references therein, that the electricity generation cost can be approximated by the quadratic form,

$$c(D_t) = \frac{1}{2}\mathsf{a}D_t^2 + \mathsf{b}D_t + \mathsf{c}, \tag{3.36}$$

with parameters a, b and c which reflect system conditions. The marginal generation cost, which is the derivative of generation cost, therefore varies linearly with the total demand, $p(D_t) \triangleq c'(D_t) = \mathsf{a}D_t + \mathsf{b}$.

Centralized PEV charging coordination is formulated as the optimization problem below:

Problem 3.2

$$\min_{\mathbf{u} \in \mathscr{U}} \ J(\boldsymbol{u}). \tag{3.37}$$

The objective is to implement a socially optimal collection of charging strategies for all PEVs, denoted by $\boldsymbol{u}^{**}$, that minimizes the system cost (3.32) or its equivalent form (3.35). ■

The following assumption will apply throughout the chapter for $g_{nt}(\cdot)$:

Assumption 3.3 $g_{nt}(\cdot)$, for all $n \in \mathscr{N}$, $t \in \mathscr{T}$, is monotonically increasing, strictly convex and differentiable. ■

When the benefit function takes the form (3.33), the solution obtained by minimizing $J(\boldsymbol{u})$ subject only to (3.1a) always satisfies (3.1b), and hence is also the solution for Optimization Problem 3.2. To see this, define the set of charging strategies that satisfy (3.1a) for PEV n,

$$\mathscr{S}_n \triangleq \left\{\boldsymbol{u}_n \equiv (u_{nt};\, t \in \mathscr{T});\ \text{s.t. constraint (3.1a)}\right\},$$

and let $\mathscr{S}$ denote the collection of such sets for all $n \in \mathscr{N}$. Consider the optimization problem $\min_{\boldsymbol{u} \in \mathscr{S}} J(\boldsymbol{u})$ rather than (3.37).

Based on Assumptions 3.2, 3.1 and 3.3, the efficient (socially optimal) charging behavior is unique and can be characterized by its associated KKT conditions [28]. The optimal solution $\boldsymbol{u}^{**}$ is therefore given by:

$$\frac{\partial}{\partial u_{nt}} J(\boldsymbol{u}) \geq 0, \quad u_{nt} \geq 0, \quad \frac{\partial}{\partial u_{nt}} J(\boldsymbol{u}) u_{nt} = 0, \tag{3.38}$$

for all $n \in \mathscr{N}$ and $t \in \mathscr{T}_n$, where:

$$\frac{\partial}{\partial u_{nt}} J(\boldsymbol{u}) = c' \left(d_t + \sum_{n \in \mathscr{N}} u_{nt} \right) - \frac{\partial}{\partial u_{nt}} v_n(\boldsymbol{u}_n).$$

It follows from (3.38) that the efficient charging behavior $\boldsymbol{u}^{**}$ is uniquely specified by:

$$p_t^{**} \begin{cases} = \frac{\partial}{\partial u_{nt}} v_n(\boldsymbol{u}_n^{**}), & \text{when } u_{nt}^{**} > 0, \\ \geq \frac{\partial}{\partial u_{nt}} v_n(\boldsymbol{u}_n^{**}), & \text{when } u_{nt}^{**} = 0, \end{cases} \tag{3.39}$$

where $p_t^{**} = c'\left(d_t + \sum_{n \in \mathscr{N}} u_{nt}^{**}\right)$ is the generation marginal cost over the charging horizon with respect to the efficient allocation $\boldsymbol{u}^{**}$.

Notice that if $\|\boldsymbol{u}_n^{**}\|_1 \geq \Gamma_n$ then (3.33) together with Assumptions 3.1 and 3.3, ensure that $\frac{\partial}{\partial u_{nt}} v_n(\boldsymbol{u}_n^{**}) < 0$. But $p_t^{**} > 0$ according to Assumption 3.2, so (3.39) implies that $\boldsymbol{u}_n^{**} = \mathbf{0}$. Hence a contradiction. Accordingly, $\|\boldsymbol{u}_n^{**}\|_1 < \Gamma_n$ and (3.1b) is always satisfied.

3.4.2 Numerical Examples

This example considers coordinated charging of a population of 5000 PEVs over a common charging interval from noon on one day to noon on the next. In accordance with (3.36), the generation cost function has the quadratic form,

$$c(D_t) = 2.9 \times 10^{-7} D_t^2 + 0.06 D_t, \tag{3.40}$$

where $D_t = d_t + \sum_{n \in \mathscr{N}} u_{nt}$. The base demand $\boldsymbol{d}$, which is shown in Fig. 3.8, is representative of a typical hot summer day.

The battery pack of each PEV is composed of $LiFePO_4$ lithium-ion cells which have a nominal voltage of 3.3 V and energy capacity of 2.5 Ah (Amp×Hour). These are typical values for batteries that are used in PEVs. Assume the price of battery cell capacity is \$10/Wh. Furthermore, let all PEVs have battery capacity of 40 kWh. Then the battery degradation cost (3.6) for each PEV is given (approximately) by,

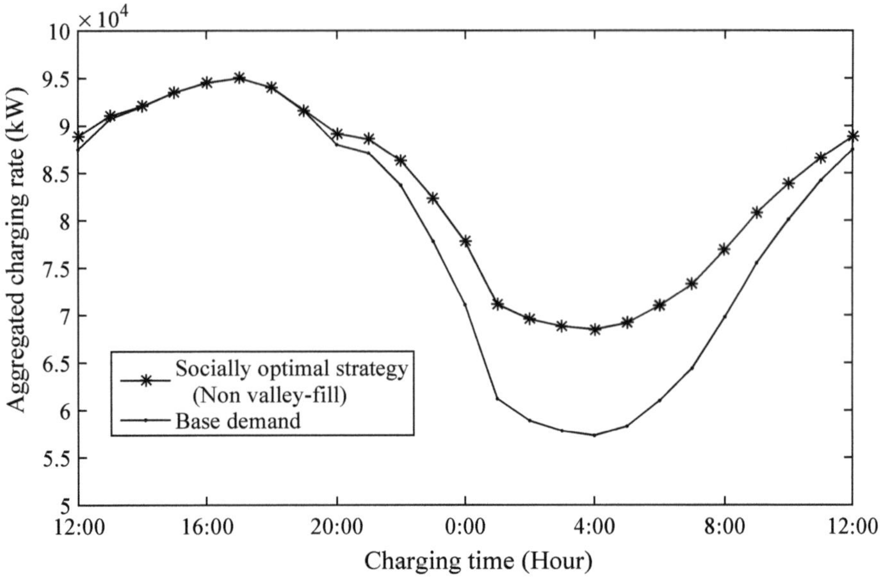

Fig. 3.8 Aggregate demand due to the efficient charging strategies

$$f_n(u_{nt}) = 0.0012u_{nt}^2 + 0.11u_{nt} - 0.02.$$

Each PEV is subject to a quadratic demand charge $f_{demand,nt}(u_{nt}) = 0.0018u_{nt}^2$. Thus, the local costs incurred by each PEV at time t amount to,

$$g_{nt}(u_{nt}) = 0.003u_{nt}^2 + 0.11u_{nt} - 0.02. \tag{3.41}$$

Initially all PEVs use the same weighting factor $\delta_n = 0.03$ in their utility function $v_n(\boldsymbol{u}_n)$ given by (3.33) and (3.34). This will be relaxed in later investigations.

For simplicity, assume all PEVs have identical minimum and maximum SOC, with $soc_{min} = 15\%$ and $soc_{max} = 90\%$. The upper limit (3.1b) on the energy that can be delivered to each PEV is given by,

$$\Gamma_n = 40(soc_{max} - soc_{n0}),$$

which equals 30 kWh if $soc_{n0} = soc_{min} = 15\%$ for all n.

The efficient (socially optimal) charging strategies given by (3.37), $\boldsymbol{u}_n^{**}$ for $n \in \mathcal{N}$, result in the aggregate demand shown in Fig. 3.8. The evolution of the SOC of one of the PEVs is shown in Fig. 3.9. As a comparison, the aggregate demand of the valley-filling strategy $\boldsymbol{u}^{vf}$ given by [15] is also shown in Fig. 3.8. Note that the algorithm developed in [15] enforces an equality constraint on the energy delivered, $\|\boldsymbol{u}_n^{vf}\|_1 = \Gamma_n^{vf}$, rather than incorporating a benefit function of the form (3.33). Therefore, to

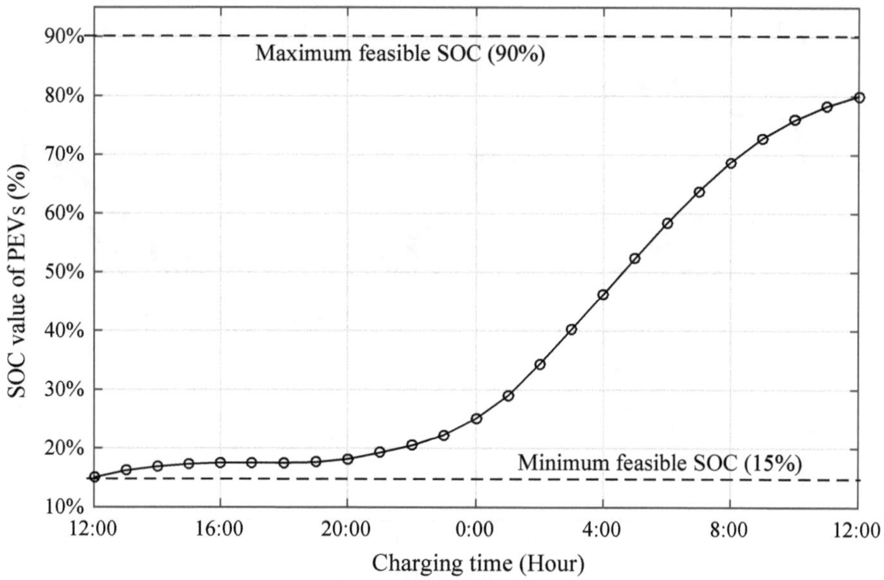

Fig. 3.9 Evolution of the SOC of an individual PEV

ensure a meaningful comparison, the total charge requirement in the valley-filling case is set equal to the energy delivered in the socially optimal solution, $\Gamma_n^{vf} = \|\boldsymbol{u}_n^{**}\|_1$.

The socially optimal charging strategy given by (3.37) establishes a tradeoff between the total generation cost of the system and the local costs (demand charge and battery degradation cost) of the PEV population. As illustrated in Fig. 3.8, this results in an outcome that differs quite considerably from the valley-filling strategy which is solely concerned with minimizing total generation cost. This distortion away from valley-filling increases as higher weighting is given to the local costs of the PEV population.

This difference between the socially optimal and valley-filling strategies can be quantified by considering,

$$\sum_{t\in\mathscr{T}} c\left(d_t + \sum_{n\in\mathscr{N}} u_{nt}^{**}\right) - \sum_{t\in\mathscr{T}} c\left(d_t + \sum_{n\in\mathscr{N}} u_{nt}^{vf}\right) = 211.7$$

$$\sum_{n\in\mathscr{N}}\sum_{t\in\mathscr{T}} g_{nt}\left(u_{nt}^{**}\right) - \sum_{n\in\mathscr{N}}\sum_{t\in\mathscr{T}} g_{nt}\left(u_{nt}^{vf}\right) = -671.3.$$

where the constraint $\Gamma_n^{vf} = \|\boldsymbol{u}_n^{**}\|_1$ has been taken into account. It can be seen that incorporating the local costs resulted in an increase in generation cost of \$211.7/day. However this is more than offset by a reduction in the local costs of \$671.3/day, resulting in an overall saving of \$459.6/day. As a further comparison, if $\Gamma_n^{vf} = \Gamma_n$,

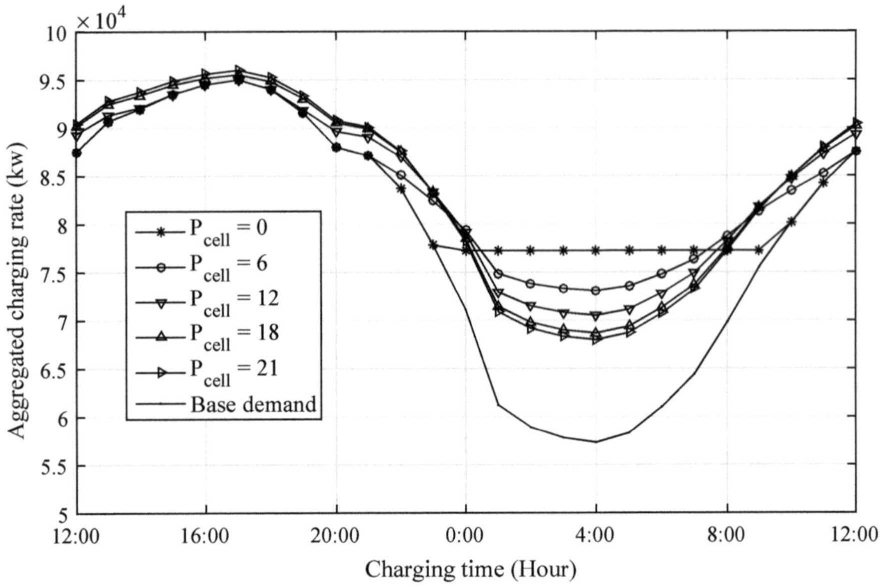

Fig. 3.10 Aggregate demand for efficient charging strategies as battery price P_{cell} varies

i.e. PEVs must fully charge rather than settle for the reduced energy delivery of $\|\boldsymbol{u}_n^{**}\|_1$, then adopting $\boldsymbol{u}^{**}$ instead of $\boldsymbol{u}^{vf}$ would result in a cost saving of \$1640.2/day.

The following investigations consider the effects of variations in the battery degradation cost f_n and the benefit function h_n on the optimal charging strategies of a population of PEVs.

Figure 3.10 shows the evolution of the (efficient) aggregate demand as the battery price P_{cell} is varied. For this study, the demand charge (3.31) is set to zero and $\delta_n = 0.03$ in all cases. It can be seen that the efficient aggregate demand approaches valley-filling as P_{cell} decreases, and becomes exactly valley-filling when $P_{cell} = 0$. Figure 3.11 shows that the total delivered energy $\|\boldsymbol{u}_n\|_1$ decreases as P_{cell} increases.

Figure 3.12 shows the variation in the efficient aggregate demand as the benefit function parameter δ_n is varied. The general shape of the aggregate demand remains largely unchanged. It is shown in Fig. 3.13 that the total delivered energy $\|\boldsymbol{u}_n\|_1$ increases with δ_n, approaching the energy capacity limit Γ_n.

Centralized coordination is only possible when the system operator has complete information, including the characteristics of PEV batteries and the valuation functions of individual PEVs. It is unlikely, however, that individuals would be willing to share such private information. Also, for a large population, centralized coordination may be computationally infeasible. Thus, the remainder of the chapter is devoted to the development of a decentralized coordination process where each PEV updates its charging strategy with respect to a common electricity price profile, and then the price profile is updated based on the latest charging strategies for the population.

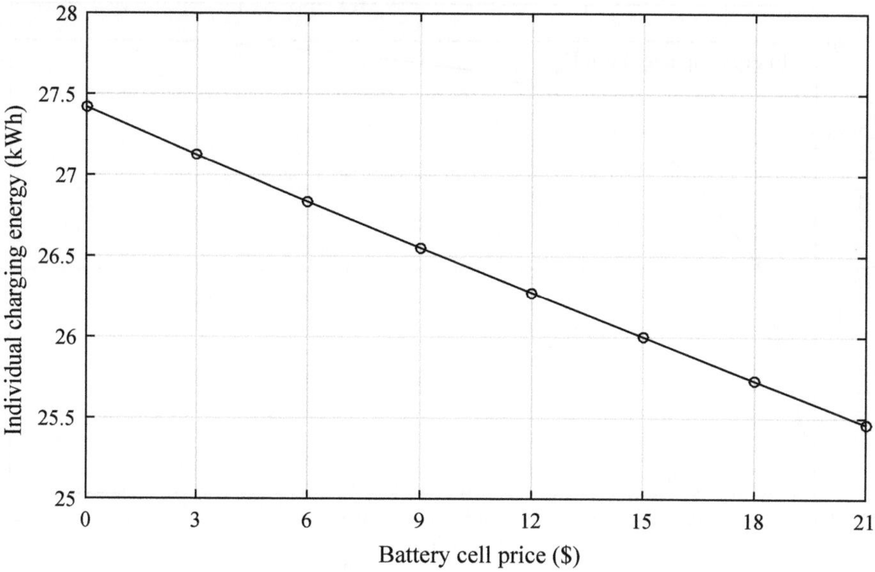

Fig. 3.11 Total delivered energy $\|\boldsymbol{u}_n\|_1$ for an individual PEV as battery price P_{cell} varies

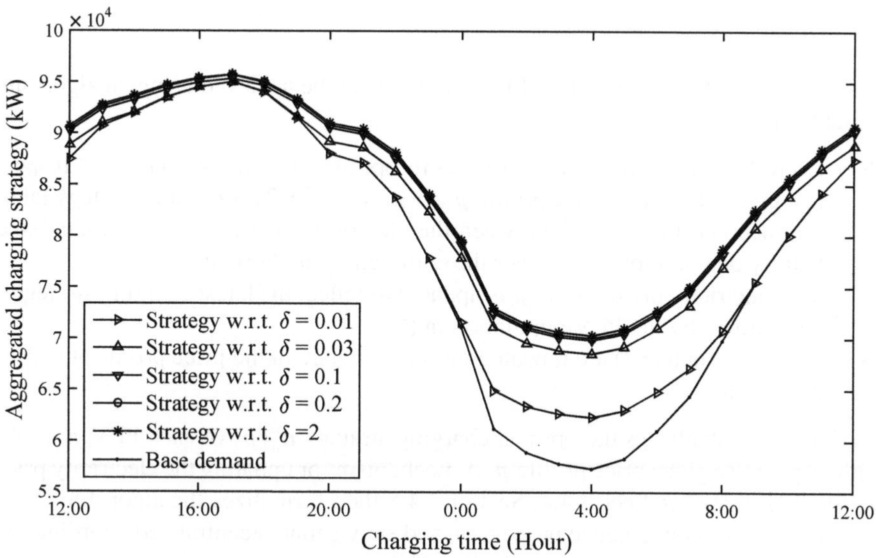

Fig. 3.12 Aggregate demand for efficient charging strategies as the benefit function parameter δ_n varies

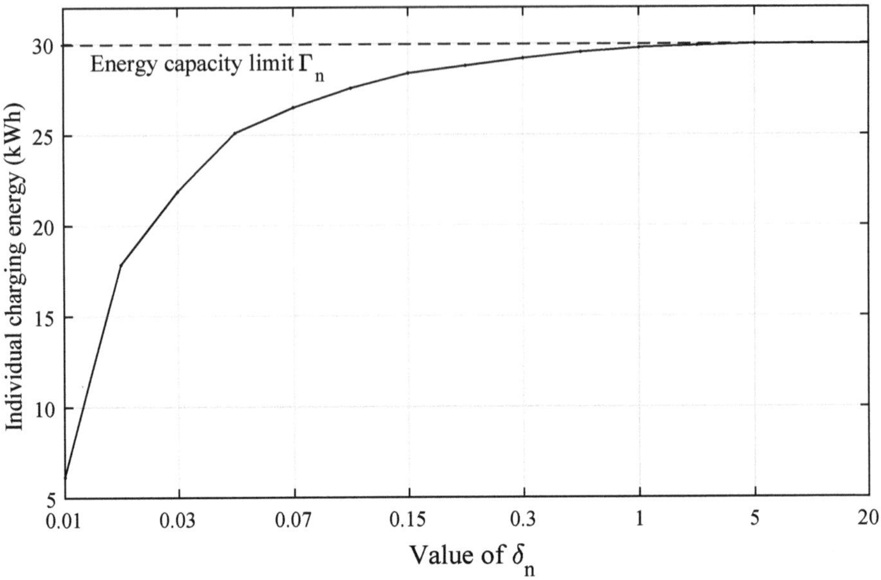

Fig. 3.13 Total delivered energy $\|\boldsymbol{u}_n\|_1$ for an individual PEV as the benefit function parameter δ_n varies

Decentralized coordination of PEV charging can be achieved using an algorithm of the form:

(S1) Each PEV autonomously determines its optimal charging strategy with respect to a given electricity price profile $\boldsymbol{p} \equiv (p_t, t \in \mathscr{T})$. This optimal strategy takes into account the tradeoff between the electricity cost and local (demand and battery degradation) costs over the entire charging horizon.
(S2) The electricity price profile $\boldsymbol{p}$ is updated to reflect the latest charging strategies determined by the PEV population in (S1).
(S3) Steps (S1) and (S2) are repeated until the change in the price profile at (S2) is negligible.

Section 3.4.3 establishes the optimal charging strategy $\boldsymbol{u}_n^*(\boldsymbol{p})$ of each PEV, $n \in \mathscr{N}$, with respect to a given price profile $\boldsymbol{p}$. A mechanism for updating the electricity price profile is designed in Sect. 3.4.4. Section 3.4.5 then formalizes the algorithm (S1)–(S3), establishes convergence properties, and shows that decentralized coordination gives the socially optimal (economically efficient) charging strategy.

3.4.3 Optimal Response of Each PEV w.r.t. a Fixed Price Profile

The individual cost function of PEV n, under charging strategy $\boldsymbol{u}_n \in \mathscr{U}_n$ and with respect to the price profile $\boldsymbol{p}$, can be written:

$$J_n(\boldsymbol{u}_n; \boldsymbol{p}) \triangleq \sum_{t\in\mathscr{T}} p_t u_{nt} - v_n(\boldsymbol{u}_n). \tag{3.42}$$

Alternatively, using (3.34), this cost function can be expressed in the form:

$$J_n(\boldsymbol{u}_n; \boldsymbol{p}) = \sum_{t\in\mathscr{T}} \Big\{ p_t u_{nt} + g_{nt}(u_{nt}) \Big\} - h_n \left(\sum_{t\in\mathscr{T}} u_{nt} \right),$$

where it becomes clear that the cost is composed of the total electricity cost $\sum_{t\in\mathscr{T}} p_t u_{nt}$, the total local cost $\sum_{t\in\mathscr{T}} g_{nt}(u_{nt})$, and the benefit derived from the total energy delivered over the charging horizon $h_n\big(\sum_{t\in\mathscr{T}} u_{nt}\big)$.

The optimal charging strategy of PEV n, with respect to $\boldsymbol{p}$, is obtained by minimizing the cost function (3.42),

$$\boldsymbol{u}_n^*(\boldsymbol{p}) = \operatorname*{argmin}_{\boldsymbol{u}_n\in\mathscr{U}_n} J_n(\boldsymbol{u}_n; \boldsymbol{p}). \tag{3.43}$$

It will be shown in Theorem 3.3 that the optimal response of PEV n has the form:

$$u_{nt}(\boldsymbol{p}, A_n) = \begin{cases} \max\big\{0, [g'_{nt}]^{-1}(A_n - p_t)\big\}, & t \in \mathscr{T}_n \\ 0, & t \in \mathscr{T} \setminus \mathscr{T}_n \end{cases} \tag{3.44}$$

for some A_n, where g'_{nt} is the derivative of g_{nt}, and $[g'_{nt}]^{-1}$ is the corresponding inverse function. Since the total charging energy is elastic, the value of A_n is dependent upon the PEV specifications and the price $\boldsymbol{p}$, with this dependence established in Theorem 3.3.

Determining the optimal charging strategy (3.43) proceeds as follows. Lemma 3.4 addresses the restricted problem of finding the optimal charging strategy when the total delivered energy takes a specified value, $\|\boldsymbol{u}_n\|_1 = \omega$, where $0 \le \omega \le \Gamma_n$. Lemma 3.5 considers the charging strategy in the form (3.44) and establishes the relationship between the value of A_n and the total energy delivered $\|\boldsymbol{u}_n(\boldsymbol{p}, A_n)\|_1$. Lemma 3.6 shows that the sensitivity of charging cost to changes in ω is given by A_n. Finally, Theorem 3.3 brings all the results together and determines the unique value of A_n that ensures (3.44) is optimal in the sense of (3.43).

To begin with, consider the cost for PEV n, but excluding h_n. This can be written,

$$F_n(\boldsymbol{u}_n; \boldsymbol{p}) \triangleq \sum_{t\in\mathscr{T}} \Big\{ p_t u_{nt} + g_{nt}(u_{nt}) \Big\}. \tag{3.45}$$

This function will be used to examine scenarios where the total charging is constant, $\|\boldsymbol{u}_n\|_1 = \omega$. In such cases, the cost $h_n(\|\boldsymbol{u}_n\|_1)$ can be neglected since it is equal across all such scenarios. Accordingly, define the set of charging strategies where total energy ω is delivered to PEV n as,

$$\mathscr{U}_n(\omega) \triangleq \left\{\boldsymbol{u}_n \in \mathscr{U}_n;\ \text{s.t. } \|\boldsymbol{u}_n\|_1 = \omega\right\}. \tag{3.46}$$

Lemma 3.4 *Consider a fixed ω, with $0 \le \omega \le \Gamma_n$, and a fixed $\boldsymbol{p}$. Then $\boldsymbol{u}_n(\boldsymbol{p}, A_n)$, defined in (3.44), is the unique charging strategy minimizing the cost function* (3.45) *subject to the set of admissible charging strategies $\mathscr{U}_n(\omega)$.*

Proof Define the Lagrangian function,

$$L_n(\boldsymbol{u}_n, A_n; \boldsymbol{p}) \triangleq F_n(\boldsymbol{u}_n; \boldsymbol{p}) + A_n(\omega - \|\boldsymbol{u}_n\|_1), \tag{3.47}$$

with $u_{nt} \ge 0$ for $t \in \mathscr{T}$, and A_n the Lagrangian multiplier associated with the constraint on total delivered energy (3.46). The desired optimal strategy must satisfy the KKT conditions [28]:

(i) $\frac{\partial L_n}{\partial A_n} = 0$.
(ii) $\frac{\partial L_n}{\partial u_{nt}} \ge 0$, $u_{nt} \ge 0$, with complementary slackness.

The equality (i) recovers the constraint on total energy $\|\boldsymbol{u}_n\|_1 = \omega$, while the inequalities given in (ii) can be expressed in the form,

$$p_t + g'_{nt}(u_{nt}) - A_n \begin{cases} = 0, & \text{when } u_{nt} > 0 \\ \ge 0, & \text{otherwise,} \end{cases} \tag{3.48}$$

which is equivalent to (3.44). Moreover, since $F_n(\boldsymbol{u}_n; \boldsymbol{p})$ is convex with respect to $\boldsymbol{u}_n$, the optimal charging strategy defined by (3.44) must be unique for a given $\mathscr{U}_n(\omega)$. ■

This lemma establishes the minimum (3.43) when $\boldsymbol{u}_n$ is restricted to $\mathscr{U}_n(\omega)$, for a specified value of ω. Relaxing that restriction, by allowing $\boldsymbol{u}_n \in \mathscr{U}_n$, is achieved in Theorem 3.3. Before reaching that point, it is necessary to establish some notation and intermediate results.

For any PEV, $n \in \mathscr{N}$, the pair of values $A_n^-(\boldsymbol{p})$ and $A_n^+(\boldsymbol{p})$ are defined as:

$$A_n^-(\boldsymbol{p}) = \max\left\{A,\ \text{ such that } \|\boldsymbol{u}_n(\boldsymbol{p}, A)\|_1 = 0\right\} \tag{3.49}$$

$$A_n^+(\boldsymbol{p}) = A,\ \text{ such that } \|\boldsymbol{u}_n(\boldsymbol{p}, A)\|_1 = \Gamma_n. \tag{3.50}$$

Note that the subscript n is included on A_n^- for consistency, though it is independent of n.

The following lemma establishes a few basic relationships between the Lagrangian multiplier A_n and the minimizing charging strategy.

Lemma 3.5 *Consider a fixed price profile* $\boldsymbol{p}$*. Then:*

(i) Every $u_{nt}(\boldsymbol{p}, A_n)$*,* $t \in \mathcal{T}$*, is non-decreasing with* $A_n \in \mathbb{R}$*, and hence* $\|\boldsymbol{u}_n(\boldsymbol{p}, A_n)\|_1$ *is non-decreasing with* A_n*. Furthermore,* $\|\boldsymbol{u}_n(\boldsymbol{p}, A_n)\|_1$ *is strictly increasing for* $A_n \geq A_n^-(\boldsymbol{p})$*.*

(ii) $\boldsymbol{u}_n(\boldsymbol{p}, A_n)$ *is admissible for* $A_n^-(\boldsymbol{p}) \leq A_n \leq A_n^+(\boldsymbol{p})$*, but not admissible for any* $A_n > A_n^+(\boldsymbol{p})$*.*

Proof Property (i) holds by the specification of $u_{nt}(\boldsymbol{p}, A_n)$ given in (3.44) and verified by Lemma 3.4, keeping in mind Assumptions 3.1 and 3.3. From (3.49) and (3.50), $\|\boldsymbol{u}_n(\boldsymbol{p}, A_n^-(\boldsymbol{p}))\|_1 = 0$ and $\|\boldsymbol{u}_n(\boldsymbol{p}, A_n^+(\boldsymbol{p}))\|_1 = \Gamma_n$. Therefore, since $\|\boldsymbol{u}_n(\boldsymbol{p}, A_n)\|_1$ is strictly increasing for $A_n \geq A_n^-(\boldsymbol{p})$, it follows that:

- $u_{nt}(\boldsymbol{p}, A_n) \geq 0$ for all $t \in \mathcal{T}$, and $\|\boldsymbol{u}_n(\boldsymbol{p}, A_n)\|_1 \leq \Gamma_n$ for $A_n^-(\boldsymbol{p}) \leq A_n \leq A_n^+(\boldsymbol{p})$, so $\boldsymbol{u}_n(\boldsymbol{p}, A_n)$ is admissible.
- $\|\boldsymbol{u}_n(\boldsymbol{p}, A_n)\|_1 > \Gamma_n$ when $A_n > A_n^+(\boldsymbol{p})$, so $\boldsymbol{u}_n(\boldsymbol{p}, A_n)$ is not admissible.

This establishes property (ii). ■

Lemma 3.5 guarantees that for a fixed $\boldsymbol{p}$, $\|\boldsymbol{u}_n(\boldsymbol{p}, A_n)\|_1$ strictly increases from 0 to Γ_n on the interval $A_n \in [A_n^-(\boldsymbol{p}), A_n^+(\boldsymbol{p})]$. This implies that $\|\boldsymbol{u}_n(\boldsymbol{p}, A)\|_1$ is invertible on $[A_n^-(\boldsymbol{p}), A_n^+(\boldsymbol{p})]$, with the inverse denoted:

$$\mathscr{A}_n(\boldsymbol{p}, \cdot) : [0, \Gamma_n] \to [A_n^-(\boldsymbol{p}), A_n^+(\boldsymbol{p})]. \tag{3.51}$$

It follows that $\mathscr{A}_n(\boldsymbol{p}, \omega)$ is strictly increasing with ω and,

$$\mathscr{A}_n(\boldsymbol{p}, \omega) = A_n \iff \|\boldsymbol{u}_n(\boldsymbol{p}, A_n)\|_1 = \omega. \tag{3.52}$$

The charging strategy that satisfies (3.44) and delivers total energy of ω will be denoted by $\boldsymbol{u}_n(\boldsymbol{p}, \mathscr{A}_n(\boldsymbol{p}, \omega))$, and therefore $\|\boldsymbol{u}_n(\boldsymbol{p}, \mathscr{A}_n(\boldsymbol{p}, \omega))\|_1 = \omega$.

Because of the non-negativity constraint on u_{nt} and the corresponding complementary slackness requirement from Lemma 3.4, it is not straightforward to determine a closed form expression for the function $\mathscr{A}_n(\boldsymbol{p}, \omega)$. A valuable property of $\mathscr{A}_n(\boldsymbol{p}, \omega)$ is, however, established in the following lemma.

Lemma 3.6 *For any fixed price profile* $\boldsymbol{p}$*,*

$$\frac{d}{d\omega} F_n^*(\boldsymbol{p}, \omega) = \mathscr{A}_n(\boldsymbol{p}, \omega), \quad \text{with } \omega \in [0, \Gamma_n], \tag{3.53}$$

where,

$$F_n^*(\boldsymbol{p}, \omega) \triangleq \min_{\boldsymbol{u}_n \in \mathscr{U}_n(\omega)} F_n(\boldsymbol{u}_n; \boldsymbol{p}). \tag{3.54}$$

Proof From (3.47), A_n is the Lagrangian multiplier associated with the constraint $\|\boldsymbol{u}_n\|_1 = \omega$. Based on duality theory [28], the sensitivity of the minimum value $F_n^*(\boldsymbol{p}, \omega)$ with respect to changes in ω is therefore given by A_n. The result follows from (3.52). ■

It is now possible to establish the optimal charging strategy for a given $\boldsymbol{p}$. This is achieved in the following theorem, which implicitly determines the optimal value for ω in the process.

Theorem 3.3 *Assume $h_n(\omega)$ is continuously differentiable, increasing and concave on $0 \le \omega \le \Gamma_n$. Define,*

$$\widehat{h}_n(\boldsymbol{p}, \omega) \triangleq \mathscr{A}_n(\boldsymbol{p}, \omega) - h_n'(\omega) \tag{3.55}$$

with $\mathscr{A}_n(\boldsymbol{p}, \cdot)$ given by (3.51), and

$$A_n^*(\boldsymbol{p}) = \begin{cases} \mathscr{A}_n(\boldsymbol{p}, \Gamma_n), & \text{if } \widehat{h}_n(\boldsymbol{p}, \Gamma_n) \le 0 \\ \mathscr{A}_n(\boldsymbol{p}, 0), & \text{if } \widehat{h}_n(\boldsymbol{p}, 0) \ge 0 \\ \mathscr{A}_n(\boldsymbol{p}, \omega^*), & \text{if } \widehat{h}_n(\boldsymbol{p}, \omega^*) = 0 \end{cases} \tag{3.56}$$

where $0 < \omega^* < \Gamma_n$. Then the charging strategy $\boldsymbol{u}_n(\boldsymbol{p}, A_n^*(\boldsymbol{p}))$ defined in (3.44) uniquely minimizes the cost function (3.42) with respect to a given $\boldsymbol{p}$, i.e. $\boldsymbol{u}_n^*(\boldsymbol{p}) = \boldsymbol{u}_n(\boldsymbol{p}, A_n^*(\boldsymbol{p}))$.

Proof Recall that

$$J_n(\boldsymbol{u}_n; \boldsymbol{p}) = F_n(\boldsymbol{u}_n; \boldsymbol{p}) - h_n(\omega),$$

for all $\boldsymbol{u}_n \in \mathscr{U}_n(\omega)$. Then,

$$\begin{aligned} \min_{\boldsymbol{u}_n \in \mathscr{U}_n} J_n(\boldsymbol{u}_n; \boldsymbol{p}) &= \min_{\boldsymbol{u}_n \in \mathscr{U}_n} \left\{ F_n(\boldsymbol{u}_n; \boldsymbol{p}) - h_n(\|\boldsymbol{u}_n\|_1) \right\} \\ &= F_n^*(\boldsymbol{p}, \omega^*) - h_n(\omega^*), \end{aligned}$$

where ω^* is the total charging energy that minimizes the cost function $J_n(\cdot; \boldsymbol{p})$. Note that ω^* is constrained to $0 \le \omega^* \le \Gamma_n$. For $0 < \omega^* < \Gamma_n$, the optimal energy demand ω^* is implicitly defined by the stationarity condition,

$$\frac{d}{d\omega}\big(F_n^*(\boldsymbol{p}, \omega) - h_n(\omega)\big)\Big|_{\omega=\omega^*} = \widehat{h}_n(\boldsymbol{p}, \omega^*) = 0 \tag{3.57}$$

where Lemma 3.6 has been used to establish the first equality. Moreover, $h_n'(\omega)$ decreases on ω, since h_n is assumed to be concave, and $\mathscr{A}_n(\boldsymbol{p}, \omega)$ is strictly increasing with ω. Therefore, if a solution for (3.57) exists over $0 < \omega^* < \Gamma_n$, then it must be unique.

If (3.57) cannot be satisfied for $0 < \omega < \Gamma_n$ then no stationary point exists over that open interval. Consequently, the cost $F_n^*(\boldsymbol{p}, \omega) - h_n(\omega)$ must exhibit monotonic behavior over $0 < \omega < \Gamma_n$. If the cost strictly increases with ω, so $\mathscr{A}_n(\boldsymbol{p}, \omega) - h_n'(\omega) = \widehat{h}_n(\boldsymbol{p}, \omega) > 0$ for $0 < \omega < \Gamma_n$, then the minimum cost solution will occur at the lower end, $\omega^* = 0$. Similarly, if the cost is strictly decreasing with ω, so the derivative $\widehat{h}_n(\boldsymbol{p}, \omega) < 0$ over $0 < \omega < \Gamma_n$, then the minimum cost solution will occur at the upper end, $\omega^* = \Gamma_n$. ■

3.4.4 Price Profile Update Mechanism

If the price profile $\boldsymbol{p}$ is equal to the optimal (efficient) generation marginal cost $\boldsymbol{p}^{**}$ given by (3.39), then the collection of PEV charging strategies $\boldsymbol{u}^*(\boldsymbol{p}) \equiv \big(\boldsymbol{u}_n(\boldsymbol{p}, \mathrm{A}_n^*(\boldsymbol{p})), n \in \mathscr{N}\big)$ given by (3.44) and (3.56) would be efficient. However, this optimal price $\boldsymbol{p}^{**}$ cannot be determined a priori. Hence there is a need for an update mechanism that guarantees convergence of the price profile to the efficient marginal cost $\boldsymbol{p}^{**}$. Consider the scheme,

$$p_t^+(\boldsymbol{p}) = p_t + \eta \left(c' \left(d_t + \sum_{n \in \mathscr{N}} u_{nt}^*(\boldsymbol{p}) \right) - p_t \right), \quad t \in \mathscr{T} \tag{3.58}$$

where $\eta > 0$ is a fixed parameter, and $\boldsymbol{u}_n^*(\boldsymbol{p})$, defined in (3.43), is the optimal charging strategy for PEV n with respect to $\boldsymbol{p}$.

Given a system price profile $\boldsymbol{p}$ over the charging horizon $\mathscr{T}$, if p_t is lower than the generation marginal cost $c'\big(d_t + \sum_{n \in \mathscr{N}} u_{nt}^*(\boldsymbol{p})\big)$ at time t, the system will set a higher price p_t^+ to encourage PEVs to reduce their charging demand at that time. Likewise, if p_t is higher than the marginal cost c', the system will set a lower system price p_t^+ to encourage PEVs to increase their charging demand at that time.

Notice that the price update mechanism (3.58) can be written in the form,

$$\boldsymbol{p}^+ = (1 - \eta)\boldsymbol{p} + \eta \mathscr{P}(\boldsymbol{p}),$$

where

$$\mathscr{P}_t(\boldsymbol{p}) = c' \left(d_t + \sum_{n \in \mathscr{N}} u_{nt}^*(\boldsymbol{p}) \right), \quad t \in \mathscr{T}.$$

This price update iteration takes the form of the Krasnoselskij iteration [35, 36], and is therefore guaranteed to converge to a fixed point of $\mathscr{P}(\cdot)$ for any $\eta \in (0, 1)$ if $\mathscr{P}(\cdot)$ is non-expansive. Corollary 3.2 establishes a more general sufficient condition under which the system converges to the unique price profile $\boldsymbol{p}^{**}$ which is the efficient marginal cost.

3.4.5 Decentralized Coordination of PEV Charging

It is now possible to formalize a decentralized coordination algorithm for determining the optimal charging strategy for a population of PEVs.

Algorithm 3.2 (*Decentralized coordination method*)

- Specify the aggregate base demand $\boldsymbol{d}$;
- Define an $\varepsilon_{\text{stop}}$ to terminate iterations;

- Initialize $\varepsilon > \varepsilon_{\text{stop}}$ and an initial price profile $\boldsymbol{p}^{(0)}$;
- Set $k = 0$;
- While $\varepsilon > \varepsilon_{\text{stop}}$
  - Determine the optimal charging profile $\boldsymbol{u}_n^{(k+1)}$ w.r.t. $\boldsymbol{p}^{(k)}$ for all $n \in \mathscr{N}$ PEVs simultaneously by minimizing the individual cost function (3.42),

$$\boldsymbol{u}_n^{(k+1)}(\boldsymbol{p}^{(k)}) \triangleq \underset{\boldsymbol{u}_n \in \mathscr{U}_n}{\operatorname{argmin}} \left\{ \sum_{t \in \mathscr{T}} p_t^{(k)} u_{nt} - v_n(\boldsymbol{u}_n) \right\};$$

  - Determine $\boldsymbol{p}^{(k+1)}$ from $\boldsymbol{p}^{(k)}$ and $\boldsymbol{u}^{(k+1)}(\boldsymbol{p}^{(k)})$ using (3.58),

$$p_t^{(k+1)} = p_t^{(k)} + \eta \left(c' \left(d_t + \sum_{n \in \mathscr{N}} u_{nt}^{(k+1)} \right) - p_t^{(k)} \right),$$

    for all $t \in \mathscr{T}$;
  - Update $\varepsilon := \|\boldsymbol{p}^{(k+1)} - \boldsymbol{p}^{(k)}\|_1$;
  - Update $k := k + 1$. ■

If Algorithm 3.2 converges, this decentralized process will achieve the efficient solution. Iterations could, however, be oscillatory or even divergent. In order to establish convergence, it is useful to define ν_{nt} as the Lipschitz constant for the function $[g'_{nt}]^{-1}(\cdot)$ over the interval $[g'_{nt}(0), g'_{nt}(\Gamma_n)]$, with

$$\nu = \max_{n \in \mathscr{N}, t \in \mathscr{T}} \nu_{nt}, \tag{3.59}$$

and to define κ as the Lipschitz constant for $c'(\cdot)$ over the typical range in the total demand. The following intermediate result is also required.

Lemma 3.7 Assume the terminal valuation function h_n is increasing and strictly concave. Then,

$$\|\boldsymbol{u}_n^*(\boldsymbol{p}) - \boldsymbol{u}_n^*(\boldsymbol{\rho})\|_1 \leq 2\nu \|\boldsymbol{p} - \boldsymbol{\rho}\|_1 \tag{3.60}$$

where $\|.\|_1$ denotes the l_1 norm of the associated vector.

The proof of Lemma 3.7 is given in Appendix 3.7.1. ■

It is now possible to establish the convergence properties of Algorithm 3.2, and hence of the decentralized coordination process.

Corollary 3.2 (Convergence of Algorithm 3.2) *Suppose* $\alpha \equiv |1 - \eta| + 2N\kappa\nu\eta < 1$ *and consider any initial charging price* $\boldsymbol{p}^{(0)}$. *Then Algorithm 3.2 converges to the efficient solution* $\boldsymbol{u}^{**}$ *which is specified in (3.39). Moreover, for any* $\varepsilon > 0$, *the system converges to a price profile* $\mathbf{p}$, *such that* $\|\boldsymbol{p} - \boldsymbol{p}^{**}\|_1 \leq \varepsilon$, *in* $K(\varepsilon)$ *iterations, with*

$$K(\varepsilon) = \left\lceil \frac{1}{\ln(\alpha)} \Big(\ln(\varepsilon) - \ln(T) - \ln(\varrho_{max}) \Big) \right\rceil, \tag{3.61}$$

where ϱ_{max} denotes the maximum possible price, and $\lceil x \rceil$ represents the minimal integer value larger than or equal to x.

Remarks:

(i) In practice, convergence to the desired tolerance ε requires many fewer iterations than the upper bound established in (3.61). Typical convergence behavior is illustrated in Sect. 3.5.

(ii) The upper bound on the iteration count, $K(\varepsilon)$, is of order $O(|\ln(\varepsilon)|)$, and is independent of the size of the PEV population. The choice of α is influenced by the PEV population size, with the condition $\alpha < 1$ requiring that $N < \frac{1}{2\kappa\nu}$. Notice from (3.36), though, that $\kappa(N) \in O(\frac{1}{N^2})$ because the generation cost $c(\cdot)$ must remain finite as the PEV population N grows. Therefore this necessary condition on N is not restrictive.

Proof of Corollary 3.2.

Consider a pair of price profiles $\boldsymbol{p}$ and $\boldsymbol{\rho}$, and the respective updated price profiles $\boldsymbol{p}^+$ and $\boldsymbol{\rho}^+$ given by (3.58). Then,

$$
\begin{aligned}
&\|\boldsymbol{p}^+ - \boldsymbol{\rho}^+\|_1 \\
&= \sum_{t\in\mathcal{T}} \left| \left\{ p_t + \eta \left(c'\left(d_t + \sum_{n\in\mathcal{N}} u^*_{nt}(\boldsymbol{p}) \right) - p_t \right) \right\} \right. \\
&\qquad \left. - \left\{ \varrho_t + \eta \left(c'\left(d_t + \sum_{n\in\mathcal{N}} u^*_{nt}(\boldsymbol{\rho}) \right) - \varrho_t \right) \right\} \right| \\
&= \sum_{t\in\mathcal{T}} \left| \eta \left(c'\left(d_t + \sum_{n\in\mathcal{N}} u^*_{nt}(\boldsymbol{p}) \right) - c'\left(d_t + \sum_{n\in\mathcal{N}} u^*_{nt}(\boldsymbol{\rho}) \right) \right) \right. \\
&\qquad \left. + (1-\eta)(p_t - \varrho_t) \right| \\
&\le \eta \sum_{t\in\mathcal{T}} \left| c'\left(d_t + \sum_{n\in\mathcal{N}} u^*_{nt}(\boldsymbol{p}) \right) - c'\left(d_t + \sum_{n\in\mathcal{N}} u^*_{nt}(\boldsymbol{\rho}) \right) \right| \\
&\qquad + |1-\eta| \times \|\boldsymbol{p} - \boldsymbol{\rho}\|_1 .
\end{aligned}
\tag{3.62}
$$

Also, given the definition of κ, it follows that,

$$
\begin{aligned}
&\sum_{t\in\mathcal{T}} \left| c'\left(d_t + \sum_{n\in\mathcal{N}} u^*_{nt}(\boldsymbol{p}) \right) - c'\left(d_t + \sum_{n\in\mathcal{N}} u^*_{nt}(\boldsymbol{\rho}) \right) \right| \\
&\qquad \le \kappa \sum_{t\in\mathcal{T}} \left| \sum_{n\in\mathcal{N}} u^*_{nt}(\boldsymbol{p}) - \sum_{n\in\mathcal{N}} u^*_{nt}(\boldsymbol{\rho}) \right| \\
&\qquad \le \kappa \sum_{t\in\mathcal{T}} \sum_{n\in\mathcal{N}} |u^*_{nt}(\boldsymbol{p}) - u^*_{nt}(\boldsymbol{\rho})|
\end{aligned}
$$

$$
\begin{aligned}
&= \kappa \sum_{n \in \mathcal{N}} \sum_{t \in \mathcal{T}} \left| u^*_{nt}(\boldsymbol{p}) - u^*_{nt}(\boldsymbol{\rho}) \right| \\
&= \kappa \sum_{n \in \mathcal{N}} \| \boldsymbol{u}^*_n(\boldsymbol{p}) - \boldsymbol{u}^*_n(\boldsymbol{\rho}) \|_1 \\
&\leq 2\kappa\nu \sum_{n \in \mathcal{N}} \| \boldsymbol{p} - \boldsymbol{\rho} \|_1 \qquad (3.63) \\
&= 2N\kappa\nu \| \boldsymbol{p} - \boldsymbol{\rho} \|_1 \qquad (3.64)
\end{aligned}
$$

where the inequality (3.63) invokes Lemma 3.7. Inequalities (3.62) and (3.64) together imply,

$$
\| \boldsymbol{p}^+ - \boldsymbol{\rho}^+ \|_1 \leq (|1 - \eta| + 2N\kappa\nu\eta) \| \boldsymbol{p} - \boldsymbol{\rho} \|_1. \qquad (3.65)
$$

If $|1 - \eta| + 2N\kappa\nu\eta < 1$, then,

$$
\| \boldsymbol{p}^+ - \boldsymbol{\rho}^+ \|_1 < \| \boldsymbol{p} - \boldsymbol{\rho} \|_1, \qquad (3.66)
$$

so the price update operator $\boldsymbol{p}^+(\boldsymbol{p})$ specified in (3.58) is a contraction map. Hence by the contraction mapping theorem [30], the system price $\boldsymbol{p}^{(k)}$ converges to a unique price profile $\boldsymbol{p}^*$ from any initial price profile $\boldsymbol{p}^{(0)}$. At the converged price, $\boldsymbol{p}^+(\boldsymbol{p}^*) = \boldsymbol{p}^*$, and so (3.58) implies,

$$
c'\left(d_t + \sum_{n \in \mathcal{N}} u^*_{nt}(\boldsymbol{p}^*) \right) = p^*_t, \quad \text{for all } t \in \mathcal{T}. \qquad (3.67)
$$

The price converges to the generation marginal cost over the charging horizon.

It will now be shown that $\boldsymbol{u}^*(\boldsymbol{p}^*)$ is the efficient (socially optimal) charging strategy that minimizes the central optimization problem (3.37). Let $\omega^* = \|\boldsymbol{u}^*(\boldsymbol{p}^*)\|_1$. If $0 < \omega^* < \Gamma_n$ then by Theorem 3.3, $A^*_n(\boldsymbol{p}^*) = h'_n(\omega^*)$. According to Lemma 3.4,

$$
p^*_t + g'_{nt}(u^*_{nt}(\boldsymbol{p}^*)) - h'_n(\omega^*) \begin{cases} = 0, & \text{if } u^*_{nt}(\boldsymbol{p}^*) > 0 \\ \geq 0, & \text{otherwise} \end{cases}
$$

so (3.39) is satisfied.

If $\omega^* = 0$ then by Theorem 3.3, $A^*_n(\boldsymbol{p}^*) \geq h'_n(\omega^*)$. Also, in this case, $u^*_{nt}(\boldsymbol{p}^*) = 0$ for all $t \in \mathcal{T}$. Therefore, from Lemma 3.4,

$$
p^*_t + g'_{nt}(u^*_{nt}(\boldsymbol{p}^*)) \geq A^*_n(\boldsymbol{p}^*) \geq h'_n(\omega^*)
$$

which again satisfies (3.39).

Suppose $\omega^* = \Gamma_n$, then by Theorem 3.3, $A_n^*(\boldsymbol{p}^*) \leq h'_n(\omega^*) = 0$. There must exist at least one instant $t \in \mathcal{T}_n$ where $u_{nt}^* > 0$. At such an instant, Lemma 3.4 implies,

$$p_t^* + g'_{nt}(u_{nt}^*(\boldsymbol{p}^*)) - A_n^*(\boldsymbol{p}^*) = 0$$
$$\Rightarrow \quad p_t^* + g'_{nt}(u_{nt}^*(\boldsymbol{p}^*)) \leq 0$$

and so $p_t^* < 0$ because $g'_{nt}(u_{nt}^*(\boldsymbol{p}^*)) > 0$ according to Assumptions 3.1 and 3.3. However this is not possible, as Assumption 3.2 ensures that $p_t^* > 0$ for all $t \in \mathcal{T}$. Hence, $\omega^* < \Gamma_n$, which is consistent with the efficient solution.

Therefore the collection of optimal PEV charging strategies is efficient, $\boldsymbol{u}^* = \boldsymbol{u}^{**}$, with respect to the converged price $\boldsymbol{p}^* = \boldsymbol{p}^{**}$.

Given $\alpha \equiv |1 - \eta| + 2N\kappa\nu\eta < 1$, (3.65) implies,

$$\|\boldsymbol{p}^{(k)} - \boldsymbol{p}^{**}\|_1 \leq \alpha^k \|\boldsymbol{p}^{(0)} - \boldsymbol{p}^{**}\|_1.$$

With $p_t^{(0)}, p_t^{**} \in [0, \varrho_{max}]$ for all $t \in \mathcal{T}$, this gives,

$$\|\boldsymbol{p}^{(k)} - \boldsymbol{p}^{**}\|_1 \leq \alpha^k T \varrho_{max},$$

which implies that $\|\boldsymbol{p}^{(k)} - \boldsymbol{p}^{**}\|_1 \leq \varepsilon$ for k satisfying (3.61). ■

3.5 Numerical Illustrations

3.5.1 Convergence

This section provides an illustration of Algorithm 3.2 using parameter values that match those of Sect. 3.4.2. For the generation cost function (3.40), the marginal cost is given by,

$$p(D_t) = c'(D_t) = 5.8 \times 10^{-7} D_t + 0.06.$$

The corresponding Lipschitz constant is $\kappa = 5.8 \times 10^{-7}$. From (3.41), the Lipschitz constant for $[g'_{nt}]^{-1}(\cdot)$ is $\nu = 1/0.006 = 166.7$. Given these values,

$$\alpha \equiv |1 - \eta| + 2N\kappa\nu\eta = |1 - \eta| + 0.967\eta,$$

which is less than 1 whenever $0 < \eta < 1.017$. Thus by Corollary 3.2, the system is guaranteed to converge to the efficient solution for all $\eta \in (0, 1.017)$. It is straightforward to show that α is a minimum when $\eta = 1$, giving $\alpha = 0.967 < 1$.

With $\alpha = 0.967$, $T = 24$ and $\varrho_{max} = 0.3$, Algorithm 3.2 will converge to a price profile $\boldsymbol{p}^*$, such that $\|\boldsymbol{p}^* - \boldsymbol{p}^{**}\|_1 < \varepsilon = 0.0001$, in less than $K = 334$ iterations, according to (3.61). Figure 3.14 shows the evolution of $\|\boldsymbol{p}^{(k)} - \boldsymbol{p}^{**}\|_1$, with $\eta = 1$, from an initial price profile $p_t^{(0)} = c'(d_t)$ for all $t \in \mathcal{T}$, i.e. zero charging load. It can

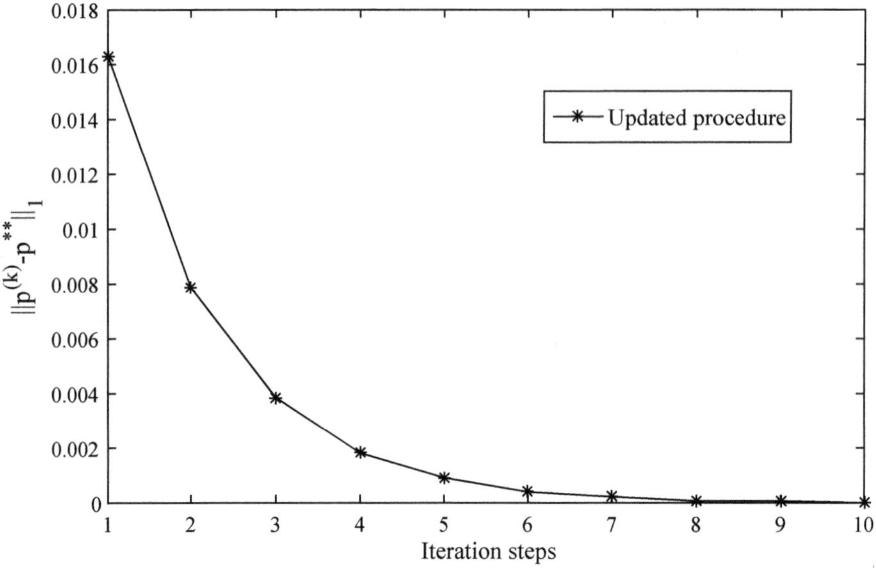

Fig. 3.14 Convergence of $\|\boldsymbol{p}^{(k)} - \boldsymbol{p}^{**}\|_1$ for Algorithm 3.2, with $\eta = 1$

be seen that convergence to the desired tolerance is achieved in about 10 iterations, which is much less than the theoretical upper bound of 334.

As η increases over the range $0 < \eta \leq 1$, the value of α decreases, with (3.65) suggesting faster convergence of Algorithm 3.2. Figure 3.15 shows this to be the case. Further increasing η results in α increasing, with convergence only guaranteed while $\eta < 1.017$. Nevertheless, as shown in Fig. 3.16, the algorithm may still converge even when this sufficient condition is violated. It is also apparent, however, that larger values of η result in non-convergence, with oscillations occurring when $\eta = 2$.

The price profile updates achieved by Algorithm 3.2 are shown in Fig. 3.17, while Fig. 3.18 shows the corresponding total aggregate demand at each iteration. Note that the converged case in Fig. 3.18 corresponds exactly to the socially optimal strategy in Fig. 3.8.

3.5.2 Heterogeneity

To consider the effect of heterogeneity in the PEV population, suppose that the initial value of each PEV's SoC, soc_{n0}, satisfies a Gaussian distribution $N(\hat{\mu}, \hat{\gamma})$ where $\hat{\mu} = 50\%$ and $\hat{\gamma} = 0.1$, which is consistent with [31, 32]. The updates of total aggregate demands are shown in Figs. 3.19 and 3.20 for charging interval in summer and spring seasons respectively.

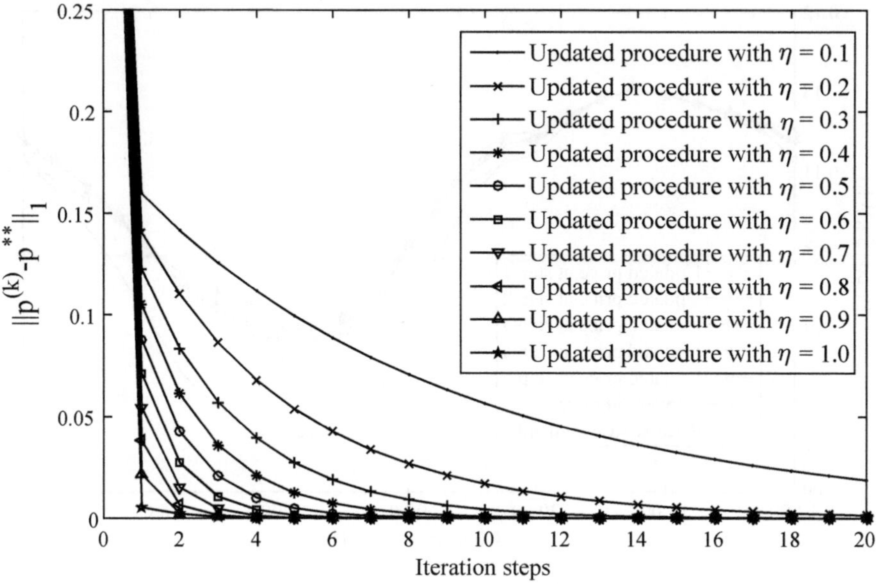

Fig. 3.15 Evolution of $\|\boldsymbol{p}^{(k)} - \boldsymbol{p}^{**}\|_1$ for values of η that satisfy the sufficient condition of Corollary 3.2

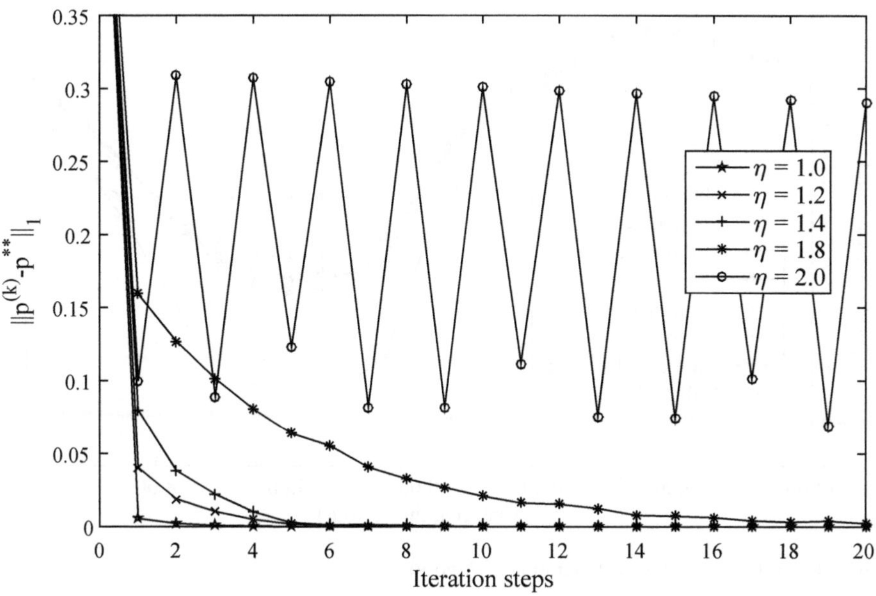

Fig. 3.16 Evolution of $\|\boldsymbol{p}^{(k)} - \boldsymbol{p}^{**}\|_1$, with the sufficient condition of Corollary 3.2 not satisfied for $\eta > 1.017$

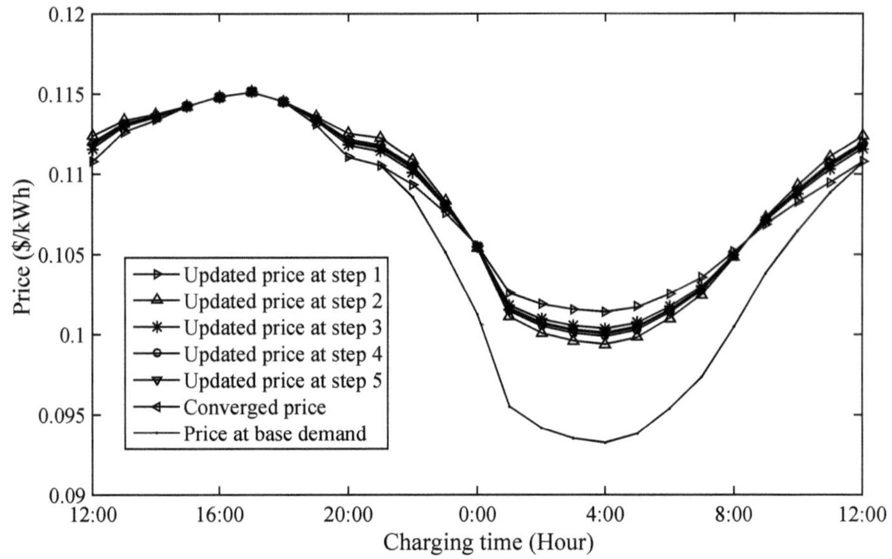

Fig. 3.17 Price profile updates achieved by Algorithm 3.2

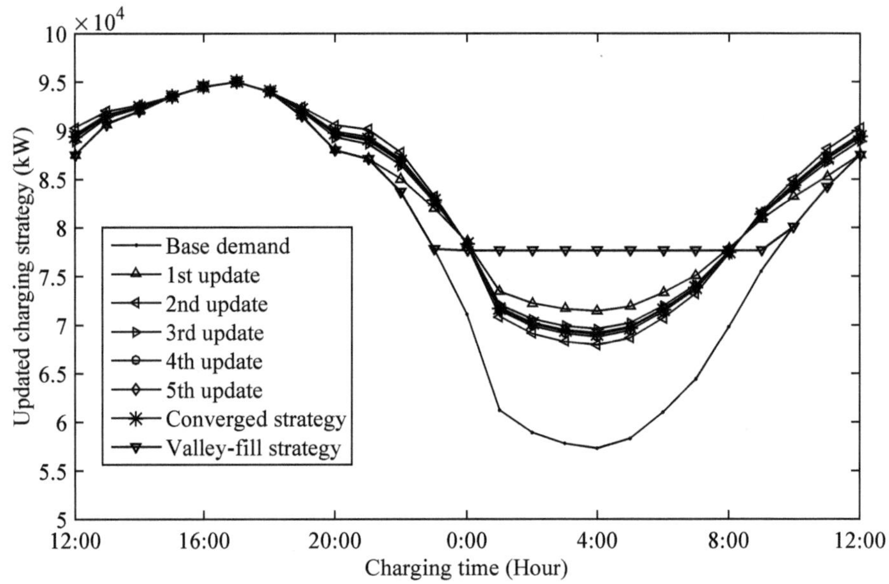

Fig. 3.18 Total demand at each iteration of Algorithm 3.2

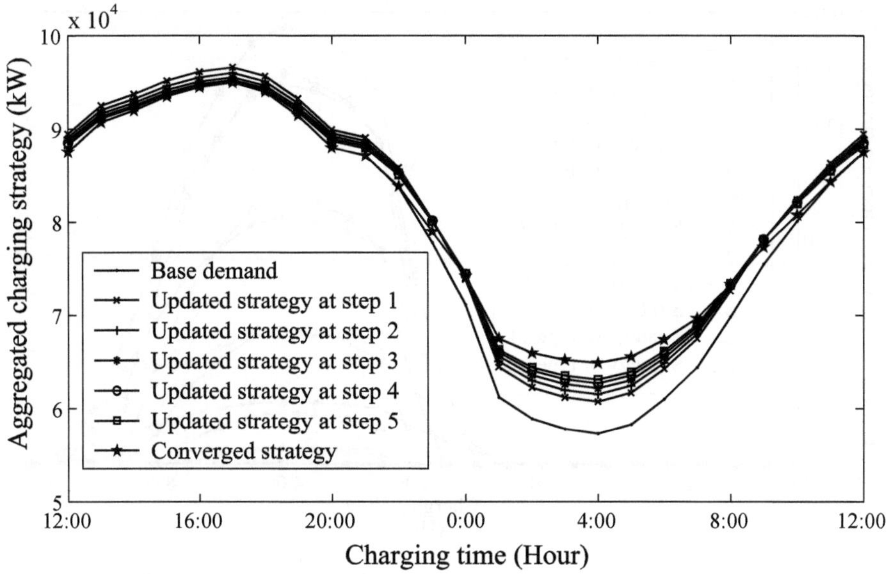

Fig. 3.19 Total demand at each iteration of Algorithm 3.2 for a heterogeneous PEV population

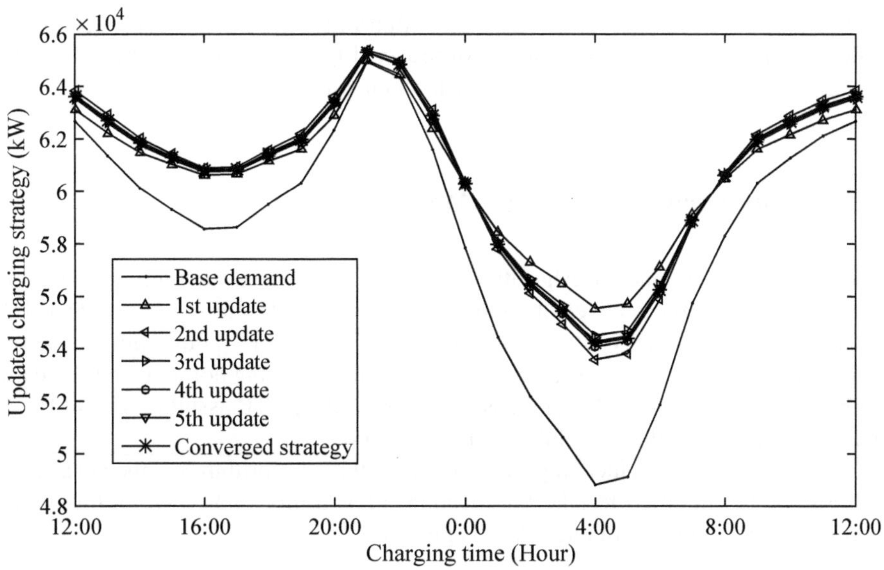

Fig. 3.20 Total demand at each iteration of Algorithm 3.2 for a heterogeneous PEV population with base demand typical of a spring day

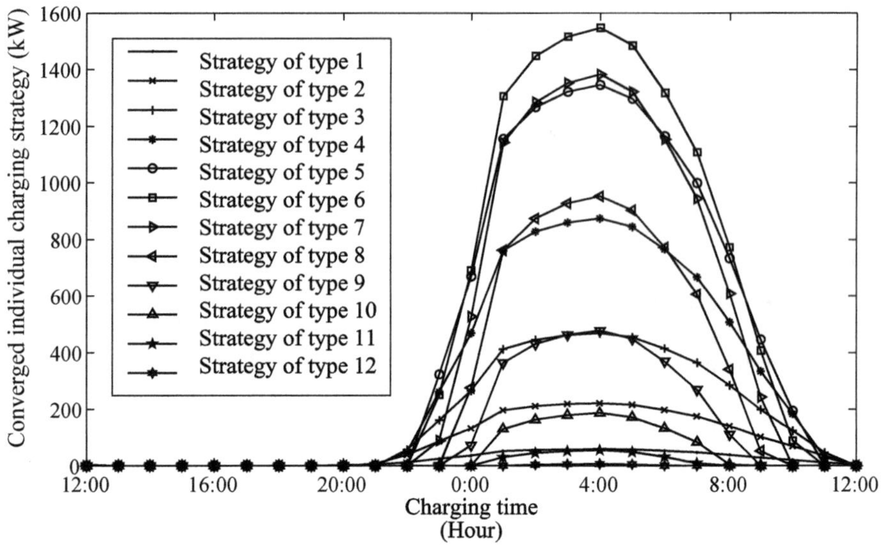

Fig. 3.21 Optimal charging strategies for individuals within a heterogeneous PEV population

By adopting the proposed decentralized algorithm, the process converges to the efficient solution in a few iterations. Moreover, Fig. 3.21 illustrates the converged charging strategies for a sample of the heterogeneous PEVs.

3.5.3 Comparison with Other Methods

The relative performance of Algorithm 3.2 will be illustrated through a comparison with the optimal decentralized charging algorithm of [16] (referred to as GTL). The optimal charging strategy attained by the GTL algorithm is valley filling since the objective is to minimize the electricity cost over the charging horizon, and there are no battery degradation or demand costs involved. Also, GTL assumes that all PEVs are fully charged by the end of the charging period.

In order to provide a meaningful comparison with Algorithm 3.2, the GTL algorithm must be modified to take into account PEV utility (3.34). Accordingly, the optimal charging profile of each PEV is given by,

$$\begin{aligned} \boldsymbol{u}_n^{(k+1)}(\boldsymbol{u}_n^{(k)}, \boldsymbol{p}^{(k)}) \triangleq \underset{\boldsymbol{u}_n \in \mathscr{U}_n}{\operatorname{argmin}} \sum_{t \in \mathscr{T}} \left\{ p_t^{(k)} u_{nt} + g_{nt}(u_{nt}) \right\} \\ + \frac{1}{2} \|\boldsymbol{u}_n - \boldsymbol{u}_n^{(k)}\|^2 - h_n \left(\sum_{t \in \mathscr{T}} u_{nt} \right). \end{aligned} \tag{3.68}$$

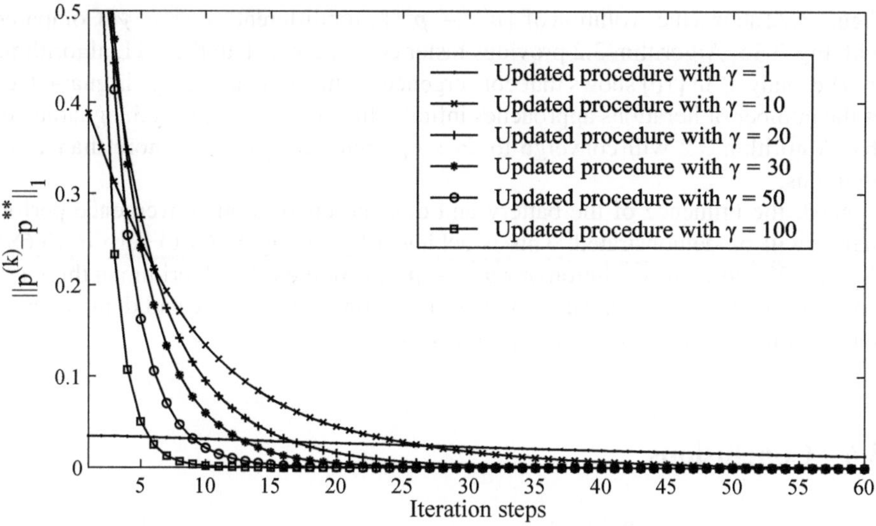

Fig. 3.22 Evolution of $\|\boldsymbol{p}^{(k)} - \boldsymbol{p}^{**}\|_1$ for the GTL algorithm for different values of γ

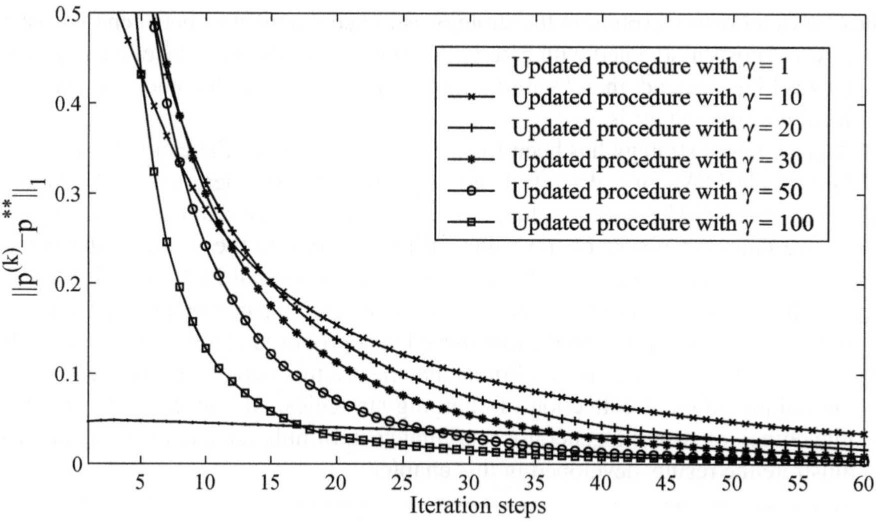

Fig. 3.23 Evolution of $\|\boldsymbol{p}^{(k)} - \boldsymbol{p}^{**}\|_1$ for the GTL algorithm without battery and demand costs

The parameter values from Sect. 3.5.1 have again been used for generation cost and PEV charging characteristics.

The GTL algorithm defines a parameter γ which governs the update process at each iteration,

$$p_t^{(k+1)} = \gamma c'\Big(d_t + \sum_{n\in\mathcal{N}} u_{nt}^{(k+1)}\Big).$$

Figure 3.22 shows the evolution of $\|\boldsymbol{p}^{(k)} - \boldsymbol{p}^{**}\|_1$ for different values of γ. Compared with Fig. 3.15, Algorithm 3.2 provides faster convergence than the GTL algorithm.

The analysis in [16] shows that convergence to the optimal strategy is guaranteed as the number of iterations approaches infinity. In contrast, Corollary 3.2 guarantees that Algorithm 3.2 will converge to an ε-optimal strategy in no more than $K(\varepsilon)$ iterations.

Next, the influence of the battery and demand charges on convergence performance will be demonstrated. This is achieved by setting $g_{nt}(\cdot)$ to zero in (3.68). Figure 3.23 shows the evolution of $\|\boldsymbol{p}^{(k)} - \boldsymbol{p}^{**}\|_1$ for the GTL algorithm in this case. Comparison with Fig. 3.22 suggests that inclusion of the battery and demand costs tends to improve the convergence performance.

3.6 Conclusions

With the increasing of PEVs, they may have significant impacts on the power grid. And some key characteristics of chemical batteries, like the state of health, the growth of resistance, and the cycle life, are effected by the charging behaviors. In this chapter, the cost function underpinning the strategy establishes a tradeoff between the cost of energy and costs associated with battery degradation. It also introduces a charge that penalizes high demand, thereby mitigating occurrences of high coincident charging on local distribution grids.

A price-based strategy has been formulated to coordinate the charging strategies of large-scale PEVs in a decentralized way with the consideration of the battery degradation cost. The optimal charging strategy avoids the damage caused by the excessive charging power of the valley filling strategy to the grid and the batteries of PEVs. A decentralized scheme is proposed where all PEVs simultaneously update their optimal charging strategies with respect to a common price profile, and then the price profile is updated using these latest proposed charging strategies. The chapter establishes sufficient conditions that ensure this iterative process converges to the unique efficient collection of charging strategies. At convergence, the price profile coincides with the generator marginal cost. Simulation examples are studied to illustrate the results developed in this chapter.

3.7 Appendices

3.7.1 Proof of Lemma 3.7

Recall that $\boldsymbol{u}_n^*(\boldsymbol{p}), \boldsymbol{u}_n^*(\boldsymbol{\rho}) \in \mathcal{U}_n$ represent the optimal response of PEV n with respect to price profiles $\boldsymbol{p}$ and $\boldsymbol{\rho}$ respectively,

$$\boldsymbol{u}_n^*(\boldsymbol{p}) \triangleq \min_{\boldsymbol{u}_n \in \mathscr{U}_n} J_n(\boldsymbol{u}_n; \boldsymbol{p}), \quad \boldsymbol{u}_n^*(\boldsymbol{\rho}) \triangleq \min_{\boldsymbol{u}_n \in \mathscr{U}_n} J_n(\boldsymbol{u}_n; \boldsymbol{\rho}).$$

Also, by Theorem 3.3,

$$\boldsymbol{u}_n^*(\boldsymbol{p}) = \boldsymbol{u}_n(\boldsymbol{p}, A_n^*(\boldsymbol{p})), \quad \boldsymbol{u}_n^*(\boldsymbol{\rho}) = \boldsymbol{u}_n(\boldsymbol{\rho}, A_n^*(\boldsymbol{\rho})).$$

For later analysis, it is useful to define another charging strategy for PEV n,

$$\boldsymbol{\vartheta}_n(\boldsymbol{p}, \boldsymbol{\rho}) \equiv \boldsymbol{u}_n(\boldsymbol{\rho}, A_n^*(\boldsymbol{p})), \tag{3.69}$$

which is the charging strategy satisfying (3.44) with respect to $\boldsymbol{\rho}$ and $A_n^*(\boldsymbol{p})$. Note that $\boldsymbol{\vartheta}_n(\boldsymbol{p}, \boldsymbol{\rho})$ may not satisfy the admissibility constraint (3.1a). However, $\boldsymbol{\vartheta}_n(\boldsymbol{p}, \boldsymbol{\rho})$ will only be used as a medium to establish a relationship between the pair of admissible charging strategies $\boldsymbol{u}_n^*(\boldsymbol{p})$ and $\boldsymbol{u}_n^*(\boldsymbol{\rho})$.

It is sufficient to show (3.60) if the following inequalities hold:

$$|u_{nt}^*(\boldsymbol{p}) - \vartheta_{nt}(\boldsymbol{p}, \boldsymbol{\rho})| \le \nu |p_t - \varrho_t|, \quad \forall t \in \mathscr{T}, \tag{3.70}$$

$$\|\boldsymbol{u}_n^*(\boldsymbol{p}) - \boldsymbol{u}_n^*(\boldsymbol{\rho})\|_1 \le 2\|\boldsymbol{u}_n^*(\boldsymbol{p}) - \boldsymbol{\vartheta}_n(\boldsymbol{p}, \boldsymbol{\rho})\|_1. \tag{3.71}$$

These relationships will be verified below respectively.

Verifying (3.70)

Firstly, consider the case where $u_{nt}^*(\boldsymbol{p}), \vartheta_{nt}(\boldsymbol{p}, \boldsymbol{\rho}) > 0$. It follows from (3.44), Theorem 3.3 and (3.69) that:

$$\begin{aligned} &|u_{nt}^*(\boldsymbol{p}) - \vartheta_{nt}(\boldsymbol{p}, \boldsymbol{\rho})| \\ &\quad = \left|[g_{nt}']^{-1}(A_n^*(\boldsymbol{p}) - p_t) - [g_{nt}']^{-1}(A_n^*(\boldsymbol{p}) - \varrho_t)\right| \\ &\quad \le \nu |p_t - \varrho_t|, \end{aligned}$$

where the inequality holds by the specification of ν given in (3.59). Following similar analysis, (3.70) holds for the other cases.

Verifying (3.71)

The proof of (3.71) will be established by considering three cases.
Case 1: $\|\boldsymbol{\vartheta}_n(\boldsymbol{p}, \boldsymbol{\rho})\|_1 = \|\boldsymbol{u}_n^*(\boldsymbol{p})\|_1$
Let $\|\boldsymbol{u}_n^*(\boldsymbol{p})\|_1 = \bar{\omega}$ so that,

$$\|\boldsymbol{\vartheta}_n(\boldsymbol{p}, \boldsymbol{\rho})\|_1 = \|\boldsymbol{u}_n(\boldsymbol{\rho}, A_n^*(\boldsymbol{p}))\|_1 = \bar{\omega}.$$

Then (3.52) implies $\mathscr{A}(\boldsymbol{\rho}, \bar{\omega}) = A_n^*(\boldsymbol{p})$. From (3.55),

$$\widehat{h}_n(\boldsymbol{\rho}, \bar{\omega}) = \mathscr{A}(\boldsymbol{\rho}, \bar{\omega}) - h_n'(\bar{\omega}) = A_n^*(\boldsymbol{p}) - h_n'(\bar{\omega}) = \widehat{h}_n(\boldsymbol{p}, \bar{\omega}).$$

It may be concluded from (3.56) that $A_n^*(\boldsymbol{\rho}) = A_n^*(\boldsymbol{p})$, and so $\boldsymbol{\vartheta}_n(\boldsymbol{p}, \boldsymbol{\rho}) = \boldsymbol{u}_n(\boldsymbol{\rho}, A_n^*(\boldsymbol{\rho})) = \boldsymbol{u}_n^*(\boldsymbol{\rho})$. Consequently,

$$\|\boldsymbol{u}_n^*(\boldsymbol{p}) - \boldsymbol{u}_n^*(\boldsymbol{\rho})\|_1 = \|\boldsymbol{u}_n^*(\boldsymbol{p}) - \boldsymbol{\vartheta}_n(\boldsymbol{p}, \boldsymbol{\rho})\|_1 \leq 2\|\boldsymbol{u}_n^*(\boldsymbol{p}) - \boldsymbol{\vartheta}_n(\boldsymbol{p}, \boldsymbol{\rho})\|_1.$$

Case 2: $\|\boldsymbol{\vartheta}_n(\boldsymbol{p}, \boldsymbol{\rho})\|_1 > \|\boldsymbol{u}_n^*(\boldsymbol{p})\|_1$

The first step is to show that,

$$\|\boldsymbol{\vartheta}_n(\boldsymbol{p}, \boldsymbol{\rho})\|_1 \geq \|\boldsymbol{u}_n^*(\boldsymbol{\rho})\|_1 \geq \|\boldsymbol{u}_n^*(\boldsymbol{p})\|_1. \tag{3.72}$$

This is achieved using proof by contradiction. Three subcases will be considered: (2A) $\|\boldsymbol{u}_n^*(\boldsymbol{p})\|_1 = \Gamma_n$, (2B) $0 < \|\boldsymbol{u}_n^*(\boldsymbol{p})\|_1 < \Gamma_n$, and (2C) $\|\boldsymbol{u}_n^*(\boldsymbol{p})\|_1 = 0$.

Case (2A): $\|\boldsymbol{u}_n^*(\boldsymbol{p})\|_1 = \Gamma_n$.

The charging strategy $\boldsymbol{\vartheta}_n(\boldsymbol{p}, \boldsymbol{\rho})$ is not admissible since the total charging demand $\|\boldsymbol{\vartheta}_n(\boldsymbol{p}, \boldsymbol{\rho})\|_1 > \Gamma_n$. Because $\boldsymbol{u}_n^*(\boldsymbol{\rho})$ must be admissible, $\|\boldsymbol{u}_n^*(\boldsymbol{\rho})\|_1$ cannot exceed $\|\boldsymbol{u}_n^*(\boldsymbol{p})\|_1 = \Gamma_n$. Therefore $\|\boldsymbol{u}_n^*(\boldsymbol{\rho})\|_1 < \|\boldsymbol{\vartheta}_n(\boldsymbol{p}, \boldsymbol{\rho})\|_1$ which implies, by Lemma 3.5, that

$$A_n^*(\boldsymbol{\rho}) < A_n^*(\boldsymbol{p}). \tag{3.73}$$

To show that $\|\boldsymbol{u}_n^*(\boldsymbol{\rho})\|_1 = \|\boldsymbol{u}_n^*(\boldsymbol{p})\|_1$, assume $\|\boldsymbol{u}_n^*(\boldsymbol{\rho})\|_1 < \|\boldsymbol{u}_n^*(\boldsymbol{p})\|_1$. Then by (3.56) and the assumed condition $\|\boldsymbol{u}_n^*(\boldsymbol{p})\|_1 = \Gamma_n$, it follows that,

$$h_n'(\omega)|_{\omega=\|\boldsymbol{u}_n^*(\boldsymbol{\rho})\|_1} \leq A_n^*(\boldsymbol{\rho}) \tag{3.74a}$$

$$h_n'(\omega)|_{\omega=\|\boldsymbol{u}_n^*(\boldsymbol{p})\|_1} \geq A_n^*(\boldsymbol{p}). \tag{3.74b}$$

From (3.73) and (3.74),

$$h_n'(\omega)|_{\omega=\|\boldsymbol{u}_n^*(\boldsymbol{\rho})\|_1} < h_n'(\omega)|_{\omega=\|\boldsymbol{u}_n^*(\boldsymbol{p})\|_1},$$

which implies, by the concavity of h_n, that $\|\boldsymbol{u}_n^*(\boldsymbol{\rho})\|_1 > \|\boldsymbol{u}_n^*(\boldsymbol{p})\|_1$. However this contradicts the assumption $\|\boldsymbol{u}_n^*(\boldsymbol{\rho})\|_1 < \|\boldsymbol{u}_n^*(\boldsymbol{p})\|_1$. Hence,

$$\|\boldsymbol{\vartheta}_n(\boldsymbol{p}, \boldsymbol{\rho})\|_1 > \|\boldsymbol{u}_n^*(\boldsymbol{\rho})\|_1 = \|\boldsymbol{u}_n^*(\boldsymbol{p})\|_1 = \Gamma_n.$$

Case (2B): $0 < \|\boldsymbol{u}_n^*(\boldsymbol{p})\|_1 < \Gamma_n$.

The desired result will be achieved by showing $\|\boldsymbol{\vartheta}_n(\boldsymbol{p}, \boldsymbol{\rho})\|_1 \geq \|\boldsymbol{u}_n^*(\boldsymbol{\rho})\|_1$ in (2B.1), and $\|\boldsymbol{u}_n^*(\boldsymbol{\rho})\|_1 \geq \|\boldsymbol{u}_n^*(\boldsymbol{p})\|_1$ in (2B.2).

(2B.1) Suppose that,

$$\|\boldsymbol{\vartheta}_n(\boldsymbol{p}, \boldsymbol{\rho})\|_1 < \|\boldsymbol{u}_n^*(\boldsymbol{\rho})\|_1. \tag{3.75}$$

Then $\|\boldsymbol{u}_n^*(\boldsymbol{\rho})\|_1 > 0$ because $\|\boldsymbol{\vartheta}_n(\boldsymbol{p}, \boldsymbol{\rho})\|_1 > 0$. Also, Lemma 3.5 implies $A_n^*(\boldsymbol{p}) < A_n^*(\boldsymbol{\rho})$. Together with $0 < \|\boldsymbol{u}_n^*(\boldsymbol{p})\|_1 < \Gamma_n$ and (3.56), this gives,

$$\begin{aligned} h_n'(\omega)|_{\omega=\|\boldsymbol{u}_n^*(\boldsymbol{p})\|_1} &= A_n^*(\boldsymbol{p}) \\ &< A_n^*(\boldsymbol{\rho}) \le h_n'(\omega)|_{\omega=\|\boldsymbol{u}_n^*(\boldsymbol{\rho})\|_1}. \end{aligned} \tag{3.76}$$

However, (3.75) and $\|\boldsymbol{\vartheta}_n(\boldsymbol{p},\boldsymbol{\rho})\|_1 > \|\boldsymbol{u}_n^*(\boldsymbol{p})\|_1$ together imply $\|\boldsymbol{u}_n^*(\boldsymbol{\rho})\|_1 > |\boldsymbol{u}_n^*(\boldsymbol{p})\|_1$, and so,

$$h_n'(\omega)|_{\omega=\|\boldsymbol{u}_n^*(\boldsymbol{\rho})\|_1} \le h_n'(\omega)|_{\omega=\|\boldsymbol{u}_n^*(\boldsymbol{p})\|_1} \tag{3.77}$$

by the concavity of h_n. This contradicts (3.76), indicating that the original assumption (3.75) is incorrect. Hence, $\|\boldsymbol{\vartheta}_n(\boldsymbol{p},\boldsymbol{\rho})\|_1 \ge \|\boldsymbol{u}_n^*(\boldsymbol{\rho})\|_1$.

(2B.2) Suppose that

$$\|\boldsymbol{u}_n^*(\boldsymbol{\rho})\|_1 < \|\boldsymbol{u}_n^*(\boldsymbol{p})\|_1. \tag{3.78}$$

Then $\|\boldsymbol{u}_n^*(\boldsymbol{\rho})\|_1 < \Gamma_n$, and $\|\boldsymbol{u}_n^*(\boldsymbol{\rho})\|_1 < \|\boldsymbol{\vartheta}_n(\boldsymbol{p},\boldsymbol{\rho})\|_1$ because $\|\boldsymbol{\vartheta}_n(\boldsymbol{p},\boldsymbol{\rho})\|_1 > \|\boldsymbol{u}_n^*(\boldsymbol{p})\|_1$. Hence, by Lemma 3.5, $A_n^*(\boldsymbol{\rho}) < A_n^*(\boldsymbol{p})$. Together with $0 < \|\boldsymbol{u}_n^*(\boldsymbol{p})\|_1 < \Gamma_n$ and (3.56), this gives,

$$\begin{aligned} h_n'(\omega)|_{\omega=\|\boldsymbol{u}_n^*(\boldsymbol{\rho})\|_1} &\le A_n^*(\boldsymbol{\rho}) \\ &< A_n^*(\boldsymbol{p}) = h_n'(\omega)|_{\omega=\|\boldsymbol{u}_n^*(\boldsymbol{p})\|_1}. \end{aligned} \tag{3.79}$$

However, given the concavity of h_n, (3.78) implies,

$$h_n'(\omega)|_{\omega=\|\boldsymbol{u}_n^*(\boldsymbol{\rho})\|_1} > h_n'(\omega)|_{\omega=\|\boldsymbol{u}_n^*(\boldsymbol{p})\|_1}, \tag{3.80}$$

which contradicts (3.79). Hence, $\|\boldsymbol{u}_n^*(\boldsymbol{\rho})\|_1 \ge \|\boldsymbol{u}_n^*(\boldsymbol{p})\|_1$.

Case (2C): $\|\boldsymbol{u}_n^*(\boldsymbol{p})\|_1 = 0$.

Because $\boldsymbol{u}_n^*(\boldsymbol{\rho})$ is an admissible strategy, $\|\boldsymbol{u}_n^*(\boldsymbol{\rho})\|_1 \ge \|\boldsymbol{u}_n^*(\boldsymbol{p})\|_1 = 0$. Suppose that,

$$\|\boldsymbol{\vartheta}_n(\boldsymbol{p},\boldsymbol{\rho})\|_1 < \|\boldsymbol{u}_n^*(\boldsymbol{\rho})\|_1. \tag{3.81}$$

Then $\|\boldsymbol{u}_n^*(\boldsymbol{\rho})\|_1 > 0$. Also, Lemma 3.5 implies $A_n^*(\boldsymbol{p}) < A_n^*(\boldsymbol{\rho})$. Together with $\|\boldsymbol{u}_n^*(\boldsymbol{p})\|_1 = 0$ and (3.56), this gives,

$$\begin{aligned} h_n'(\omega)|_{\omega=\|\boldsymbol{u}_n^*(\boldsymbol{p})\|_1} &\le A_n^*(\boldsymbol{p}) \\ &< A_n^*(\boldsymbol{\rho}) \le h_n'(\omega)|_{\omega=\|\boldsymbol{u}_n^*(\boldsymbol{\rho})\|_1}. \end{aligned} \tag{3.82}$$

However, (3.81) together with $\|\boldsymbol{\vartheta}_n(\boldsymbol{p},\boldsymbol{\rho})\|_1 > \|\boldsymbol{u}_n^*(\boldsymbol{p})\|_1$ imply $\|\boldsymbol{u}_n^*(\boldsymbol{\rho})\|_1 > \|\boldsymbol{u}_n^*(\boldsymbol{p})\|_1$, giving,

$$h_n'(\omega)|_{\omega=\|\boldsymbol{u}_n^*(\boldsymbol{p})\|_1} \ge h_n'(\omega)|_{\omega=\|\boldsymbol{u}_n^*(\boldsymbol{\rho})\|_1}, \tag{3.83}$$

by the concavity of h_n. This contradicts (3.82), indicating that the assumption (3.81) is incorrect. Hence, $\|\boldsymbol{\vartheta}_n(\boldsymbol{p},\boldsymbol{\rho})\|_1 \ge \|\boldsymbol{u}_n^*(\boldsymbol{\rho})\|_1$.

Case 2 summary: It has been shown in Cases (2A)–(2C) that (3.72) holds. Because $\|\boldsymbol{\vartheta}_n(\boldsymbol{p},\boldsymbol{\rho})\|_1 \ge \|\boldsymbol{u}_n^*(\boldsymbol{\rho})\|_1$, Lemma 3.5 implies that $A_n^*(\boldsymbol{\rho}) \le A_n^*(\boldsymbol{p})$, and therefore that,

$$0 \le u^*_{nt}(\boldsymbol{\rho}) \le \vartheta_{nt}(\boldsymbol{p}, \boldsymbol{\rho}), \tag{3.84}$$

for all $t \in \mathscr{T}$. Hence,

$$\begin{aligned} 0 &\le \|\boldsymbol{\vartheta}_n(\boldsymbol{p}, \boldsymbol{\rho}) - \boldsymbol{u}^*_n(\boldsymbol{\rho})\|_1 \\ &= \|\boldsymbol{\vartheta}_n(\boldsymbol{p}, \boldsymbol{\rho})\|_1 - \|\boldsymbol{u}^*_n(\boldsymbol{\rho})\|_1 \qquad \text{by (3.84)} \\ &\le \|\boldsymbol{\vartheta}_n(\boldsymbol{p}, \boldsymbol{\rho})\|_1 - \|\boldsymbol{u}^*_n(\boldsymbol{p})\|_1 \qquad \text{by (3.72)} \\ &\le \|\boldsymbol{\vartheta}_n(\boldsymbol{p}, \boldsymbol{\rho}) - \boldsymbol{u}^*_n(\boldsymbol{p})\|_1 \end{aligned} \tag{3.85}$$

where the final inequality is due to the reverse triangle inequality. Therefore,

$$\begin{aligned} &\|\boldsymbol{u}^*_n(\boldsymbol{p}) - \boldsymbol{u}^*_n(\boldsymbol{\rho})\|_1 \\ &\quad \le \|\boldsymbol{\vartheta}_n(\boldsymbol{p}, \boldsymbol{\rho}) - \boldsymbol{u}^*_n(\boldsymbol{p})\|_1 + \|\boldsymbol{\vartheta}_n(\boldsymbol{p}, \boldsymbol{\rho}) - \boldsymbol{u}^*_n(\boldsymbol{\rho})\|_1 \\ &\quad \le \|\boldsymbol{\vartheta}_n(\boldsymbol{p}, \boldsymbol{\rho}) - \boldsymbol{u}^*_n(\boldsymbol{p})\|_1 + \|\boldsymbol{\vartheta}_n(\boldsymbol{p}, \boldsymbol{\rho}) - \boldsymbol{u}^*_n(\boldsymbol{p})\|_1 \\ &\quad = 2\|\boldsymbol{u}^*_n(\boldsymbol{p}) - \boldsymbol{\vartheta}_n(\boldsymbol{p}, \boldsymbol{\rho})\|_1, \end{aligned}$$

where the first inequality is due to the triangle inequality and the second is from (3.85).

Case 3: $\|\boldsymbol{\vartheta}_n(\boldsymbol{p}, \boldsymbol{\rho})\|_1 < \|\boldsymbol{u}^*_n(\boldsymbol{p})\|_1$

Following a similar approach to that adopted for Case 2, it can again be shown that (3.71) holds. ■

References

1. S. Bashash, S.J. Moura, J.C. Forman, H.K. Fathy, Plug-in hybrid electric vehicle charge pattern optimization for energy cost and battery longevity. J. Power Sour. **196**, 541–549 (2011)
2. J. Wang, P. Liu, J. Hicks-Garner, E. Sherman, S. Soukiazian, M. Verbrugge, H. Tataria, J. Musser, P. Finamore, Cycle-life model for graphite-LiFePO4 cells. J. Power Sour. **194**(8), 3942–3948 (2011)
3. L. Gan, U. Topcu, S.H. Low, Stochastic distributed protocol for electric vehicle charging with discrete charging rate, in *IEEE Power and Energy Society General Meeting* (2012), pp. 1–8
4. K.W.E. Cheng, B.P. Divakar, H. Wu, K. Ding, H.F. Ho, Battery-management system (BMS) and SOC development for electrical vehicles. IEEE Trans. Veh. Technol. **60**(1), 76–88 (2011)
5. J. Forman, J. Stein, H. Fathy, Optimization of dynamic battery paramter characterization experiments via differential evolution. In *American Control Conference (ACC)*, pp. 867–874, Washington, DC, USA, 17–19 June 2013
6. J. Forman, S. Moura, J. Stein, H. Fathy, Optimal experimental design for modeling battery degradation, in *Proceedings of Dynamic Systems and Control Conference* (2012), pp. 309–318
7. S. Moura, J. Forman, S. Bashash, J. Stein, H. Fathy, Optimal control of film growth in lithium-ion battery packs via relay switches. IEEE Trans. Ind. Electr. **58**(8), 3555–3566 (2011)
8. D. Callaway, I. Hiskens, Achieving controllability of electric loads. Proc. IEEE **99–1**, 184–199 (2011)

9. A.-H. Mohsenian-Rad, A. Leon-Garcia, Optimal residential load control with price prediction in real-time electricity pricing environments. IEEE Trans. Smart Grid **1**(2), 120–133 (2010)
10. P. Samadi, A. Mohsenian-Rad, R. Schober, V. Wong, J. Jatskevich, Optimal real-time pricing algorithm based on utility maximization for smart grid, in *Proceedings of 1st IEEE International Conference on Smart Grid Communications*, Gaithersburg, 4–6 Oct 2010
11. R.A. Waraich, M. Galus, C. Dobler, M. Balmer, G. Andersson, K. Axhausen, Plug-in hybrid electric vehicles and smart grid: investigations based on a micro-simulation. Technical Report 10.3929/ethza-005916811 (Institute for Transport Planning and Systems, ETH Zurich, 2009)
12. C. Wu, H. Mohsenian-Rad, J. Huang, Vehicle-to-aggregator interaction game. IEEE Trans. Smart Grid **3**(1), 434–442 (2012)
13. Z. Fan, A distributed demand response algorithm and its application to PHEV charging in smart grids. IEEE Trans. Smart Grid **3**(3), 1280–1290 (2012)
14. L. Gan, N. Chen, A. Wierman, U. Topcu, S.H. Low, Real-time deferrable load control: handling the uncertainties of renewable generation, in *Fourth International Conference on Future Energy Systems*, ACM (2013)
15. Z. Ma, D. Callaway, I. Hiskens, Decentralized charging control of large populations of plug-in electric vehicles. IEEE Trans. Control Syst. Technol. **21**(1), 67–78 (2013)
16. L. Gan, U. Topcu, S. Low, Optimal decentralized protocol for electric vehicle charging. IEEE Trans. Power Syst. **28**(2), 940–951 (2013)
17. F. Kelly, A. Maulloo, D. Tan, Rate control for communication networks: shadow prices, proportional fairness and stability. J. Oper. Res. Soc. **49**(3), 237–252 (1998)
18. T. Lee, Z. Bareket, T. Gordon, Stochastic modeling for studies of real-world PHEV usage: driving schedule and daily temporal distributions. IEEE Trans. Veh. Technol. **61**(4), 1493–1502 (2012)
19. P. Denholm, W. Short, An evaluation of utility system impacts and benefits of optimally dispatched plug-in hybrid electric vehicles. Technical Report NREL/TP-620-40293 (National Renewable Energy Laboratory, 2006)
20. P. Prosini, Modeling the voltage profile for LiFePO4. J. Electrochem. Soc. **152**(10), A1925–A1929 (2005)
21. J. Kim, G.-S. Seo, C. Chun, B.-H. Cho, S. Lee, OCV hysteresis effect-based SOC estimation in extended Kalman filter algorithm for an LiFePO4/C cell, in *IEEE International Electric Vehicle Conference (IEVC)*, pp. 1–5, Greenville, SC, 4–8 March 2012
22. A. Singh, A. Izadian, S. Anwar, Fault diagnosis of li-ion batteries using multiple-model adaptive estimation, in *39th Annual Conference of the IEEE Industrial Electronics Society (IECON)*, pp. 3524–3529, Vienna, 10–13 November 2013
23. M.D. Galus, G. Andersson. Demand management of grid connected plug-in hybrid electric vehicles (PHEV), in *IEEE Energy 2030*, Atlanta, Georgia, 17–18 Nov 2008
24. Y. He, B. Venkatesh, L. Guan, Optimal scheduling for charging and discharging of electric vehicles. IEEE Trans. Smart Grid **3**(3), 1095–1105 (2012)
25. E. Bompard, Y. Ma, R. Napoli, G. Abrate, The demand elasticity impacts on the strategic bidding behavior of the electricity producers. IEEE Trans. Power Syst. **22**(1), 188–197 (2007)
26. V.P. Gountis, A.G. Bakirtzis, Bidding strategies for electricity producers in a competitive electricity marketplace. IEEE Trans. Power Syst. **19**(1), 356–365 (2004)
27. F.S. Wen, A.K. David, Strategic bidding for electricity supply in a day-ahead energy market. Electr. Power Syst. Res. **59**, 197–206 (2001)
28. S. Boyd, L. Vandenberghe, *Convex Optimization* (Cambridge University Press, Cambridge, 2004)
29. O. Sundstrom, C. Binding, Planning electric-drive vehicle charging under constrained grid conditions. Technical Report (IBM - Zurich, Switzerland, 2010)
30. D.R. Smart, *Fixed Point Theorems* (Cambridge University Press, London, 1974)
31. Z. Luo, Z. Hu, Y. Song, Z. Xu, H. Lu, Optimal coordination of plug-in electric vehicles in power grids with cost-benefit analysis - part I: enabling techniques. IEEE Trans. Power Syst. **28**(4), 3546–3555 (2013)

32. Z. Luo, Z. Hu, Y. Song, Z. Xu, H. Lu, Optimal coordination of plug-in electric vehicles in power grids with cost-benefit analysis - part II: a case study in China. IEEE Trans. Power Syst. **28**(4), 3556–3565 (2013)
33. R. Deliso, Understanding peak demand charges. EnerNOC EnergySMART (2013)
34. S. Han, S. Han, K. Sezaki, Development of an optimal vehicle-to-grid aggregrator for frequency regulation. IEEE Trans. Smart Grid **1**(1), 65–72 (2010)
35. S. Grammatico, F. Parise, M. Colombino, J. Lygeros. Decentralized convergence to Nash equilibria in constrained deterministic mean field control. IEEE Trans. Autom. Control (cond. accepted) (2015), arXiv:1410.4421v2
36. V. Berinde, *Iterative Approximation of Fixed Points* (Springer, Berlin, 2007)

Chapter 4
Decentralized Charging Coordination of Large-Population PEVs Under a Hierarchical Structure

Abstract Centralized or decentralized charging schedules of large-scale PEVs coordinated by a system operator usually require significant management, computation and communication capabilities on the system. Alternatively in this chapter, it constructs a hierarchical model for the PEV charging coordination problems where a collection of agents are introduced between the system operator and individual vehicles, and proposes an off-line decentralized method for the constructed hierarchical optimization problems. Under the decentralized method, each PEV implements its best behavior with respect to a given local charging price curve set by its agent. Each agent submits the collected aggregated charging behaviors under this agent to the system operator who then updates the electricity generation price and broadcasts it to PEVs via agents. To reimburse the transaction operation costs on agents, the charging price on the PEVs under an agent comprises both the generation price broadcasted from the system operator and the operation price set by this agent. It is shown that, under certain conditions, the proposed dynamical procedure converges to the efficient (or socially optimal) solution. The proposed method under the hierarchical structure presents the advantage of the autonomy of the individual PEVs and the low computation and communication capability requirements on the system.

4.1 Introduction

Compared with charging strategies in centralized ways e.g. [1–4] and the references therein. decentralized charging methods proposed in Chaps. 2 and 3, and other methods studied in the literature, e.g., [5], have several benefits such that the individual authority remains and only limited individual charging information is required.

However, the system under decentralized methods mentioned above has to directly deal with the transactions with all of the individuals. As a consequence, it is still challenging and impractical to implement the decentralized charging behaviors for large-population PEVs, since it requires complicated management, computation and heavy communications between the system operator and PEVs. Actually, in the realistic power market, the electricity is always traded under hierarchical structure, from the generator individual end-users and intermediaries, say utilities.

Z. Ma, *Decentralized Charging Coordination of Large-scale Plug-in Electric Vehicles in Power Systems*, https://doi.org/10.1007/978-981-13-7652-8_4

Much research work, studied in [6–9], has presented hierarchical charging systems for PEVs by organizing the PEVs into several groups, and setting a collection of intermediaries to manage the charging behaviors of the PEVs belonging to them. More specifically, in [6], the power grid is divided into multiple areas with a charging management center setting up the management of PEVs in a certain geographic area, providing communication interfaces between the PEVs and the dispatching centers. [7] focused on the PEVs charging coordination scheme for multiple parking decks in multifamily dwellings accounting for every user's satisfaction in a centralized way and distributed method, while in [8], it presents a hierarchical and zonal architecture for dispatching PEVs which aims to optimize the objective composed of electricity generation cost, battery degradation cost and penalty cost between the upper and lower layers. [9] introduced a hierarchical network resource allocation model. In the above work, the agents need to collect the charging characteristics of individual PEVs and coordinate the charging behaviors of these PEVs. As a result, agents take implement complicated optimization computing and heavy information transmission missions.

In this part, following the exact individual cost model given in Chap. 3, say each of the individual PEVs deals with the tradeoff between the battery degradation cost and the charging pricing cost; then the system can improve the longevity of the batteries and may mitigate the oscillation caused by the greedy behavior for the cheap electricity resource by individual PEVs. As mentioned before, this kind of cost plays the same role as the congestion pricing which has been widely used in Internet traffic control, see [10], and has been adopted in [11] to schedule PEV charging strategies in the power grid.

To deal with the charging coordination of large-population of PEVs here, it formulates a class of decentralized PEV charging coordination problems under a hierarchical structure such that the electricity is traded from the generators to load units via (intermediate) agents. To reimburse the transaction operation cost on agent, an agent operation price is charged to all of the individual vehicles under this agent. Each of the agents sets a charging price which is composed of the generation price and the operation price, and broadcasts it to the collection of individuals under its supervision. Each of the PEVs under an agent implements the best charging behavior with respect to the charging price by minimizing individual charging cost. The agents collect the implemented charging behaviors, updates the operation price and submits the aggregated charging behaviors to the central system operator; the system operator updates the marginal generation cost with respect to the collected updated aggregated charging strategies from collection of agents. The dynamic system continues until the charging price does not update any longer. It is further shown that the hierarchical charging system can converge to the efficient (or socially optimal) solution by applying the proposed decentralized method under certain mild conditions. Case studies are specified to illustrate the results developed in the work. Our decentralized method under the hierarchical structure presents the advantage of the autonomy of the individual PEVs and the low computation and communication capability requirements on the system.

This chapter is organized as follows. Section 4.2 formulates a class of charging control problems for a population of PEVs under a hierarchial structure. In Sect. 4.3, it proposes a decentralized method for the underlying hierarchical charging coordination problems such that the system converges to the efficient solution under certain mild conditions. Case studies are illustrated in Sect. 4.4 to verify the results developed in the work. In Sect. 4.5, it makes conclusions of this work.

4.2 Formulation of PEV Hierarchical Charging Coordination Problems

Differen from previous chapters, here suppose there exists a collection of intermediate agents, between the system operator and the individual vehicles, denoted by $\mathscr{M}$, each of which is responsible for the charging coordination of a specific group of PEVs $\mathscr{N}_m$, with $m \in \mathscr{M}$, such that $\dot{\bigcup}_{m\in\mathscr{M}} \mathscr{N}_m = \mathscr{N}$ with $\dot{\bigcup}$ representing the disjoint union operator. Denote by N and N_m the cardinality of $\mathscr{N}$ and $\mathscr{N}_m$ respectively.

The structure of the hierarchical charging system described above is illustrated in Fig. 4.1.

Call $\boldsymbol{u}_n^m \equiv (u_{nt}^m; t \in \mathscr{T})$, with $m \in \mathscr{M}$ and $n \in \mathscr{N}$, is an admissible charging behavior of PEV n under agent m, if

$$u_{nt}^m \begin{cases} \in [0, \gamma_{nm}], & \text{in case } t \in \mathscr{T}_{nm} \\ = 0, & \text{otherwise} \end{cases}; \quad \sum_{t\in\mathscr{T}} u_{nt}^m = \Gamma_{nm}, \tag{4.1}$$

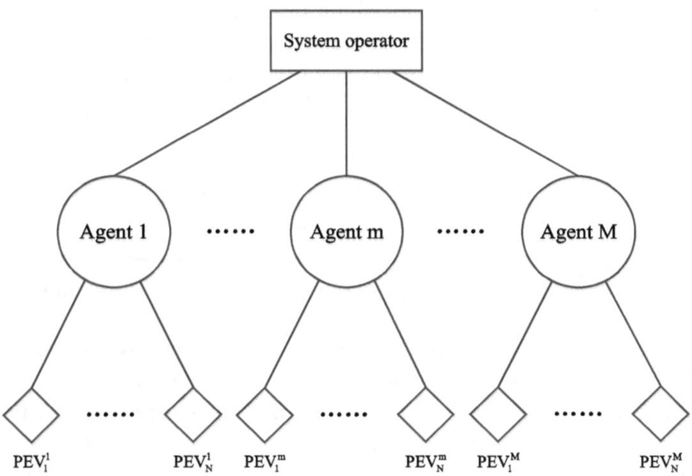

Fig. 4.1 The structure of the hierarchical charging system for a large-scale of PEVs

with $\mathcal{T}_{nm} \subset \mathcal{T}$ and where γ_{nm} represents a uniform maximum charging rate over $\mathcal{T}$, and Γ_{nm} represents the total required charging energy of PEV n under agent m.

Denote by $\mathcal{U}_n^m$ and $\mathcal{U}$ the set of admissible charging behaviors of PEV n under agent m, and the whole PEV populations respectively, such that

$$\mathcal{U} \triangleq \left\{ \boldsymbol{u} \equiv \left(\boldsymbol{u}_n^m; n \in \mathcal{N}_m; m \in \mathcal{M} \right); \text{ s.t. } \boldsymbol{u}_n^m \in \mathcal{U}_n^m \right\}.$$

Remark: It is assumed that $|\mathcal{T}_{nm}|\gamma_{nm} > \Gamma_{nm}$ to guarantee the set of admissible controls is nonempty, where $|\mathcal{T}_{nm}|$ denotes the cardinality of the set of charging intervals of PEV n.

It formulates a class of charging coordination problems for PEV populations under the hierarchical structure in Sect. 4.2.1 below.

4.2.1 Formulation of PEV Charging Coordination Problems

The research work in the literature mainly deals with how to minimize the impact on the power grid by scheduling the charging behaviors of high-penetration PEVs, and ignores the negative effects of the charging behavior of a PEV on itself. In fact it has been widely studied, some key characteristics of battery, like the state of health, the resistance impedance growth and the cycle life, which are strongly related to the energy capacity of battery, are effected by the charging behavior, e.g., [12, 13]. Some research work in the literature, e.g., [12, 14], studies the optimal charging behavior for a single PEV in a way that takes into account both total energy cost and state of health of battery. In this chapter, under the hierarchical structure, it studies the PEV charging coordination in power grid which deals with the total generation, agent operation and accumulated battery degradation costs over multi-time interval.

Following the above discussions, subject to a collection of admissible charging behaviors $\boldsymbol{u} \in \mathcal{U}$, the system cost function, denoted by $J(\boldsymbol{u})$, is specified below:

$$J(\boldsymbol{u}) \triangleq \sum_{t \in \mathcal{T}} \left\{ c(D_t) + \sum_{m \in \mathcal{M}} H_m(\boldsymbol{u}_t^m) \right\}, \tag{4.2}$$

with $H_m(\boldsymbol{u}_t^m) \equiv B_m(D_{mt}) + \sum_{n \in \mathcal{N}_m} f_{nm}\left(u_{nt}^m\right)$, where $c(D_t)$, $B_m(D_{mt})$ and $f_{nm}\left(u_{nt}^m\right)$ represent the electricity power generation cost, the operation cost of agent m and the battery degradation cost of PEV n under agent m respectively.

Interpretation of the system cost function $J(\boldsymbol{u})$: The system deals with the trade-off among the total electricity generation cost, the operation cost from agents and the accumulated battery degradation cost over the whole charging interval $\mathcal{T}$. The specification of generation cost $c(\cdot)$ has been discussed in Assumption 3.2, Chap. 3. And the specifcations of the battery degradation cost of PEV $f_{nm}(\cdot)$ and the operation cost of agent $B_m(\cdot)$ will be given in Assumptions 4.1 and 4.2 respectively.

A class of centralized PEV charging coordination problems is formulated as $\min_{u\in\mathscr{U}} \{J(u)\}$, i.e., the objective of system is to implement a collection of efficient charging behaviors for PEVs, denoted by u^{**}, minimizing the system cost (4.2).

Assumption 4.1 $f_{nm}(x)$, for all $n \in \mathscr{N}_m$, $m \in \mathscr{M}$, is monotonic increasing, strictly convex and differentiable on x. ■

Remark: Assumption 4.1 above and Assumption 3.1 given in Chap. 3 specify the same specifications for each of individual PEVs. Actually as analyzed in Sect. 3.2.2 in Chap. 3, the degradation cost for typical $LiFePO_4$-type batteries is consistent with these two assumptions and is in a quadratic form as below:

$$f_{nm}\left(u_{nt}^{m}\right) = \mathfrak{a}_{nm}[u_{nt}^{m}]^2 + \mathfrak{b}_{nm}u_{nt}^{m} + \mathfrak{c}_{nm}. \tag{4.3}$$

Assumption 4.2 The operation cost of agent m, $B_m(\cdot)$, for all $m \in \mathscr{M}$, is monotonically increasing, strictly convex and continuously differentiable. ■

Remarks on Assumption 4.2: For demonstration simplicity, in the simulations specified later in this chapter, it also consider a quadratic form for the agent operation cost B_m, that is to say $B_m(D_{mt}) = \frac{1}{2}\mathsf{p}_m D_{mt}^2 + \mathsf{q}_m D_{mt} + \mathsf{l}_m$, for some properly positive valued parameters p_m, q_m and l_m.

By applying the Lagrange multiplier method, see [15], and under Assumptions 3.2, 4.1 and 4.2, it can be obtained that the efficient charging behavior, denoted by u^{**}, is unique and can be characterized by its associated KKT conditions. Firstly, the Lagrangian function can be written:

$$L(u; \lambda) \triangleq J(u) + \sum_{m\in\mathscr{M}} \sum_{n\in\mathscr{N}_m} \lambda_{nm} \left(\Gamma_{nm} - \sum_{t\in\mathscr{T}} u_{nt}^{m}\right), \tag{4.4}$$

where λ_{nm} is a Lagrangian multiplier associated with the constraint specified in (4.1), say $\sum_{t\in\mathscr{T}} u_{nt}^{m} = \Gamma_{nm}$. The centralized optimization problem that minimizes $J(u)$, subject to $u \in U$, must satisfy,

- (I). $\dfrac{\partial L(u; \lambda)}{\partial \lambda_{nm}} = 0$,
- (II). $\dfrac{\partial L(u; \lambda)}{\partial u_{nt}^{m}} \le 0$, $u_{nt}^{m} \ge 0$, with complementary slackness.

Condition (I) contains the constraint $\sum_{t\in\mathscr{T}} u_{n}^{m} = \Gamma_{nm}$. It could be derived from condition (II) that

$$f'_{nm}\left(u_n^{m,**}\right) + c'\left(D_t^{**}\right) + B'_m\left(D_{mt}^{**}\right) \begin{cases} = \lambda_{nm}, & \text{in case } u_{nt}^{m,**} > 0, \\ \leq \lambda_{nm}, & \text{in case } u_{nt}^{m,**} = 0, \end{cases} \tag{4.5}$$

with $D_{mt}^{**} = \sum_{n \in \mathcal{N}_m} u_{nt}^{m,**} + d_{mt}$ and $D_t^{**} = \sum_{m \in \mathcal{M}} D_{mt}^{**}$.

4.2.2 *Case Studies for Centralized Charging Coordination Problems*

It considers the PEV charging coordination problems with four agents each of which coordinates the charging schedules of a specific collection of PEVs with size of 10,000. That is to say, the total number of PEVs is 40,000. The multi-time charging interval begins from 12:00AM on one day and terminates at 12:00AM on the next day, and the length of each interval is one hour. The 24-hour inelastic base demand under each agent, $\boldsymbol{d}_m$ as illustrated in Fig. 4.2 for the case studies, is taken as a scaled demand in a typical day in the region of the MISO (Midwest Independent System Operator) during a hot summer season.

As stated earlier, it adopts the quadratic form for generation cost $c(\cdot)$ and agent operation cost $B_m(\cdot)$ for all $m \in \mathcal{M}$, such that $c(D_t) = 4 \times 10^{-8} D_t^2 + 0.03 D_t$ and $B_m(D_{mt}) \triangleq 2.5 \times 10^{-8} D_{mt}^2 + 0.002 D_{mt}$, for all $m \in \mathcal{M}$.

For the purpose of demonstration, simply suppose that that $\mathcal{T}_{nm} = \mathcal{T}$ for all $n \in \mathcal{N}_m$ and all $m \in \mathcal{M}$, and all PEVs are supposed to share identical battery capacity size of 30 kWh, and minimum and maximum soc values equal to 15% and 90% respectively. Also suppose that the initial soc values of all of the PEVs are identical with each other, such that $soc_{n0}^m = 15\%$ for all $n \in \mathcal{N}_m$ and all $m \in \mathcal{M}$; then the total charging energy over the charging interval $\mathcal{T}$, is specified as $\Gamma_{nm} = \Phi_{nm}(soc_{max} - soc_{n0})$, which is equal to 22.5 kWh for all $n \in \mathcal{N}_m$ and all $m \in \mathcal{M}$.

It can be obtained that the battery degradation cost is a quadratic form on the charging rate for all of PEVs, such that $f_{nm} = 0.004\left(u_{nt}^m\right)^2 + 0.075 u_{nt}^m$.

The socially optimal charging behaviors of individual PEVs in each agent m are illustrated in Fig. 4.2 respectively with respect to inelastic based demands in typical hot summer season.

Figure 4.3 displays the aggregated socially optimal behaviors of the whole PEV populations in the charging system.

As studied, the centralized efficient coordination behaviors of PEV populations can be effectively implemented in case that the system has complete information, say the agent operation costs, the battery characteristics and the degradation costs of individual PEVs, and has the permission to directly schedule the charging behaviors of all PEVs. However, in practice the individuals may not be willing to share their private information with others or to allow the system to directly control their vehicles. Thus in the rest of the part, it will propose a decentralized coordination method for the underlying hierarchical coordination problems.

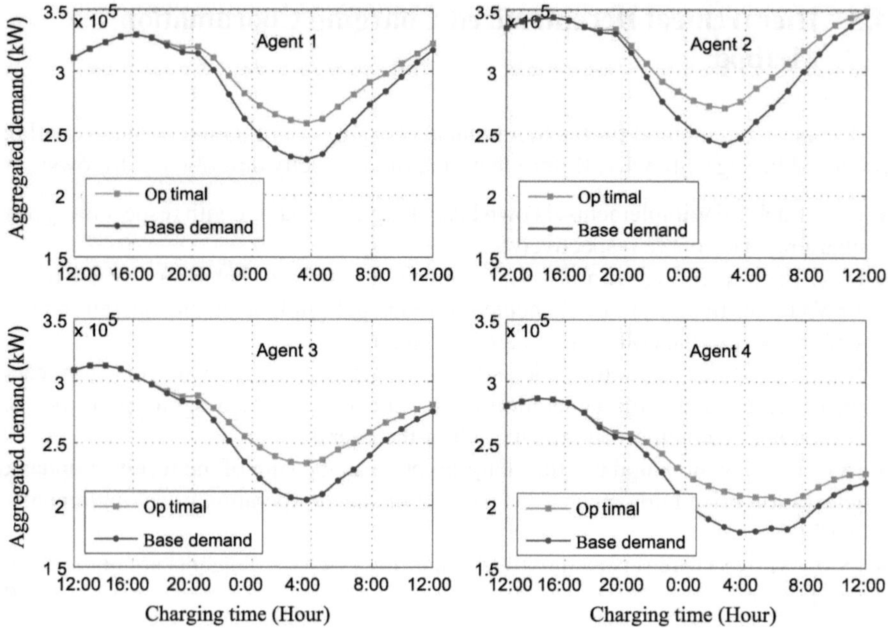

Fig. 4.2 Centralized socially optimal charging behaviors of individual PEVs under each of the agents

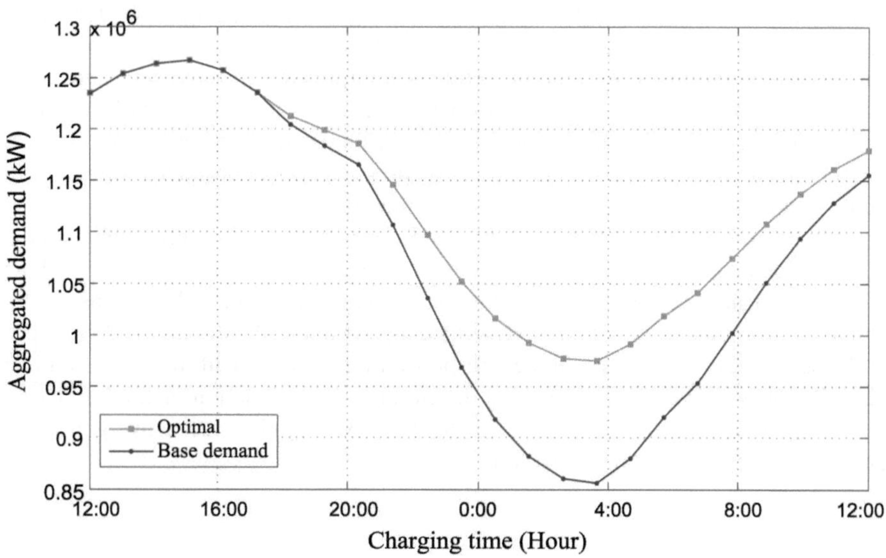

Fig. 4.3 Aggregated centralized socially optimal behaviors of the whole PEV populations

4.3 Hierarchical Decentralized Charging Coordination Method

A decentralized method for the hierarchical charging coordination problems will be proposed in Algorithm 4.1. Before that, it is firstly briefly described as follows:

- (S1). Each PEV implements its own best charging behavior with respect to a given charging price curve respectively;
- (S2). Each agent m, with $m \in \mathcal{M}$, collects the updated behaviors, of all of the PEVs under this agent, implemented in (S1), and implements the operation price with respect to the collected updated behaviors;
- (S3). The system operator collects the aggregated charging behaviors from each agent m, calculates the generation price w.r.t. the collected aggregated updated behaviors, and then broadcasts it to all of the agents;
- (S4). Each agent m updates the charging price as the sum of the received updated generation price implemented in (S3) and the agent operation price implemented in (S2);
- (S5). (S1)–(S4) are repeated until the charging price of all agents are not updated any longer. ■

The dynamic procedure (S1)–(S5) is illustrated in Fig. 4.4.

In the decentralized charging problems without any intermediate agents between the system operator and individual vehicles, see [16, 17], the system operator has to directly deal with the transactions with all of the individuals. As a consequence, it is still challenging and impractical to implement the decentralized charging behaviors for large-population PEVs, since it requires complicated management and heavy communications between the system operator and the large number of individual PEVs. While, under the hierarchical structure, the system operator only needs to communicate with M agents, and each of these agents coordinates the charging behaviors of a specific group PEVs under this agent, e.g, agent m only needs to communicate with the system operator and coordinate the charging behaviors of N_m individuals under this agent. In summary, compared with the charging problems without intermediate agents, the communication and computational missions are distributed among the system and intermediate agents under the hierarchical structure.

In Sect. 4.3.1, it analyzes the best response of each PEV with respect to a given charging price curve; then proposes a decentralized coordination method under the hierarchical structure in Sect. 4.3.2. In Sect. 4.3.3, it studies the convergence and performance of the proposed method.

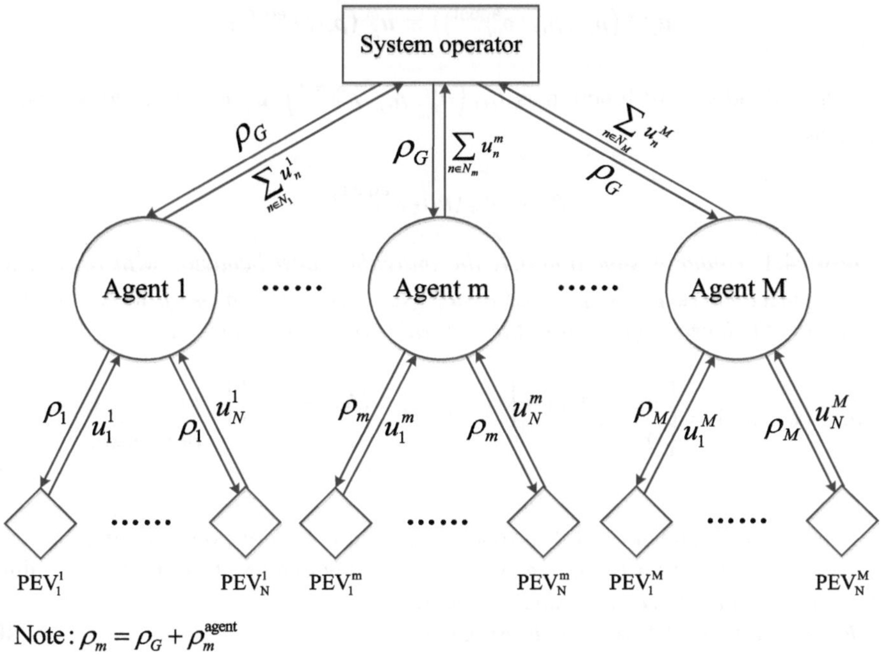

Fig. 4.4 Information exchange under the decentralized method for the hierarchical charging coordination problems

4.3.1 *Best Response of PEVs with a Given Charging Price Curve*

Subject to the charging behavior $\boldsymbol{u}_n^m$, the charging cost of an individual PEV is denoted by $J_{nm}\left(\boldsymbol{u}_n^m; \boldsymbol{\rho}_m\right)$, with respect to a given charging price curve $\boldsymbol{\rho}_m$, such that

$$J_{nm}\left(\boldsymbol{u}_n^m; \boldsymbol{\rho}_m\right) = \sum_{t \in \mathscr{T}} \left\{ f_{nm}\left(u_{nt}^m\right) + \rho_{mt} u_{nt}^m \right\}, \tag{4.6}$$

where $\boldsymbol{\rho}_m$ represents the charging price curve for individual vehicles under agent m, and is composed of the generation price and the operation price of agent m.

Denote by $\boldsymbol{u}_n^{m,*}\left(\boldsymbol{\rho}_m\right)$ the best charging behavior of individual PEV n with respect to $\boldsymbol{\rho}_m$, by minimizing the individual charging cost $J_{nm}\left(\boldsymbol{u}_n^m; \boldsymbol{\rho}_m\right)$, such that

$$\boldsymbol{u}_n^{m,*}\left(\boldsymbol{\rho}_m\right) = \underset{\boldsymbol{u}_n^m \in \mathscr{U}_n^m}{\operatorname{argmin}} \left\{ J_{nm}\left(\boldsymbol{u}_n^m; \boldsymbol{\rho}_m\right) \right\} \tag{4.7}$$

which is determined by Lemma 4.1 below.

Note: For notational simplicity, the best individual charging strategy may be rewritten as below:

$$\boldsymbol{u}_n^{m,*}\left(\boldsymbol{\rho}_m\left(\boldsymbol{\rho}_G,\boldsymbol{\rho}_m^{\text{agent}}\right)\right)\equiv\boldsymbol{u}_n^m\left(\boldsymbol{\rho}_G,\boldsymbol{\rho}_m^{\text{agent}}\right),$$

also the individual cost function as $J_{nm}\left(\boldsymbol{u}_n^m;\boldsymbol{\rho}_G,\boldsymbol{\rho}_m^{\text{agent}}\right)$. $\boldsymbol{\rho}_m$ is applied to represent as below:

$$\boldsymbol{\rho}_m\equiv\boldsymbol{\rho}_m\left(\boldsymbol{\rho}_G,\boldsymbol{\rho}_m^{\text{agent}}\right).$$

Lemma 4.1 *Under Assumption 4.1, the individual best behavior with respect to a given price curve $\boldsymbol{\rho}_m$ is specified as $\boldsymbol{u}_n^m\left(\boldsymbol{\rho}_m,A_m\right)\in\mathscr{U}_n^m$ in (4.8), where $\left[f'_{nm}\right]^{-1}$ represents the inverse operator of the derivative of the function f_{nm}.*

$$u_{nt}^{m,*}\left(\boldsymbol{\rho}_m,A_m\right)=\begin{cases}\max\left\{0,\min\left\{\gamma_{nm},\left[f'_{nm}\right]^{-1}\left(A_m-\rho_{mt}\right)\right\}\right\}, & \textit{in case } t\in\mathscr{T}_{nm}\\ 0, & \textit{otherwise},\end{cases}\tag{4.8}$$

Remark: The trajectory (4.8) provides the unique best response minimizing the individual cost function $J_{nm}(\boldsymbol{u}_n^m;\boldsymbol{\rho}_m)$ given in (4.6) with respect to a particular value of A_m, denoted by $A_m^*(\boldsymbol{\rho}_m)$, uniquely dependent upon $\boldsymbol{\rho}_m$.

Proof of Lemma 4.1 *is given in Appendix* 4.6.1. ■

4.3.2 Decentralized Method Under Hierarchical Structure

This work assumes that the individual PEV deals with the tradeoff among the electricity generation cost, the agent operation cost and battery degradation cost. In this section, a decentralized iterative charging strategy which is the update procedure in Algorithm 4.1 below is specified. The system operator and the agents will update the electricity generation price curve and the operation price curve respectively, which minimizes the local cost function (4.6) with respect to all the individual charging behaviors in the last step. The iterative update strategy will make the system converges to a unique collection.

As formulated in Sect. 4.2.1, the electricity generation price $\boldsymbol{\rho}_G$, the agent operation price $\boldsymbol{\rho}_m^{\text{agent}}$ shall be determined by the total demand via the marginal generation cost $p_G(\cdot)$, and by the demand under agent m via marginal operation cost $p_m^{\text{agent}}(\cdot)$, respectively. Then the presumed price curves $\boldsymbol{\rho}_G$, $\boldsymbol{\rho}_m^{\text{agent}}$ and $\boldsymbol{\rho}_m$ shall satisfy the following equations:

$$\rho_{mt}^{\text{agent}}=p_{mt}^{\text{agent}}\left(D_{mt}\right),\quad\text{with } D_{mt}\triangleq\sum_{n\in\mathscr{N}_m}u_{nt}^{m,*}\left(\boldsymbol{\rho}_G,\boldsymbol{\rho}_m^{\text{agent}}\right)+d_{mt},\tag{4.9a}$$

$$\rho_{Gt}=p_G\left(D_t\right),\quad\text{with } D_t\triangleq\sum_{m\in\mathscr{M}}D_{mt},\tag{4.9b}$$

Since the solution to (4.9) can not be presumably known, it will specify an off-line distributed iterative price curve update procedure prior to the charging interval $\mathscr{T}$ in Algorithm 1 below. Essentially each of individual PEVs updates its best strategy minimizing the cost function (4.6) with respect to a the signal of price curves $\boldsymbol{\rho}_G$, $\boldsymbol{\rho}_m^{\text{agent}}$ and $\boldsymbol{\rho}_m$, which will be updated with $\boldsymbol{\rho}_G^{\dagger}$, $\boldsymbol{\rho}_m^{\text{agent},\dagger}$ and $\boldsymbol{\rho}_m^{\dagger}$ respectively, such that

$$\rho_{mt}^{\text{agent},\dagger} = p_{mt}^{\text{agent}}\left(\sum_{n\in\mathscr{N}_m} u_{nt}^{m,*}\left(\boldsymbol{\rho}_G, \boldsymbol{\rho}_m^{\text{agent}}\right) + d_{mt}\right), \tag{4.10a}$$

$$\rho_{Gt}^{\dagger} = p_G\left(\sum_{m\in\mathscr{M}}\left\{\sum_{n\in\mathscr{N}_m} u_{nt}^{m,*}\left(\boldsymbol{\rho}_G, \boldsymbol{\rho}_m^{\text{agent}}\right) + d_{mt}\right\}\right). \tag{4.10b}$$

A collection of distributed charging strategies is implemented in case the proposed price curves update procedure converges. Here it is ready to formalize a hierarchical decentralized charging coordination method for PEV populations in Algorithm 4.1 below.

Here it is ready to formalize, in Algorithm 4.1 below, the decentralized charging coordination method under the hierarchical structure described in (S1)–(S5) earlier.

Algorithm 4.1 (*Hierarchical decentralized coordination method*)

- Provide an initial generation price and a collection of agent operation prices:

$$\boldsymbol{\rho}_G^{(0)} \equiv \left(\rho_{Gt}^{(0)}; t\in\mathscr{T}\right),$$
$$\boldsymbol{\rho}_m^{\text{agent},(0)} \equiv \left(\rho_{mt}^{\text{agent},(0)}; t\in\mathscr{T}\right), \text{ for all } m\in\mathscr{M};$$

- Set $\boldsymbol{\rho}_m^{(0)} = \left(\rho_{Gt}^{(0)} + \rho_{mt}^{\text{agent},(0)}; t\in\mathscr{T}\right)$;
- Set $k = 0$ and $\varepsilon = \varepsilon_0$ for some $\varepsilon_0 > 0$;
- While $\varepsilon > 0$
  - Implement a best individual charging behavior $\boldsymbol{u}_n^{m,(k+1)}$ w.r.t. $\boldsymbol{\rho}_m^{(k)}$ for all $n\in\mathscr{N}_m$ by minimizing the individual cost function (4.6), such that

$$\boldsymbol{u}_n^{m,(k+1)}\left(\boldsymbol{\rho}_m^{(k)}\right) \triangleq \underset{\boldsymbol{u}_n^m\in\mathscr{U}_n^m}{\text{argmin}} \sum_{t\in\mathscr{T}}\left\{f_{nm}\left(u_{nt}^m\right) + \rho_{mt}^{(k)} u_{nt}^m\right\};$$

  - Each agent m collects the updated behaviors of $\boldsymbol{u}_n^{m,(k+1)}$ of PEV n, $n\in\mathscr{N}_m$, and calculates the operation price $\boldsymbol{\rho}_m^{\text{agent},(k+1)}$, such that

$$\rho_{mt}^{\text{agent},(k+1)} = B'_m\left(D_{mt}^{(k+1)}\right), \tag{4.11}$$

with $D_{mt}^{(k+1)} = \sum_{n \in \mathcal{N}_m} u_{nt}^{m,(k+1)} + d_{mt}$;

- The system operator collects the aggregated charging behaviors $\sum_{n \in \mathcal{N}_m} \boldsymbol{u}_n^{m,(k+1)}$ from each agent m, and calculates the generation price $\boldsymbol{\rho}_G$, such that

$$\rho_{Gt}^{(k+1)} = c' \left(\sum_{m \in \mathcal{M}} D_{mt}^{(k+1)} \right); \tag{4.12}$$

- Each agent m collects the updated generation price $\boldsymbol{\rho}_G^{k+1}$ and calculates the charging price $\boldsymbol{\rho}_m^{k+1}$, such that

$$\rho_{mt}^{(k+1)} = \rho_{Gt}^{(k+1)} + \rho_{mt}^{\text{agent},(k+1)}; \tag{4.13}$$

- Update $\varepsilon = \sum_{m \in \mathcal{M}} \| \boldsymbol{\rho}_m^{(k+1)} - \boldsymbol{\rho}_m^{(k)} \|_1$;
- Update $k = k + 1$; ■

Remarks: In Algorithm 4.1, x' represents the derivative of the function x, and $|| \cdot ||_1$ represents the l_1-norm of a vector.

4.3.3 Convergence of the Algorithm

In this section, as the sufficient conditions of Theorem 4.1, Lemma 4.2 is introduced firstly. Then the convergence of the decentralized algorithm will be shown in Theorem 4.1. In Corollary 4.1, the efficient charging behaviors of the system are analyzed.

Lemma 4.2 $\boldsymbol{u}_n^{m,*}(\boldsymbol{\rho}_m)$ *and* $\boldsymbol{u}_n^{m,*}(\widehat{\boldsymbol{\rho}}_m)$ *satisfy the following inequality*

$$\begin{aligned} &\left\| \boldsymbol{u}_n^{m,*}(\boldsymbol{\rho}_m) - \boldsymbol{u}_n^{m,*}(\widehat{\boldsymbol{\rho}}_m) \right\|_1 \\ &\le 2 \sum_{t \in \mathcal{T}} \left| \left[f'_{nm} \right]^{-1} \left(A_m^*(\boldsymbol{\rho}_m) - \rho_{mt} \right) - \left[f'_{nm} \right]^{-1} \left(A_m^*(\boldsymbol{\rho}_m) - \widehat{\rho}_{mt} \right) \right|. \end{aligned} \tag{4.14}$$

Proof of Lemma 4.2 *is given in Appendix* 4.6.2. ■

Based on Lemma 4.2, it will show in Theorem 4.1 that, by adopting the decentralized method proposed in Algorithm 4.1, the system converges to a unique collection of charging behaviors under certain mind conditions.

Theorem 4.1 (Convergence of Algorithm) *Considering Assumptions 3.2, 4.1 and 4.2, such that*

$$\left|c'(y_1) - c'(y_2)\right| \le \widehat{\alpha}\left|y_1 - y_2\right|, \tag{4.15a}$$

$$\left|B'_m(z_1) - B'_m(z_2)\right| \le \widehat{\beta}\left|z_1 - z_2\right|, \tag{4.15b}$$

$$\left|f'_{nm}(x_1) - f'_{nm}(x_2)\right| \ge \widehat{\zeta}\left|x_1 - x_2\right|, \tag{4.15c}$$

for all $n \in \mathcal{N}_m$ and all $m \in \mathcal{M}$; then, by adopting Algorithm 4.1, the system converges to a unique collection of charging behaviors, if the following inequality holds,

$$\frac{2}{\widehat{\zeta}} N_{max}\left(\widehat{\alpha}M + \widehat{\beta}\right) < 1, \tag{4.16}$$

where $N_{max} \equiv \max\{N_m;\, m \in \mathcal{M}\}$.

Proof Under Assumption 4.1, the inverse operator of f'_{nm}, denoted by $\left[f'_{nm}\right]^{-1}$, exists; then by (4.15c), it follows:

$$\left|\left[f'_{nm}\right]^{-1}(\xi_1) - \left[f'_{nm}\right]^{-1}(\xi_2)\right| \le \frac{1}{\widehat{\zeta}}\left|\xi_1 - \xi_2\right|,$$

by which it can be obtained that

$$\begin{aligned}
&\left|\left[f'_{nm}\right]^{-1}\left(A^*_m(\boldsymbol{\rho}_m) - \rho_{mt}\right) - \left[f'_{nm}\right]^{-1}\left(A^*_m(\boldsymbol{\rho}_m) - \widehat{\rho}_{mt}\right)\right| \\
&\le \frac{1}{\widehat{\zeta}}\left|\left(A^*_m(\boldsymbol{\rho}_m) - \rho_{mt}\right) - \left(A^*_m(\boldsymbol{\rho}_m) - \widehat{\rho}_{mt}\right)\right| \\
&= \frac{1}{\widehat{\zeta}}\left|\rho_{mt} - \widehat{\rho}_{mt}\right|,
\end{aligned} \tag{4.17}$$

by which together with (4.14), the following inequality holds

$$\sum_{n \in \mathcal{N}_m}\left\|\boldsymbol{u}^{m,*}_n - \widehat{\boldsymbol{u}}^{m,*}_n\right\|_1 \le \frac{2}{\widehat{\zeta}} N_{max}\left\|\boldsymbol{\rho}_m - \widehat{\boldsymbol{\rho}}_m\right\|_1, \tag{4.18}$$

where for notational simplicity, it considers $\boldsymbol{u}^{m,*}_n \equiv \boldsymbol{u}^{m,*}_n(\boldsymbol{\rho}_m)$ and $\widehat{\boldsymbol{u}}^{m,*}_n \equiv \boldsymbol{u}^{m,*}_n(\widehat{\boldsymbol{\rho}}_m)$.

It defines $D^*_{it} = \sum_{n \in \mathcal{N}_i} u^{i,*}_{nt} + d_{it}$ and $\widehat{D}^*_{it} = \sum_{n \in \mathcal{N}_i} \widehat{u}^{i,*}_{nt} + d_{mt}$ for all $i \in \mathcal{M}$; while $D^*_t = \sum_{i \in \mathcal{M}} D^*_{it}$ and $\widehat{D}^*_t = \sum_{i \in \mathcal{M}} \widehat{D}^*_{it}$.

Now the following analysis holds for any agent $m \in \mathcal{M}$,

$$\begin{aligned}
&\left\|\boldsymbol{\rho}^\dagger_m - \widehat{\boldsymbol{\rho}}^\dagger_m\right\|_1 \\
&= \sum_{t \in \mathcal{T}}\left|\left\{G'\left(D^*_t\right) + B'_m\left(\sum_{n \in \mathcal{N}_m} D^*_{mt}\right)\right\}\right.
\end{aligned}$$

$$
\begin{aligned}
&\quad - \left\{ G'\left(\widehat{D}_t^*\right) + B_m'\left(\sum_{n\in\mathcal{N}_m} \widehat{D}_{mt}^*\right)\right\} \Bigg|, \\
&\qquad \text{by (4.11), (4.12) and (4.13)} \\
&\leq \widehat{\alpha} \sum_{i\in\mathcal{M}} \sum_{n\in\mathcal{N}_i} \left\| \boldsymbol{u}_n^{i,*} - \widehat{\boldsymbol{u}}_n^{i,*} \right\|_1 + \widehat{\beta} \sum_{n\in\mathcal{N}_m} \left\| \boldsymbol{u}_n^{m,*} - \widehat{\boldsymbol{u}}_n^{m,*} \right\|_1, \\
&\qquad \text{by (4.15a) and (4.15b)} \\
&\leq \frac{2\widehat{\alpha}}{\widehat{\zeta}} N_{max} \sum_{i\in\mathcal{M}} \left\| \boldsymbol{\rho}_i - \widehat{\boldsymbol{\rho}}_i \right\|_1 + \frac{2\widehat{\beta}}{\widehat{\zeta}} N_{max} \left\| \boldsymbol{\rho}_m - \widehat{\boldsymbol{\rho}}_m \right\|_1,
\end{aligned} \tag{4.19}
$$

where the last equality holds by (4.18).

Based upon the above analysis, it can obtain that

$$
\begin{aligned}
&\sum_{m\in\mathcal{M}} \left\| \boldsymbol{\rho}_m^\dagger - \widehat{\boldsymbol{\rho}}_m^\dagger \right\|_1 \\
&\leq \sum_{m\in\mathcal{M}} \left\{ \frac{2\widehat{\alpha}}{\widehat{\zeta}} N_{max} \sum_{i\in\mathcal{M}} \left\| \boldsymbol{\rho}_i - \widehat{\boldsymbol{\rho}}_i \right\|_1 + \frac{2\widehat{\beta}}{\widehat{\zeta}} N_{max} \left\| \boldsymbol{\rho}_m - \widehat{\boldsymbol{\rho}}_m \right\|_1 \right\}, \qquad \text{by (4.19)} \\
&= \frac{2\widehat{\alpha}}{\widehat{\zeta}} N_{max} M \sum_{m\in\mathcal{M}} \left\| \boldsymbol{\rho}_m - \widehat{\boldsymbol{\rho}}_m \right\|_1 + \frac{2\widehat{\beta}}{\widehat{\zeta}} N_{max} \sum_{m\in\mathcal{M}} \left\| \boldsymbol{\rho}_m - \widehat{\boldsymbol{\rho}}_m \right\|_1 \\
&= \frac{2}{\widehat{\zeta}} N_{max} \left(\widehat{\alpha} M + \widehat{\beta}\right) \sum_{m\in\mathcal{M}} \left\| \boldsymbol{\rho}_m - \widehat{\boldsymbol{\rho}}_m \right\|_1,
\end{aligned}
$$

which implies that, in case $\frac{2}{\widehat{\zeta}} N_{max}\left(\widehat{\alpha} M + \widehat{\beta}\right) < 1$, $\boldsymbol{\rho}^\dagger(\cdot)$ is a contraction mapping in the l_1 norm; then by the contraction mapping theorem [18], under (4.16), the system converges to a unique fix point, denoted by $\boldsymbol{\rho}_m^*$, by applying the decentralized method proposed in Algorithm 4.1. ■

Corollary 4.1 *Under Assumptions 3.2, 4.1 and 4.2, and considering (4.15) and (4.16), the charging system converges to the efficient charging behaviors* $\boldsymbol{u}^{**}(\boldsymbol{\rho}_m^*)$ *as specified in Sect. 4.2.1 by applying Algorithm 4.1.*

Proof It is firstly shown that the trajectories of $\boldsymbol{\rho}_G^{(k)}$ and $\boldsymbol{\rho}_m^{\text{agent},(k)}$ converge.

Following Theorem 4.1, the convergence of $\boldsymbol{\rho}_m$ indicates that the charging system converges to a unique collection of charging behaviors $\boldsymbol{u}_n^{m,*}\left(\boldsymbol{\rho}_m^*\right)$ under (4.15) and (4.16). Also by Algorithm 4.1,

$$
\boldsymbol{\rho}_m^{\text{agent},(k+1)} = B_m'\left(D_m^*\left(\boldsymbol{\rho}_m^{(k)}\right)\right), \tag{4.20a}
$$

$$
\boldsymbol{\rho}_G^{(k+1)} = c'\left(D^*\left(\boldsymbol{\rho}^{(k)}\right)\right), \tag{4.20b}
$$

where $D_m^*\left(\boldsymbol{\rho}_m^{(k)}\right) = \sum_{n\in\mathcal{N}_m} \boldsymbol{u}_n^{m,*}\left(\boldsymbol{\rho}_m^{(k)}\right) + d_m$ and $D^*\left(\boldsymbol{\rho}^{(k)}\right) = \sum_{m\in\mathcal{M}} D_m^*\left(\boldsymbol{\rho}_m^{(k)}\right)$.

By (4.20), if the charging system converges to a unique collection of charging behaviors $\boldsymbol{u}_n^{m,*}(\boldsymbol{\rho}_m^*)$, the price curves $\boldsymbol{\rho}_G^{(k)}$ and $\boldsymbol{\rho}_m^{\text{agent},(k)}$ are certain to converge to a unique collection of fixed trajectories $\boldsymbol{\rho}_G^*$ and $\boldsymbol{\rho}_m^{\text{agent},*}$ respectively.

By Lemma 4.1, the best response $\boldsymbol{u}_n^{m,*}$, with respect to a fixed trajectory $\boldsymbol{\rho}_G^*$ and $\boldsymbol{\rho}_m^{\text{agent},*}$, satisfies the following:

$$f'_{nm}\left(u_{nt}^{m,*}\right) + \boldsymbol{\rho}_G^* + \boldsymbol{\rho}_m^{\text{agent},*} \begin{cases} = A_m, & \text{in case } u_{nt}^{m,*} > 0 \\ \leq A_m, & \text{in case } u_{nt}^{m,*} = 0 \end{cases},$$

where $\boldsymbol{\rho}_G^*$ and $\boldsymbol{\rho}_m^{\text{agent},*}$ are defined in (4.20) respectively; then by which together with (4.5), it implies that the best response of PEVs, denoted by $\boldsymbol{u}_n^{m,*}$, with respect to the converged price curves $\boldsymbol{\rho}_G^*$ and $\boldsymbol{\rho}_m^{\text{agent},*}$, is the efficient behavior. ■

4.4 Case Studies

Apply the decentralized method proposed for the hierarchical charging coordination problems with numerical case studies. For the purpose of demonstration, it adopts the same parameters for the examples specified in this section as those considered in Sect. 4.2.2.

Following the specifications in Sect. 4.2.2 on the generation cost $c(\cdot)$, agent operation cost $B_m(\cdot)$ and battery degradation cost f_{nm}, $c'(D_t) = 8 \times 10^{-8} D_t + 0.03$, $B'_m(D_{mt}) = 5 \times 10^{-8} D_{mt} + 0.002$, where D_t and D_{mt} denote the total demand in the whole charging system and the total demand under agent m respectively, and $f'_{nm}(u_{nt}) = 0.008 u_{nt}^m + 0.075$. Hence $|c'(y_1) - c'(y_2)| \leq \widehat{\alpha}\,|y_1 - y_2|$ with $\widehat{\alpha} = 8 \times 10^{-8}$ and $\left|B'_m(\eta_1) - B'_m(\eta_2)\right| \leq \widehat{\beta}|\eta_1 - \eta_2|$, with $\widehat{\beta} = 5 \times 10^{-8}$, for all $m \in \mathcal{M}$, and

$$|f'_{nm}(x_1) - f'_{nm}(x_2)| \geq \widehat{\zeta}\,|x_1 - x_2|, \text{ with } \widehat{\zeta} = 0.008.$$

By the above analysis and with $N_{max} = 10,000$, $\frac{2}{\widehat{\zeta}} N_{max}\left(\widehat{\alpha} M + \widehat{\beta}\right) = 0.925 < 1$, that is to say, by this inequality and the above series of inequalities concerning f'_{nm}, c' and B'_m for all $m \in \mathcal{M}$ and all $n \in \mathcal{N}_m$, the hierarchical charging system can converge to the efficient solution by Theorem 4.1 and Corollary 4.1.

Figure 4.5 displays the iteration update behaviors of PEVs in case each of agents adopts the decentralized method proposed in Algorithm 4.1; while the iteration updates of the aggregated charging behaviors of the whole PEV populations are illustrated in Fig. 4.6.

As observed from Figs. 4.5 and 4.6, and compared with Figs. 4.2 and 4.3, the charging system converges to the efficient solution in 12 iteration steps by applying the proposed decentralized method. This is an expected result as stated in Theorem 4.1 and Corollary 4.1.

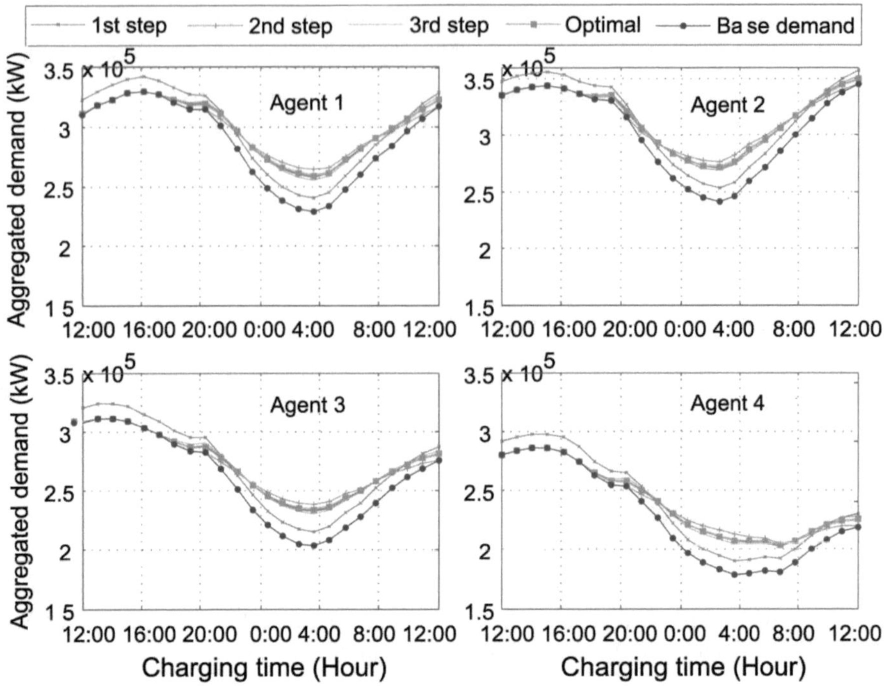

Fig. 4.5 Iteration updates of PEV charging behaviors under each agent by applying Algorithm 4.1

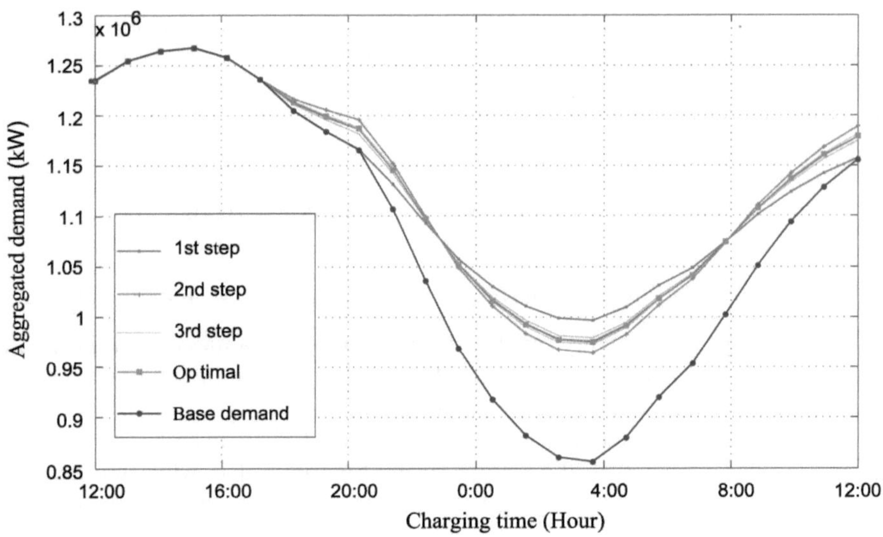

Fig. 4.6 Iteration updates of aggregated behaviors by applying Algorithm 4.1

4.5 Conclusions

In this chapter, it firstly introduces a hierarchical model for the charging coordination problems of large-scale PEVs; then proposes a decentralized method for the underlying hierarchical optimization problems such that each of vehicles implements its best charging behaviors with respect to a given charging price curve. Under the proposed method, the charging price curves of all of the intermediate agents iteratively update in the hierarchical system until no price curves update any longer. It is shown that under certain conditions, the charging system converges to the unique efficient solution. Owed to the hierarchical structure, the computation and communication complexity are largely reduced under the decentralized method.

4.6 Appendices

4.6.1 Proof of Lemma 4.1

To solve the optimization problem, a Lagrangian function is defined as follow $L_{nm}\left(\boldsymbol{u}_n^m; \boldsymbol{\rho}_m\right) \triangleq J_{nm}\left(\boldsymbol{u}_n^m; \boldsymbol{\rho}_m\right) + A_m\left(\Gamma_{nm} - \sum_{t\in\mathscr{T}} \boldsymbol{u}_n^m\right)$, where A_m is Lagrange multiplier. Since $J_{nm}\left(\boldsymbol{u}_n^m; \boldsymbol{\rho}_m\right)$ is convex with respect to $\boldsymbol{u}_n^m$ due to the convexity of f_{nm}, the charging control that minimizes $J_{nm}\left(\boldsymbol{u}_n^m; \boldsymbol{\rho}_m\right)$, subject to $\boldsymbol{u}_n^m \in \mathscr{U}_n^m$, must satisfy,

- (I). $\dfrac{\partial L_{nm}}{\partial A_m} = 0$,
- (II). $\dfrac{\partial L_{nm}}{\partial u_{nt}^m} \leq 0$, $u_{nt}^m \geq 0$, with complementary slackness.

Condition (I) contains the constraint $\sum_{t\in\mathscr{T}} \left(\boldsymbol{u}_n^m\right) = \Gamma_{nm}$. It can be derived from condition (II) that

$$\left[f_{nm}'\right]^{-1}\left(u_{nt}^m\right) + \rho_{mt} - A_m \begin{cases} = 0, & \text{in case } u_{nt}^m > 0, \\ < 0, & \text{otherwise,} \end{cases}$$

which is equivalent to (4.8).

The dependence of $u_{nt}^m\left(\boldsymbol{\rho}_m, A_m\right)$ on A_m expressed in (4.8) ensures that, for any fixed $\boldsymbol{\rho}_m$,

- There exists an A_{m-} such that for $A_m \leq A_{m-}$, $\sum_{t\in\mathscr{T}} u_{nt}^m\left(\boldsymbol{\rho}_m, A_m\right) = 0$.

- For $A_m > A_{m-}$, $\sum_{t \in \mathcal{T}} u_{nt}^m(\rho_m, A_m)$ is strictly increasing with A_m, with the relationship continuous, though not smooth. Hence, $\sum_{t \in \mathcal{T}} u_{nt}^m(\rho_m, A_m)$ is invertible over this domain.

Therefore a constraint $\sum_{t \in \mathcal{T}} u_{nt}^m(\rho_m, A_m) = K > 0$ defines a unique $A_m > A_{m-}$ for each fixed $\boldsymbol{\rho}_m$, which may be written as $A_m(\boldsymbol{\rho}_m)$. The particular value of A_m that ensures satisfaction of the constraint $\sum_{t \in \mathcal{T}} u_{nt}^m = \Gamma_{nm}$ shall be denoted by $A_m^*(\boldsymbol{\rho}_m)$. The resulting control trajectory can be written $\boldsymbol{u}_n^m(\boldsymbol{\rho}_m, A_m^*(\boldsymbol{\rho}_m)) = \boldsymbol{u}_n^{m,*}(\boldsymbol{\rho}_m)$.

Since $J_{nm}(\boldsymbol{u}_n^m; \boldsymbol{\rho}_m)$ is convex with respect to $\boldsymbol{u}_n^m$, the minimizing control defined by (4.8) must be unique. ■

4.6.2 Proof of Lemma 4.2

Recall that $\boldsymbol{u}_n^{m,*}(\boldsymbol{\rho}_m) \triangleq \underset{\boldsymbol{u}_n^m \in \mathcal{U}_n^m}{\operatorname{argmin}} \{J_{nm}(\boldsymbol{u}_n^m; \boldsymbol{\rho}_m)\}$, it means that $\boldsymbol{u}_n^{m,*}(\boldsymbol{\rho}_m)$ represents the charging behavior that minimizes the individual cost function (4.6) with respect to a given price curve $\boldsymbol{\rho}_m$.

By the same token, $\boldsymbol{u}_n^{m,*}(\widehat{\boldsymbol{\rho}}_m) \in \mathcal{U}_n^m$ represents the best response of PEV n with respect to $\widehat{\boldsymbol{\rho}}_m$, such that $\boldsymbol{u}_n^{m,*}(\widehat{\boldsymbol{\rho}}_m) \triangleq \min_{\boldsymbol{u}_n \in \mathcal{U}_n} J_n^m(\boldsymbol{u}_n^m; \widehat{\boldsymbol{\rho}}_m)$. Then, by Lemma 4.1, it obtains that $\boldsymbol{u}_n^{m,*}(\boldsymbol{\rho}_m) = \boldsymbol{u}_n^m(\boldsymbol{\rho}_m, A_m(\boldsymbol{\rho}_m))$ and $\boldsymbol{u}_n^{m,*}(\widehat{\boldsymbol{\rho}}_m) = \boldsymbol{u}_n^m(\widehat{\boldsymbol{\rho}}_m, A_m(\widehat{\boldsymbol{\rho}}_m))$.

Another behavior $\boldsymbol{v}_n^m(\boldsymbol{\rho}_m, \widehat{\boldsymbol{\rho}}_m)$ is defined as follow

$$v_{nt}^m(\boldsymbol{\rho}_m, \widehat{\boldsymbol{\rho}}_m) = \begin{cases} u_{nt}^m(\widehat{\boldsymbol{\rho}}_m, A_m^*(\boldsymbol{\rho}_m)), & \text{in case } t \in \mathcal{T}, \\ 0, & \text{otherwise} \end{cases} \tag{4.21}$$

Remark: As specified in (4.21), the charging behavior $v_{nt}^m(\boldsymbol{\rho}_m, \widehat{\boldsymbol{\rho}}_m)$ describes an individual charging control satisfying (4.8) with respect to $\widehat{\boldsymbol{\rho}}_m$ and $A_m^*(\boldsymbol{\rho}_m)$. There is no guarantee that $v_{nt}^m(\boldsymbol{\rho}_m, \widehat{\boldsymbol{\rho}}_m)$ is an admissible charging behavior, such that $\sum v_{nt}^m(\boldsymbol{\rho}_m, \widehat{\boldsymbol{\rho}}_m) = \Gamma_{nm}$.

The inequality (4.14) will be testified in (I) and (II) below.

(I) It will be show that:

$$\begin{aligned} &\left|u_{nt}^{m,*}(\boldsymbol{\rho}_m) - v_{nt}^m(\boldsymbol{\rho}_m, \widehat{\boldsymbol{\rho}}_m)\right| \\ &\leq \left| [f_{nm}']^{-1}(A_m^*(\boldsymbol{\rho}_m) - \rho_{mt}) - [f_{nm}']^{-1}(A_m^*(\boldsymbol{\rho}_m) - \widehat{\rho}_{mt}) \right|. \end{aligned} \tag{4.22}$$

Obviously, (4.22) holds for $t \notin \mathcal{T}_n^m$, since $u_{nt}^{m,*}(\boldsymbol{\rho}_m) = v_{nt}^m(\boldsymbol{\rho}_m, \widehat{\boldsymbol{\rho}}_m) = 0$
It is shown that (4.22) holds for all $t \in \mathcal{T}_n^m$.

Note: For notational simplicity, $\boldsymbol{v}_n^m \equiv \boldsymbol{v}_n^m\left(\boldsymbol{\rho}_m, \widehat{\boldsymbol{\rho}}_m\right)$ will be adopted throughout the following proof.

(I.a) $v_{nt}^m = u_{nt}^{m,*} = 0$. It follows immediately that $v_{nt}^m - u_{nt}^{m,*} = 0$

(I.b) $v_{nt}^m > 0$ and $u_{nt}^{m,*} = 0$. By (4.8), $v_{nt}^m > 0$ implies $\left[f_{nm}'\right]^{-1}\left(A_m^*\left(\boldsymbol{\rho}_m\right) - \widehat{\rho}_{mt}\right) > 0$, and $u_{nt}^{m,*} = 0$ implies $\left[f_{nm}'\right]^{-1}\left(A_m^*\left(\boldsymbol{\rho}_m\right) - \rho_{mt}\right) \leq 0$. Together these give:

$$\begin{aligned}
&0 < v_{nt}^m - u_{nt}^{m,*} \\
&\leq \left[f_{nm}'\right]^{-1}\left(A_m^*\left(\boldsymbol{\rho}_m\right) - \widehat{\rho}_{mt}\right) - \left[f_{nm}'\right]^{-1}\left(A^*\left(\boldsymbol{\rho}_m\right) - \rho_{mt}\right).
\end{aligned}$$

(I.c) $v_{nt}^m = 0$ and $u_{nt}^{m,*} > 0$. Similar to (I.b),

$$\begin{aligned}
&0 < u_{nt}^{m,*} - v_{nt}^m \\
&\leq \left[f_{nm}'\right]^{-1}\left(A^*\left(\boldsymbol{\rho}_m\right) - \rho_{mt}\right) - \left[f_{nm}'\right]^{-1}\left(A^*\left(\boldsymbol{\rho}_m\right) - \widehat{\rho}_{mt}\right).
\end{aligned}$$

(I.d) $v_{nt}^m > 0$ and $u_{nt}^{m,*} > 0$. By (4.8) the following holds:

$$\begin{aligned}
&\left|v_{nt}^m - n_{nt}^{m,*}\right| \\
=&\Big| \min\left\{\gamma_{nm}, \left[f_{nm}'\right]^{-1}\left(A_m^*\left(\boldsymbol{\rho}_m\right) - \widehat{\rho}_{mt}\right)\right\} \\
&\quad - \min\left\{\gamma_{nm}, \left[f_{nm}'\right]^{-1}\left(A_m^*\left(\boldsymbol{\rho}_m\right) - \rho_{mt}\right)\right\}\Big| \\
\leq&\left|\left[f_{nm}'\right]^{-1}\left(A_m^*\left(\boldsymbol{\rho}_m\right) - \widehat{\rho}_{mt}\right) - \left[f_{nm}'\right]^{-1}\left(A_m^*\left(\boldsymbol{\rho}_m\right) - \rho_{mt}\right)\right|
\end{aligned}$$

(II) The following inequality below will be verified:

$$\left\|\boldsymbol{u}_n^{m,*}\left(\boldsymbol{\rho}_m\right) - \boldsymbol{u}_n^{m,*}\left(\widehat{\boldsymbol{\rho}}_m\right)\right\|_1 \leq 2\left\|\boldsymbol{u}_n^{m,*}\left(\boldsymbol{\rho}_m\right) - \boldsymbol{v}_n^m\left(\boldsymbol{\rho}_m, \widehat{\boldsymbol{\rho}}_m\right)\right\|_1. \tag{4.23}$$

For notational simplicity, $\sum \boldsymbol{u}_n^m \equiv \sum_{t\in\mathscr{T}} u_{nt}^m$ is applied in the following (II.a) to (II.c).

(II.a) $\sum \boldsymbol{v}_n^m = \sum \boldsymbol{u}_n^{m,*}\left(\boldsymbol{\rho}_m\right)$. This equality ensures $\boldsymbol{v}_n^m \in \mathscr{U}_n^m$. Also, charging control $\boldsymbol{v}_n^m$ has the form (4.8) with $A_m = A_m\left(\boldsymbol{\rho}_m\right)$. Therefor, by Lemma 4.1, $\boldsymbol{v}_n^m$ is the local optimal control with respect to $\widehat{\boldsymbol{\rho}}_m$. Hence, $\boldsymbol{u}_n^{m,*}\left(\widehat{\boldsymbol{\rho}}_m\right) = \boldsymbol{v}_n^m$. It follows that:

$$\left\|\boldsymbol{u}_n^{m,*}\left(\widehat{\boldsymbol{\rho}}_m\right) - \boldsymbol{u}_n^m\left(\boldsymbol{\rho}_m\right)\right\|_1 = \left\|\boldsymbol{v}_n^m - \boldsymbol{u}_n^m\left(\boldsymbol{\rho}_m\right)\right\|_1 \leq 2\left\|\boldsymbol{v}_n^m - \boldsymbol{u}_n^m\left(\boldsymbol{\rho}_m\right)\right\|_1.$$

(II.b) $\sum \boldsymbol{v}_n^m < \sum \boldsymbol{u}_n^{m,*}\left(\boldsymbol{\rho}_m\right)$. By (4.1) and the definition of $\boldsymbol{u}_n^m\left(\widehat{\boldsymbol{\rho}}_m\right)$, $\sum \boldsymbol{u}_n^m\left(\boldsymbol{\rho}_m\right) = \sum \boldsymbol{u}_n^m\left(\widehat{\boldsymbol{\rho}}_m\right) = \Gamma_{nm}$.
Therefor $\sum \boldsymbol{v}_n^m < \sum \boldsymbol{u}_n^{m,*}\left(\widehat{\boldsymbol{\rho}}_m\right)$, together with (4.8) and the definition of $\boldsymbol{v}_n^m$

and $\boldsymbol{u}_n^{m,*}\left(\widehat{\boldsymbol{\rho}}_m\right)$ implies,

$$A_m^*\left(\widehat{\boldsymbol{\rho}}_m\right) > A_m^*\left(\boldsymbol{\rho}_m\right), \quad u_{nt}^{m,*}\left(\widehat{\boldsymbol{\rho}}_m\right) > v_{nt}^m, \quad \text{for all } t,$$

Hence,

$$\begin{aligned} 0 &\leq \left\|\boldsymbol{v}_n^m - \boldsymbol{u}_n^{m,*}\left(\widehat{\boldsymbol{\rho}}_m\right)\right\|_1 \\ &= \sum \boldsymbol{u}_n^{m,*}\left(\widehat{\boldsymbol{\rho}}_m\right) - \sum \boldsymbol{v}_n^m = \sum \boldsymbol{u}_n^{m,*}\left(\boldsymbol{\rho}_m\right) - \sum \boldsymbol{v}_n^m \\ &\leq \left\|\boldsymbol{v}_n^m - \boldsymbol{u}_n^{m,*}\left(\boldsymbol{\rho}_m\right)\right\|_1 \end{aligned}$$

where the last inequality is a consequence of the triangle inequality for norms, taking into account that $\sum(\cdot) = \|\cdot\|_1$ for all admissible non-negative control trajectories; then:

$$\begin{aligned} &\left\|\boldsymbol{u}_n^{m,*}\left(\widehat{\boldsymbol{\rho}}_m\right) - \boldsymbol{u}_n^{m,*}\left(\boldsymbol{\rho}_m\right)\right\|_1 \\ \leq &\left\|\boldsymbol{v}_n^m - \boldsymbol{u}_n^{m,*}\left(\boldsymbol{\rho}_m\right)\right\|_1 + \left\|\boldsymbol{v}_n^m - \boldsymbol{u}_n^{m,*}\left(\widehat{\boldsymbol{\rho}}_m\right)\right\|_1 \\ \leq &2\left\|\boldsymbol{v}_n^m - \boldsymbol{u}_n^{m,*}\left(\boldsymbol{\rho}_m\right)\right\|_1 \end{aligned}$$

(II.c) $\sum \boldsymbol{v}_n^m > \sum \boldsymbol{u}_n^{m,*}\left(\boldsymbol{\rho}_m\right)$. A similar argument to (II.b) can be used to show the inequality above holds in this case.

Together with (I) and (II), the conclusion of (4.14) in Lemma 4.2 can be obtained. ■

References

1. O. Sundstrom, C. Binding, Planning electric-drive vehicle charging under constrained grid conditions. Technical Report (IBM - Zurich, Switzerland, 2010)
2. K. Clement-Nyns, E. Haesen, J. Driesen, The impact of charging plug-in hybrid electric vehicles on a residential distribution grid. IEEE Trans. Power Syst. **25**(1), 371–380 (2010)
3. R. Hermans, M. Almassalkhi, I.A. Hiskens. Incentive-based coordinated charging control of plug-in electric vehicles at the distribution-transformer level, in *American Control Conference (ACC)*, Montreal, Canada (2012), pp. 264–269
4. M.D. Galus, G. Andersson, Demand management of grid connected plug-in hybrid electric vehicles (PHEV), in *IEEE Energy 2030*, Atlanta, Georgia, 17–18 November 2008
5. Z. Li, Q. Guo, H. Sun, S. Xin, and J. Wang. A new real-time smart-charging method considering expected electric vehicle fleet connections. *IEEE Trans. Power Syst.*, **pp**(99), 1–2 (2014)
6. S. Wang, L. Han, D. Wang, M. Shahidehpour, Z. Li, Hierarchical charging management strategy of plug-in hybrid electric vehicles to provide regulation service, in *2012 3rd IEEE PES International Conference and Exhibition on Innovative Smart Grid Technologies (ISGT Europe)* (IEEE, Berlin, 2012), pp. 1–6
7. W. Qi, Z. Xu, Z.J.M. Shen, Z. Hu, Y. Song, Hierarchical coordinated control of plug-in electric vehicles charging in multifamily dwellings. IEEE Trans. Smart Grid **5**(3), 1465–1474 (2014)

8. W. Yao, J. Zhao, F. Wen, Y. Xue, G. Ledwich, A hierarchical decomposition approach for coordinated dispatch of plug-in electric vehicles. IEEE Trans. Power Syst. **28**(3), 2768–2778 (2013)
9. W. Tang, R. Jain, Hierarchical auction mechanisms for network resource allocation. IEEE J. Sel. Areas Commun. **30**(11), 2117–2125 (2012)
10. F. Kelly, A. Maulloo, D. Tan, Rate control for communication networks: shadow prices, proportional fairness and stability. J. Oper. Res. Soc. **49**(3), 237–252 (1998)
11. Zhong Fan, A distributed demand response algorithm and its application to PHEV charging in smart grids. IEEE Trans. Smart Grid **3**(3), 1280–1290 (2012)
12. S. Bashash, S.J. Moura, J.C. Forman, H.K. Fathy, Plug-in hybrid electric vehicle charge pattern optimization for energy cost and battery longevity. J. Power Sources **196**(1), 541–549 (2011)
13. J. Wang, P. Liu, J. Hicks-Garner, E. Sherman, S. Soukiazian, M. Verbrugge, H. Tataria, J. Musser, P. Finamore, Cycle-life model for graphite-$LiFePO_4$ cells. J. Power Sources **194**(8), 3942–3948 (2011)
14. K.W.E. Cheng, B.P. Divakar, H. Wu, K. Ding, H.F. Ho, Battery-management system (BMS) and SOC development for electrical vehicles. IEEE Trans. Veh. Technol. **60**(1), 76–88 (2011)
15. S. Boyd, L. Vandenberghe, *Convex optimization* (Cambridge University Press, Cambridge, 2004)
16. Z. Ma, D.S. Callaway, I.A. Hiskens, Decentralized charging control of large populations of plug-in electric vehicles. IEEE Trans. Control Syst. Technol. **21**(1), 67–78 (2013)
17. L. Gan, U. Topcu, S.H. Low, Optimal decentralized protocol for electric vehicle charging. IEEE Trans. Power Syst. **28**(2), 940–951 (2013)
18. D. Bertsekas, *Dynamic Programming and Optimal Control*, vol. I (Athena Scientific, Singapore, 1995)

Chapter 5
Decentralized Charging and Discharging Coordination

Abstract Due to the energy storage property of PEV batteries, they could be considered as an energy storage system (ESS) to minimize the total costs of the power systems during coordination interval, and potentially smooth out the fluctuation of the intermittent renewable energy source (RES) and the power load profile as well. In this chapter, it firstly formulates a class of charging and discharging coordination problems of PEVs which deals with the tradeoff between total electricity cost and the accumulated battery degradation costs. It shows that the optimal solution possesses the valley-filling & peak-shift property for any pair of adjacent instants, and then proposes a coordination method by applying the gradient projection method. Due to the decoupling relationship among the admissible sets of coordination behaviors of all the PEVs in the modeling issue, a decentralized algorithm is proposed by applying the gradient projection method, such that the coordination behaviors of individual PEVs can be updated locally and simultaneously. It is proved that, the convergence and optimality of the proposed method are guaranteed in case the step-size parameter of the update procedure is in a certain region, and its performance doesn't rely on the shape of the base demand trajectory. Furthermore, a receding horizon based algorithm is presented to account for the forecast errors occurred in the predictions of PEV populations, RES productions and inelastic load profiles. The results developed in this chapter are demonstrated with numerical simulations and comparisons with other decentralized methods are also provided.

5.1 Introduction

Besides the orderly coordinated charging behaviors discussed in previous chapters, it has emerged to offer benefits to the power grid by considering both charging and discharging coordination behaviors for individual PEVs, [1, 2], since PEVs can also be potentially considered as storage units by discharging their batteries, e.g., mitigating their negative effects on the grid system, see [3–6], shaving peak load [7–9] and providing ancillary services [10, 11].

Moreover, within the operation scheme, PEVs could serve as energy storage systems (ESSs), which can promote the integration of renewable energy sources (RES)

Z. Ma, *Decentralized Charging Coordination of Large-scale Plug-in Electric Vehicles in Power Systems*, https://doi.org/10.1007/978-981-13-7652-8_5

and maintain the stability of the grid against the intermittence and randomness of the RES. Due to this, some recent work has studied the interaction between the PEVs and the RES, such as [12, 13]. In this chapter, it is also focused on how to coordinate PEVs' charging and discharging behaviors in order to minimize total system costs and reduce the effects of the RES on the power grid.

Compared with the work on how to effectively coordinate PEVs' charging behaviors, the PEV management algorithms under bidirectional scheme can achieve smaller load variance through cost minimization or benefit maximization [3, 14–17]. Specifically, in [3], the authors firstly analyze the bidirectional coordination of PEVs via a global scheme, and then propose a local scheme to reduce the burden on the communication and computation of the centralized formulation. In the local scheme, the whole population of PEVs are divided into groups according to their physical locations and each group is coordinated by a local controller which can obtain background demand information from the central controller, collect the information of all the PEVs located in its group and compute a local optimization problem independently. It can achieve a performance near to the optimal scheduling solution under the proposed hierarchical coordination method. In [14], a coordinated charging and discharging algorithm was proposed to minimize PEVs' charging cost, under the consideration of network limits. In [15], a real-time scheduling algorithm was applied to estimate the capacity of V2G and the availability of the RES within a building energy management system. Stochastic characteristics of PEV driving behaviors were taken into account in [16, 17]. The aim of [16] is to minimize the overall load variance of power grid in a centralized way, whilst [17] aims at evaluating the V2G capacity using aggregate PEV model.

The research work in [3, 14–17] deals with the charging and discharging coordination of PEVs in different scenarios under centralized schemes, which usually require complex and expensive communication infrastructure, powerful central controller with computation modular, and full information of all the PEVs. Also, decisions of all the PEVs are controlled by the central controller. In contrast, charging/discharging decisions are made and executed by the PEVs themselves in decentralized ways. The communication and computation burden can be largely reduced in the meanwhile, especially for large scale systems. A couple of decentralized methods have been designed alternatively to achieve coordination solutions. For example, in PEV charging coordination, real-time pricing [18, 19] and noncooperative games [20, 21] are leveraged. However, these developed decentralized methods cannot be directly applied to solve bidirectional coordination, since the evolution of the SOC of PEVs' batteries is no longer non-decreasing as in the charging-only scenarios.

The centralized optimal scheduling solution can be effectively implemented in the case that the system has complete information and has a permission to directly coordinate the behaviors of all of the individual PEVs. However, in practice, the PEVs may not be willing to share their private information with others, and the transmission of complete information may create heavy communication signals.

Due to the above difficulties concerning the centralized method, certain research work considers how to effectively coordinate the charging and discharging behaviors of PEVs in decentralized schemes, see [4–6, 22, 23]. Reference [22] extolled the advantage of providing services in the form of V2G power injections to the system, which was scheduled by taking into account the damage cost introduced to the battery. This study presented a scheduling algorithm for PEVs which are supposed to be a price taker, i.e. PEVs' charging behavior will not influence the price to get electricity. Hence each PEV's coordination problem is independent with each other. In [23], data communication risks were considered for the V2G coordination. The authors formulated the optimal energy cost problem for the PEV user to make optimal decisions in an online fashion, while the control variable lies in a discrete set with three elements, say charging, discharging and neither charging nor discharging. The optimal charging/discharging rates could not be provided.

The work in [4–6] designed decentralized coordination methods based on the partition of the whole coordination period into charging-only instants and discharging-only instants. In [4], the partition was implemented by comparing the base demand of each instant with the average value of the base demand trajectory over the whole period, and the total period was divided into multiple sub-intervals. At those instants whose base demand is higher than the average value, the PEVs are supposed to be discharged only; otherwise PEVs are supposed to be charged only. By doing this, the evolution of the state of charge of PEVs is largely simplified. In [5, 6], the authors proposed a decentralized water-fill algorithm such that the whole period is partitioned into two sub-intervals. The first sub-interval is discharging only interval such that all the PEVs are only allowed to be discharged, and the second one is charging only. This method mainly focuses on a kind of typical load profiles such that peak demand appears at first followed by the valley demand. The performance of these methods is highly related to the shape of the inelastic demand trajectory, which means the resulted coordination strategy may be far away from the social optimum in case the shape of the inelastic demand trajectory doesn't fulfill the shapes considered.

In this chapter, it firstly formulates a class of PEV charging and discharging coordination problems which deals with the tradeoff between total generation cost and the accumulated battery degradation cost of PEV populations, which is formulated in Chap. 3.

Considering the convexity property of generation and battery degradation costs, the optimal solution of the underlying optimization problems is *valley-fill* and *peak-shift* on any pair of adjacent time slots due to the constraints of the state of charge (SOC) values at any instants.

In this part it will propose a coordination method by applying the gradient projection method. Due to the decoupling relationship among the admissible sets of coordination behaviors of all the individual PEVs, the gradient projection on an individual admissible coordination set is independent upon the admissible sets of all the others. It shows that, following this modeling issue, the coordination behaviors of all the individual PEVs can be updated locally and simultaneously. Following the proposed decentralized method, the system converges to the optimal solution in case the step-size parameter of the update procedure is set in a certain region. It is worth to mention that, unlike those methods stated above, the performance properties, say optimality & convergence, of our proposed method don't rely on the shape of the base demand curve. The results developed in our work are demonstrated and verified with several numerical simulations.

Moreover, to address the forecast errors appeared in the predictions of PEV populations, RES productions and inelastic load profiles, a receding horizon based algorithm is constructed such that the prediction proceeds at each instant. As a result, the system can obtain a coordination strategy against those forecast errors.

The rest of the chapter is organized as follows. In Sect. 5.2, it formulates the charging and discharging coordination problems over a multi-time interval as a class of constrained optimization problems, and analyze the valley-filling and peak-shift property of the optimal solution on each pair of adjacent instants. A few of numerical simulations are studied to demonstrate the optimality property of the optimal solution and the performance of some decentralized methods proposed in the literature. In Sect. 5.3, a decentralized method based upon the gradient projection is proposed for the underlying PEV coordination problems, and then it shows the convergence and the optimality performance of the proposed method. The results developed in this chapter are demonstrated with several numerical examples in Sect. 5.4. In Sect. 5.5, a receding horizon based algorithm is also proposed to account for the forecast errors in the system, Some numerical examples and and certain comparisons to some other methods are given to demonstrate the performance of the proposed method. Finally in Sect. 5.6 it makes a brief conclusion of the work.

5.2 Formulation of PEV Charging and Discharging Coordination Problems

In this part, it studies the charging and discharging coordination of a large population of PEVs with most of their specifications given in Sect. 2.2.1 of Chap. 2, except that, instead of the common charging horizon $\mathscr{T}$, the charging horizon of PEV n is $\mathscr{T}_n$ such that $\mathscr{T}_n \subset \mathscr{T}$.

Call $\boldsymbol{u}_n \equiv (u_{nt}; t \in \mathscr{T})$ an admissible coordination strategy of PEV n, with $n \in \mathscr{N}$, if (2.2) holds for all $t \in \mathscr{T}$.

Note: In Chap. 2, it considers that $\Upsilon_n^- = 0$, that is to say, in that part, each of individual PEVs isn't allowed to be discharged during the period of time except of this chapter.

Different from other chapters, however here it considers that $\Upsilon_n^- > 0$, That is, by (2.2), the PEV n is allowed to be charged and discharged over the period of $\mathscr{T}_n$ such that at each instant the SOC value has to be in a feasible range and at the terminal instant it has to be fully charged.

Recall that denote by $\mathbf{soc}_n$ the SOC trajectory of PEV n subject to an admissible strategy $\boldsymbol{u}_n$, and its evolution is given in (2.1) in Chap. 2.

By (2.1), (2.2b) and (2.2c) are equivalent to (5.1a), (5.1b) and (5.1c) below:

$$\sum_{r\in[0,t]} u_{nr} \geq \Gamma_n(\text{soc}_{n,\min} - \text{soc}_{n0}), \text{ for all } t \in \mathscr{T}_n, \tag{5.1a}$$

$$\sum_{r\in[0,t]} u_{nr} \leq \Gamma_n(\text{soc}_{n,\max} - \text{soc}_{n0}), \text{ for all } t \in \mathscr{T}_n, \tag{5.1b}$$

$$\sum_{t\in\mathscr{T}} u_{nt} = \Gamma_n(\text{soc}_{n,\max} - \text{soc}_{n0}). \tag{5.1c}$$

Note: In order to simplify the the constraints of (5.1a) and (5.1b), in some research work in the literature, e.g., [4–6], the whole coordination interval is partitioned into a few of sub-intervals each of which is composed of a collection of adjacent instants. All the PEVs are allowed to be either charged only or discharged only over each of such sub-intervals.

The set of admissible strategies of PEV n is denoted by $\mathscr{U}_n$, such that

$$\mathscr{U}_n \triangleq \left\{ \boldsymbol{u}_n \in \mathbb{R}^T;\ \text{s.t. (2.2a) and (5.1)} \right\}, \tag{5.2}$$

and the set of admissible charging strategies for PEV populations, is denoted by $\mathscr{U}$, such that

$$\mathscr{U} = \mathscr{U}_1 \times \cdots \mathscr{U}_{N-1} \times \mathscr{U}_N. \tag{5.3}$$

In the next section, it will formulate a class of coordination problems for the PEV populations by considering the tradeoff between the power generation cost and the accumulated PEV battery degradation costs.

5.2.1 Formulations of PEV Charging and Discharging Coordination Concerning Battery Degradation Cost

The system cost over the horizon $\mathscr{T}$, denoted by $J(\boldsymbol{u})$, is defined as below:

$$J(\boldsymbol{u}) \triangleq \sum_{t\in\mathscr{T}} \left\{ \eta c(d_t + U_t) + (1-\eta) \sum_{n\in\mathscr{N}} f_n(u_{nt}) \right\}, \tag{5.4}$$

with $U_t \equiv \sum_{n\in\mathscr{N}} u_{nt}$ and $\eta \in [0, 1]$, where d_t denotes the base demand at instant t in the power system, $c(\cdot)$ and $f_n(\cdot)$ represent respectively the total generation cost and individual battery degradation cost of PEV n which have been analyzed and specified in previous chapters in details.

Note: In the system cost function $J(\cdot)$ given above, η is a weighting parameter which deals with the tradeoff between total generation cost and the aggregated battery degradation costs over the whole PEV population.

A class of centralized PEV charging and discharging coordination problems will be formally formulated as follows:

Problem 5.1 $\boldsymbol{u}^* = \text{argmin}_{\boldsymbol{u}\in\mathscr{U}} J(\boldsymbol{u})$, that is to say, the objective is to implement a collection of optimal coordination strategies $\boldsymbol{u}^*$, by minimizing the system cost in (5.4). ■

For notational simplicity, it considers $D_t(\boldsymbol{u}_t) \equiv d_t + \sum_{n\in\mathscr{N}} u_{nt}$ and $D_{-n,t}(\boldsymbol{u}_t) \equiv d_t + \sum_{m\in\mathscr{N}/\{n\}} u_{mt}$, which represents the total demand except PEV n at t.

Theorem 5.1 *Under Assumptions 3.2 and 3.6, the optimal strategy $\boldsymbol{u}^*$ satisfies the following properties, for each PEV n and a pair of adjacent instants $t, s \in \mathscr{T}$:*

$$u^*_{nt} \ge u^*_{ns}, \tag{5.5a}$$

$$D_t(\boldsymbol{u}^*) \le D_s(\boldsymbol{u}^*), \tag{5.5b}$$

in case $D_{-n,t}(\boldsymbol{u}^*) \le D_{-n,s}(\boldsymbol{u}^*)$. ■

Proof of Theorem 5.1 is given in Appendix 5.7.1.

Remark: In case the PEV owners don't allow to discharge their vehicles, the optimality properties of (5.5) stated in Theorem 5.1, holds for any $t, s \in \mathscr{T}$, as developed in [18]. That is, the valley-filling property holds for each pair of instants in case the PEVs are only allowed to be charged; while the peak-shift and valley-filling property holds for each pair of *adjacent* instants as stated in Theorem 5.1 in case PEV owners allow to charge and discharge their vehicles respectively.

In Theorem 5.2, it will study the properties of the optimal strategy under certain specific conditions. Before that, it defines a notion $\mathscr{N}_t$ with respect to $t \in \mathscr{T}$, such that

$$\mathscr{N}_t \triangleq \left\{n \in \mathscr{N};\ \text{s.t. } t \in \mathscr{T}_n\right\}, \tag{5.6}$$

i.e., $\mathscr{N}_t$ represents all the PEVs which are plugged in the grid at instant t.

Theorem 5.2 *Under Assumptions 3.2 and 3.6, the optimal solution $\boldsymbol{u}^*$ satisfies the properties of (5.5) in case $d_t \le d_s$, for each PEV and each pair of adjacent instants $t, s \in \mathscr{T}$ with $\mathscr{N}_t = \mathscr{N}_s$.* ■

Proof of Theorem 5.2 is given in Appendix 5.7.2.

In Lemma 5.1 below, it gives the optimality property of the charging and discharging problems without considering any battery degradation costs.

Lemma 5.1 *With weighting parameter $\eta = 1$, (5.5) degenerates to the following exact peak-shift and valley-filling property:*

$$D_t(\boldsymbol{u}^*) = D_s(\boldsymbol{u}^*), \tag{5.7}$$

in case $u^*_{nt}, u^*_{ns} \in (-\Upsilon^-_n, \Upsilon^+_n)$, and $soc_{nt}(\boldsymbol{u}^*_n), soc_{ns}(\boldsymbol{u}^*_n) \in (soc_{n,\min}, soc_{n,\max})$ for all $n \in \mathscr{N}$.

Proof In case $u^*_{nl} \in (-\Upsilon_n^-, \Upsilon_n^+)$, and $\mathrm{soc}_{nl}(\boldsymbol{u}_n^*) \in (\mathrm{soc}_{n,\min}, \mathrm{soc}_{n,\max})$ for all $n \in \mathscr{N}$ and $l = t, s, \gamma_{nl}^- = \gamma_{nl}^+ = \lambda_{nl}^- = \lambda_{nl}^+ = 0$ by (5.10). Then by (5.10e), it can obtain that $\eta\big(c'(D_t) - c'(D_s)\big) = 0$ which implies (5.7) under Assumptions 3.2 and 3.6. ■

Remark: By Lemma 5.1, the property of (5.7) holds for a collection of instants which are adjacent to each other, say $\{s, s+1, \ldots, s+k\}$, if at each of these instants, $u^*_{nt} \in (-\Upsilon_n^-, \Upsilon_n^+)$, and $\mathrm{soc}_{nt}(\boldsymbol{u}_n^*) \in (\mathrm{soc}_{\min}, \mathrm{soc}_{\max})$ for all $n \in \mathscr{N}$.

In Sect. 5.2.2 below, it illustrates the properties of the optimal coordination solutions with numerical simulations.

5.2.2 *Numerical Examples*

It specifies the coordination of PEVs with a population size of $N = 350$ over a multi-time interval from 20:00 on one day to 20:00 on the next day. It adopts a base demand trajectory used in [24] and considers that the time length ΔT is 1 h, see Fig. 5.1 for an illustration.

For the purpose of demonstration, here it is simply supposed that $\mathscr{T}_n = \mathscr{T}, \Gamma_n =$ 20 kWh, $\mathrm{soc}_{n,\min} = 20\%$ and $\mathrm{soc}_{n,\max} = 90\%$ for all n. Also consider that all the PEVs share a common initial SOC value of 30%; then the total required energy over the period of $\mathscr{T}$ is valued with 12 kWh, for all $n \in \mathscr{N}$. Suppose that the generation cost is $c(D_t) = 10^{-5} D_t^2 + 0.02 D_t$, and the battery degradation cost is as formulated in [19], such that $f_n(u_{nt}) = 0.004u_{nt}^2 + 0.075u_{nt} + 0.003$.

5.2.2.1 Optimal Solution

Figure 5.1 displays the resulting optimal coordination behaviors with weighting parameter η equal to 1.0 and 0.8 respectively.

As illustrated, for each pair of adjacent instants t and s, (5.5) holds in case $d_t \le d_s$, which is consistence with Theorems 5.1 and 5.2. However as observed, this property may not hold for a pair of instants which are not adjacent to each other. For example, $D^*_{21:00} < D^*_{16:00}$, but $d_{21:00} > d_{16:00}$.

Figure 5.2 displays of the associated SOC trajectories of an individual PEV $\mathrm{soc}_n(\boldsymbol{u}_n^*)$, subject to the optimal solution $\boldsymbol{u}_n^*$, with different valued parameter η respectively, which are all in the region of $[\mathrm{soc}_{n,\min}, \mathrm{soc}_{n,\max}] = [20\%, 90\%]$ over the whole interval $\mathscr{T}$.

As illustrated, each optimal coordination trajectory of an individual PEV reaches the boundaries of $\mathrm{soc}_{n,\min}$ and $\mathrm{soc}_{n,\max}$ before it is fully charged at the end of interval. More specifically, it is firstly discharged until it reaches the minimal SOC at 23:00; it is then charged from 24:00 until reaching the maximum SOC state at 8:00 on the next day; finally it is discharged until 13:00 and is charged again until reaching the maximum SOC at the terminal instant 20:00.

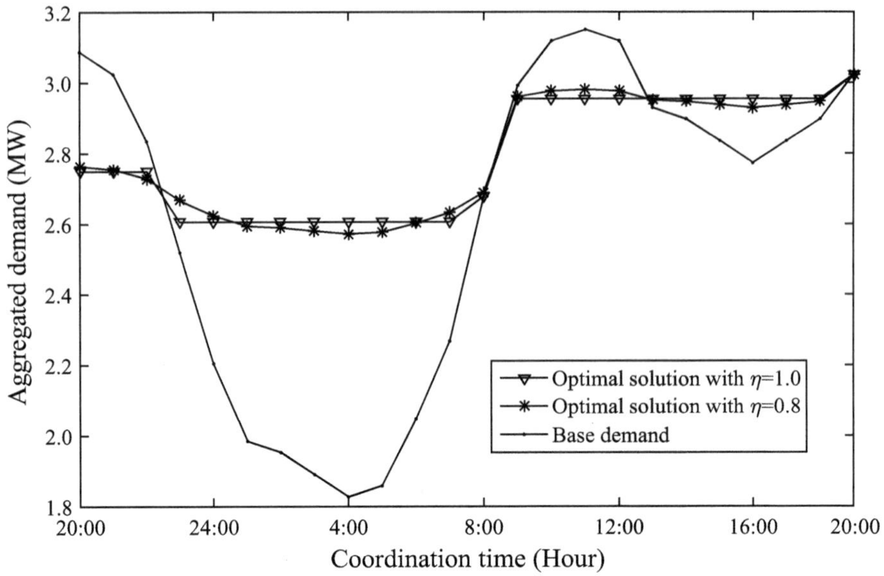

Fig. 5.1 Optimal coordination behaviors with weighting parameter $\eta = 0.8$ and $\eta = 1.0$, say without degradation costs, respectively

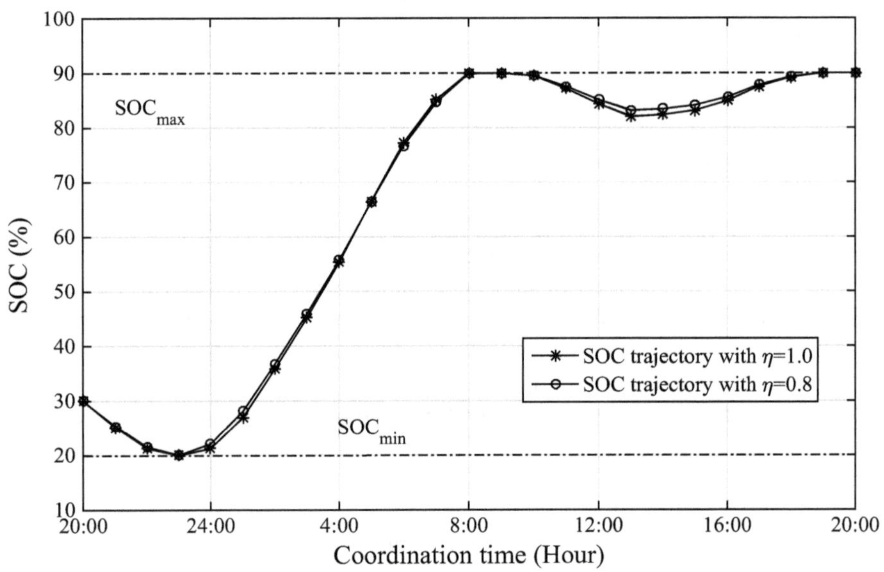

Fig. 5.2 Evolutions of the SOC values of an individual PEV subject to optimal coordination behaviors with parameter $\eta = 0.8$ and $\eta = 1.0$ respectively

Also as observed from Figs. 5.1 and 5.2, (5.7) holds for each of collections of sub-intervals, say {20:00, 21:00, 22:00}, {23:00, 0:00, ..., 7:00} and {9:00, 10:00, ..., 19:00}, respectively, such that $soc_{nt} \in (soc_{n,min}, soc_{n,max})$. As a result, it can obtain that the simulation result is consistent with Lemma 5.1.

5.2.2.2 Brief Discussions on the Implemented Solutions by Applying Some Decentralized Methods in the Literature

The centralized optimal scheduling solution can be effectively implemented in the case that the system has complete information and has a permission to directly coordinate the behaviors of all the individual PEVs. However, in practice, the PEVs may not be willing to share their private information with others, and the transmission of complete information may create heavy communication signals. Hence, the centralized method is usually impractical.

To overcome the difficulties concerning the centralized method, some research work in the literature have designed some methods to implement the coordination behaviors in decentralized ways. In [4], the individuals are supposed to be discharged at those instants at which the base demand is higher than the average value of the based demand over the whole interval and to be charged otherwise; while in [5, 6], the authors propose other decentralized methods such that the whole interval is partitioned into two sub-intervals, such that all the PEVs are only allowed to be discharged during the first sub-interval, and to be charged during the second one.

Figures 5.3, 5.4 and 5.5 display the optimal solution and the implemented solutions by applying the methods proposed in [4, 5] considering different shapes of the base demand trajectory respectively. As illustrated, the performance of the implemented solutions by applying those decentralized methods is dependent upon the shape of the base demand curve.

5.3 Decentralized Charging and Discharging Coordination Algorithm

In this section, a decentralized coordination algorithm is designed for the underlying coordination problem in Algorithm 5.1 by applying a *gradient projection* method such that all the PEVs can update their strategies locally and simultaneously.

By adopting this algorithm, PEVs don't need to share their private information with others and the computation tasks are distributed to all the PEVs. Moreover, the performance properties, such as the optimality and convergence, of the proposed algorithm are analyzed.

Gradient projection method [25] has been used extensively for a variety of constrained optimization problems, and it is successful in real-world applications due to its computational superiority and fast rates of convergence in conjunction with

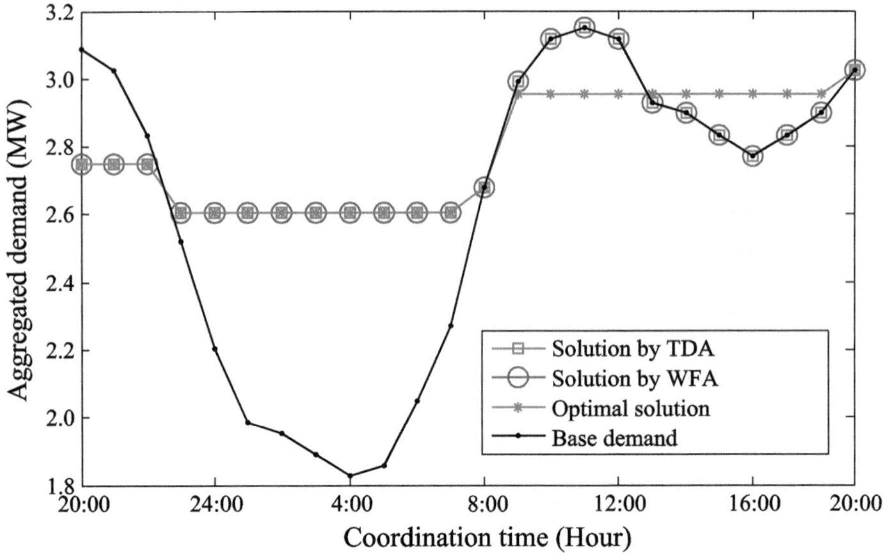

Fig. 5.3 Optimal solution versus solutions implemented by applying TDA, WFA and Algorithm 5.1 without considering degradation costs

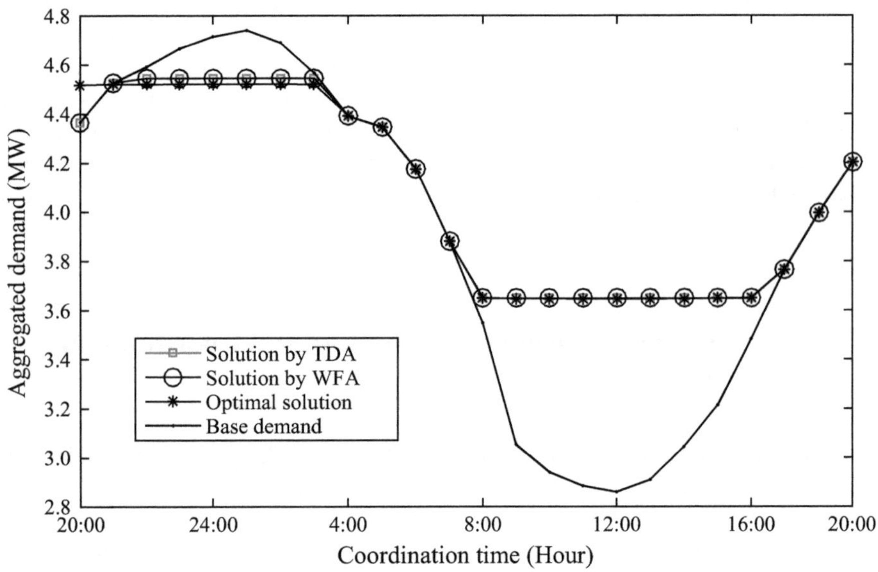

Fig. 5.4 Optimal solution versus solutions implemented by TDA, WFA and Algorithm 5.1 without considering degradation costs

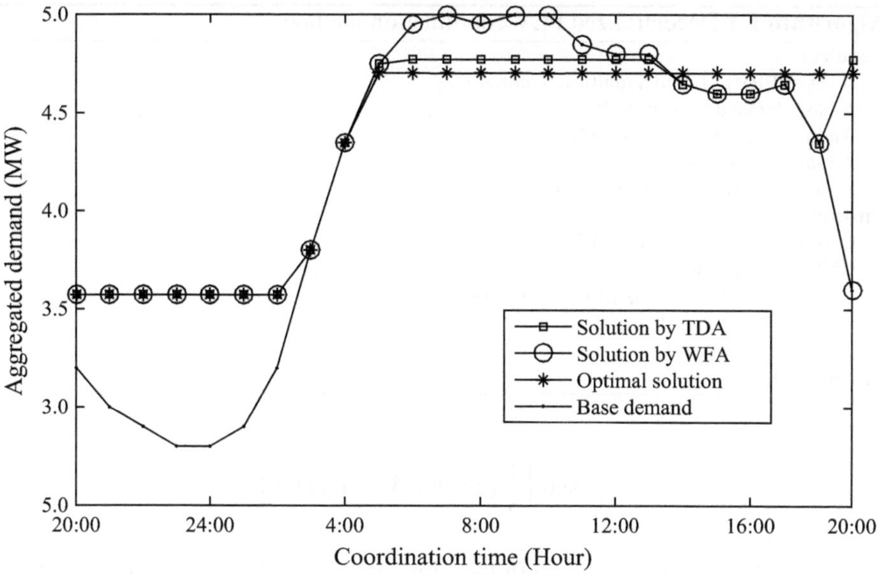

Fig. 5.5 Optimal solution versus solutions implemented by TDA, WFA and Algorithm 5.1 without considering degradation costs

other methods. Hence, an algorithm is presented based on the gradient projection method. For a given strategy $\boldsymbol{u}$, the update of the strategy $\widehat{\boldsymbol{u}}(\boldsymbol{u})$ by applying gradient projection is as below [25]:

$$\widehat{\boldsymbol{u}}(\boldsymbol{u}) = [\boldsymbol{u} - \alpha \nabla J(\boldsymbol{u})]^+_{\mathscr{U}} \tag{5.8}$$

where $[\boldsymbol{x}]^+_{\mathscr{U}}$ represents the projection of the vector $\boldsymbol{x}$ on the set $\mathscr{U}$.

The update rule in (5.8) is still a centralized form which requires a central node for computation. Nevertheless, due to the decoupling property of the admissible set defined in (5.3), say $\mathscr{U} = \mathscr{U}_1 \times \cdots \mathscr{U}_{N-1} \times \mathscr{U}_N$, it can be verified that the updated strategy $\widehat{\boldsymbol{u}}(\boldsymbol{u})$ specified in (5.8) can be implemented by the following decentralized form:

$$\widehat{\boldsymbol{u}}_n(\boldsymbol{u}) = [\boldsymbol{u}_n - \alpha[\nabla J(\boldsymbol{u})]_n]^+_{\mathscr{U}_n}, \quad \forall n \in \mathscr{N}, \tag{5.9}$$

where $[\nabla J(\boldsymbol{u})]_n$ is the nth row of $\nabla J(\boldsymbol{u})$.

In Algorithm 5.1, a decentralized update procedure is designed for the coordination using the gradient projection method given in (5.9).

Remark: Algorithm 5.1 is decentralized and can deal with the case involving large-scale PEVs, since all the decisions are updated locally and simultaneously.

The convergence of Algorithm 5.1 will be analyzed in Theorem 5.3. Before this, it defines a notion κ, such that

Algorithm 5.1 Decentralized PEV coordination method.

Require:
A set of PEVs $\mathcal{N}$ and a multi-time horizon $\mathcal{T}$;
A base demand trajectory $\boldsymbol{d}$;
An initial strategy of PEVs $\boldsymbol{u}^0$;
An initial $\varepsilon > \varepsilon_{\text{stop}}$;
Set $k := 0$, and a step size $\alpha > 0$;
Ensure:
Coordination behaviors of PEVs $\boldsymbol{u}$;'
1: **while** $\varepsilon > \varepsilon_{\text{stop}}$ **do**
2: Update $\boldsymbol{u}_n^{k+1}(\boldsymbol{u}^k) := \widehat{\boldsymbol{u}}(\boldsymbol{u}^k)$ by (5.9), $\forall n \in \mathcal{N}$;
3: Update $\varepsilon := \|\boldsymbol{u}^{k+1} - \boldsymbol{u}^k\|_1$;
4: Update $k := k + 1$;
5: **end while**

$$\kappa \geq \sup_{D,\delta} \left\{ \frac{1}{\delta} \left(c'(D+\delta) - c'(D) \right) \right\},$$

i.e., κ represents an upper bound of the Lipschitz constant of $c'(\cdot)$. Suppose that ρ_n is the Lipschitz constant of $f_n'(\cdot)$ over $[-\Upsilon_n, \Upsilon_n]$ for each $n \in \mathcal{N}$; then it defines another notion ρ, such that $\rho \geq \max_{n \in \mathcal{N}} \rho_n$.

Theorem 5.3 (Convergence and optimality of Algorithm 5.1) *By applying Algorithm 5.1, the system cost $J(\boldsymbol{u}^k)$ monotonically decreases with respect to iteration step k, and converges to the optima $J(\boldsymbol{u}^*)$ in case* $0 < \alpha < 2\left(\eta N^2 \kappa^2 + (1-\eta)\rho^2\right)^{-1/2}$.

Proof Please refer to Appendix 5.7.3.

5.4 Numerical Examples

In this section, unless specified, it studies several numerical cases by adopting the parameters considered in the example in Sect. 5.2.2.

5.4.1 Case Study I

In this part, it considers the coordination problem of a homogenous population of PEVs with their specifications specified in Sect. 5.2.2. Also consider the base demand curve illustrated in Fig. 5.1.

The behaviors of all the PEVs are updated simultaneously by adopting the decentralized coordination method proposed in Algorithm 5.1. As illustrated in Fig. 5.6 and compared with the optimal solution displayed in Fig. 5.1, the system converges

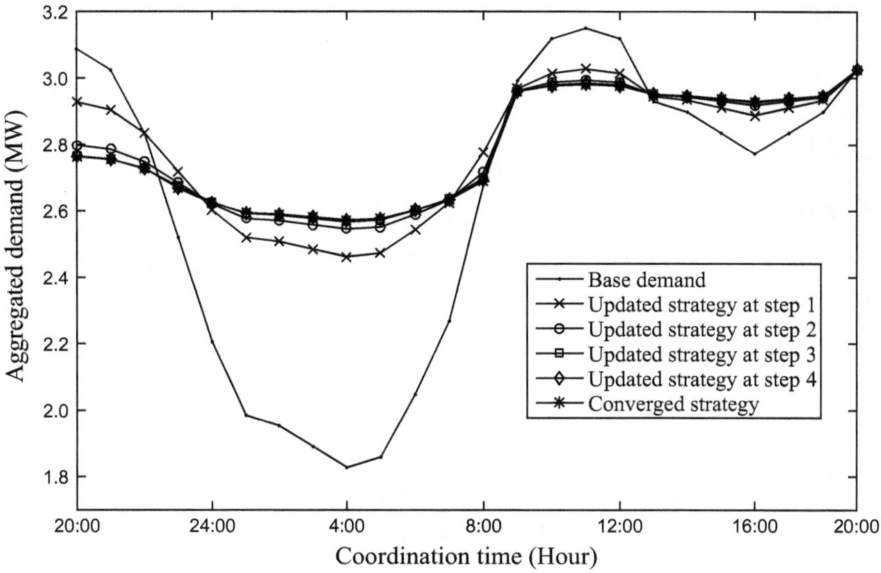

Fig. 5.6 Updates of the aggregated coordination behaviors of homogeneous PEV population with degradation costs considering $\eta = 0.8$

to the optimal solution in a few of iteration steps under the proposed decentralized method.

It also considers the coordination problem of the same population of PEVs as specified above without the degradation cost, say setting parameter $\eta = 1.0$. Figure 5.7 displays the corresponding optimal solution by applying Algorithm 5.1; then with the comparison of the optimal solution displayed in Fig. 5.1, the implemented solution is optimal as well. This is consistent with Theorem 5.3.

5.4.2 *Case Study II*

For approaching realistic characteristics of PEV populations, it is supposed that the initial SOC values of PEV populations, $\{\mathrm{SOC}_{n0}; n \in \mathcal{N}\}$, approximately satisfy a Gaussian distribution, denoted by $\mathrm{N}(\hat{\mu}, \hat{\gamma})$, see [26, 27]. For the purpose of demonstration, it considers $\hat{\mu} = 0.5$ and $\hat{\gamma} = 0.1$. Figure 5.8 illustrates the iteration updates of the aggregated individual behaviors following Algorithm 5.1. As illustrated, the system converges in a few of iteration steps.

Figure 5.9 displays the evolutions of the SOC trajectories for individual PEVs with different initial SOC values respectively.

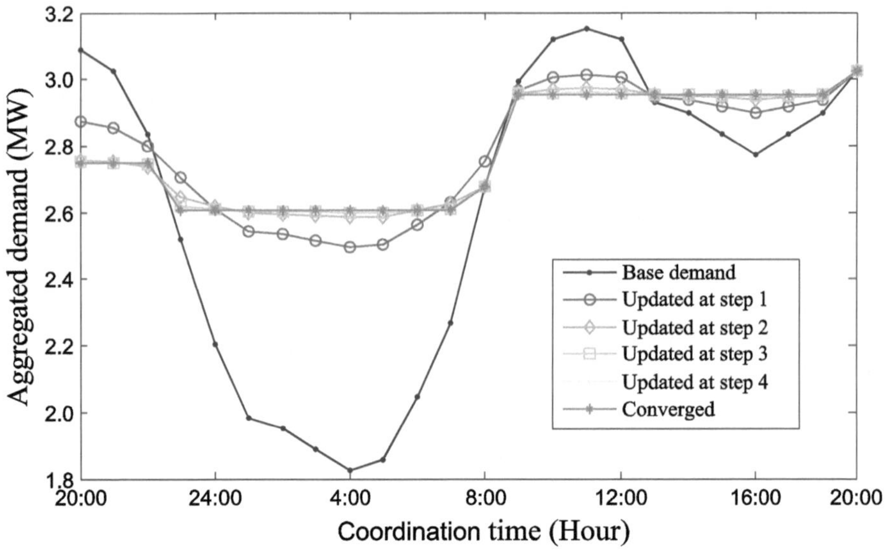

Fig. 5.7 Updates of the aggregated coordination behaviors of homogeneous PEV population without degradation costs, say $\eta = 1.0$

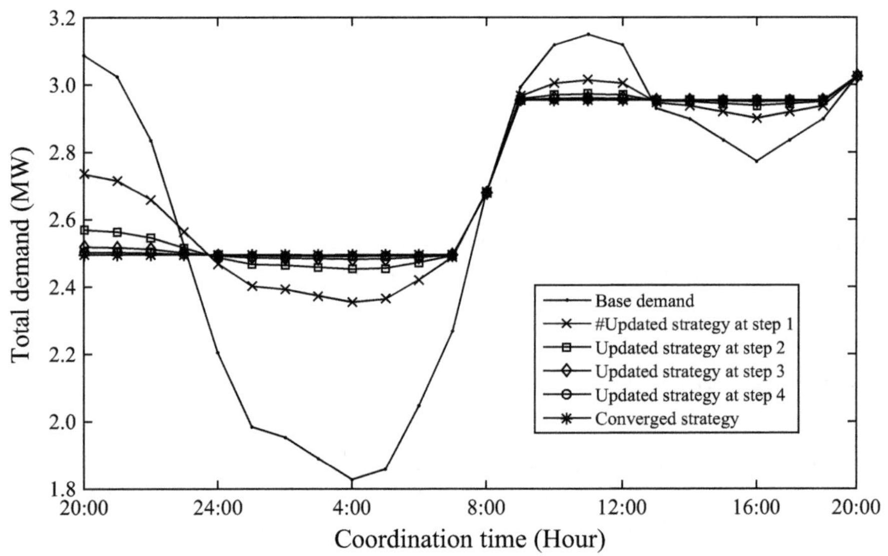

Fig. 5.8 Updates of aggregated behaviors of PEV population in case study II

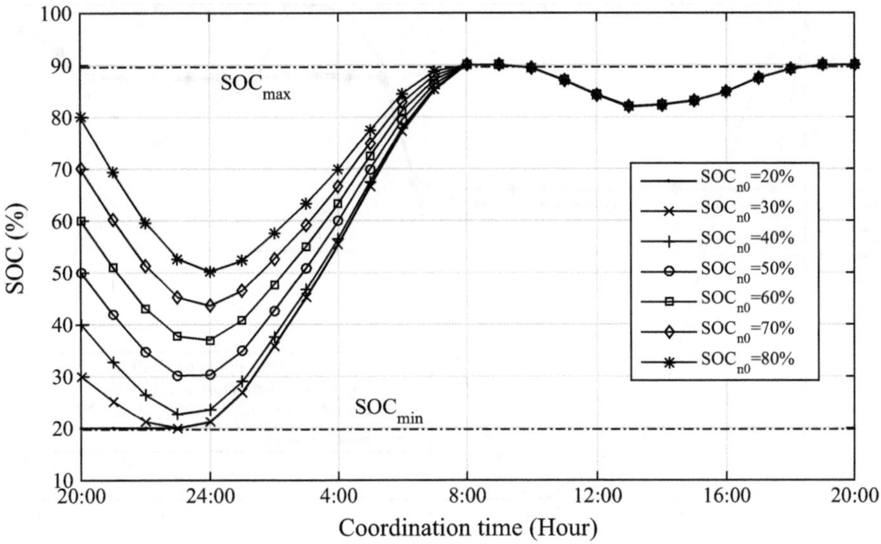

Fig. 5.9 Evolutions of SOC trajectories of PEV population subject to the implemented solution displayed in Fig. 5.8

5.4.3 *Case Study III*

Here, it considers a more complicated case. It adopts the initial SOC distribution of the PEV population specified in Sect. 5.4.2, and supposes that the PEV population arrives during the period between 20:00 on one day and 2:00 on the next day, approximately following a Poisson distribution, [28], with a parameter $\delta = 100$ per hour. For illustration simplicity, it is supposed that all the PEVs terminate at 20:00 on the next day.

Denote by $\widehat{\mathscr{N}}$ a subpopulation of PEVs each of which arrives at 2:00 on a day with a common initial SOC value of 50%. Figure 5.10 displays the aggregated behaviors of all the PEVs and all the PEVs except $\widehat{\mathscr{N}}$ over the whole interval. As illustrated, the implemented result satisfies the optimality property stated in Lemma 5.1.

Figure 5.11 displays the SOC trajectories of all the heterogeneous PEV populations. For example, the bottom-right sub-figure displays the SOC curves of the PEVs which possess a common initial SOC value of 80% and plug in the system at different instants respectively; more specifically, among all these curves, the red-colored curve represents the SOC curve of the PEVs plugging in the system at 22:00 on a day.

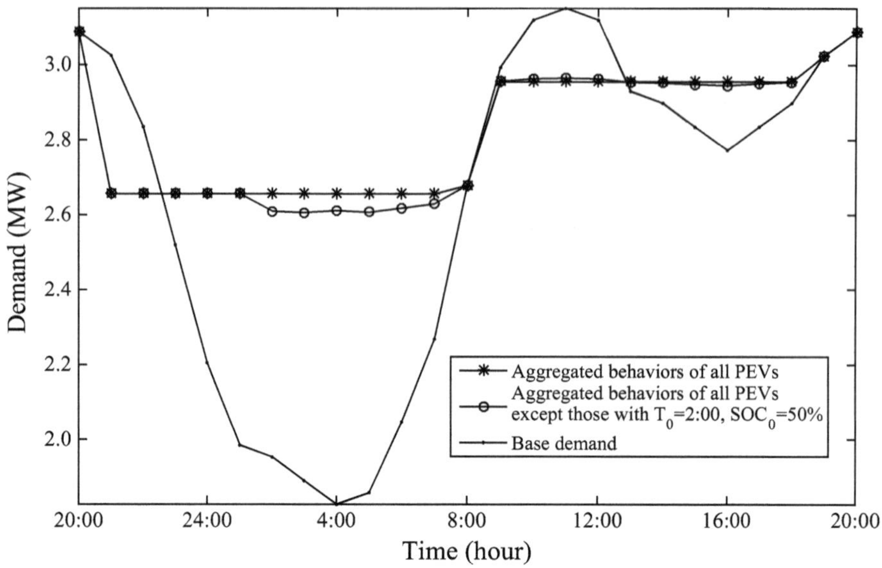

Fig. 5.10 Illustration of aggregated behaviors of all the PEVs and all the PEVs except those with $T_0 = 2:00$ and $\mathrm{soc}_0 = 50\%$, respectively

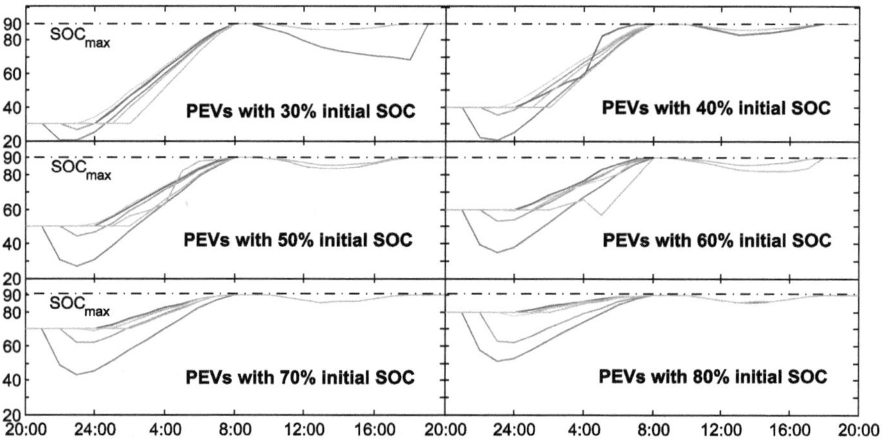

Fig. 5.11 SOC trajectories of all the heterogeneous PEV populations

5.5 Accounting for Forecast Errors Based on Receding Horizon Algorithm

Besides large-scale dispatchable PEVs, inelastic base demand and conventional generators considered in earlier parts, the smart grid system may also contain certain intermittent renewable energy sources (RES) as well, see Fig. 5.12 for an illustration.

Besides the characteristics of PEVs and base demand trajectory have been given in earlier parts, the RES units and generators are briefly stated below.

Suppose that the aggregated power generated by the RES, denoted by $\boldsymbol{R} \equiv \{R_t, t \in \mathscr{T}\}$, is given at the beginning of the operation period $\mathscr{T}$. Moreover, in order to maximize the social welfare and become as environmentally friendly as possible, as considered in the literature, in our work the power generated by the renewable generators shall be fully utilized.

Suppose that the aggregated renewable resource R_t is known prior to the operation period, and has to be completely consumed; then the net base demand is given as: $d_t - R_t$, with d_t representing the base demand of the system as stated earlier.

Denote by $\boldsymbol{G} \equiv \{G_t, t \in \mathscr{T}\}$ the power generated by conventional generators; and the power should be balanced at each instant, i.e.

$$G_t = d_t + U_t - R_t,$$

where $U_t \equiv \sum_{n \in \mathscr{N}} u_{nt}$ denotes the aggregated power consumed/supplied by PEVs.

The model, which is analyzed in earlier sections, underpinning the coordination scheme assumes that the PEV population, the power generated by the RES and the inelastic demand are known with certainty prior to the operation period. The resulting model formulation is static and the decentralized gradient projection process can be undertaken off-line ahead of actual operation. In reality, it may be impossible to accurately predict the above required information. For example, PEVs may arrive or depart without any advanced warning. Also, the predictions of RES supplies and

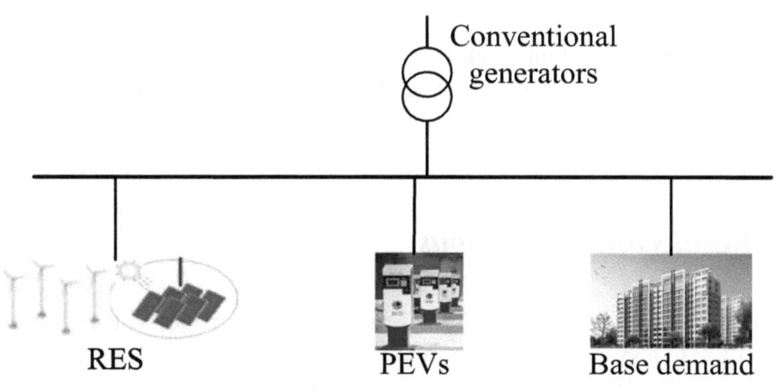

Fig. 5.12 Structure of the smart grid system

inelastic demands are imprecise in general. Consequently, the implemented solution based upon the inaccurate information predicted ahead of the operation period may be far away from the optimal one. Suppose that the prediction of the information will be updated as the operation time goes; then, to enhance the performance of the proposed method in this work, in this section, it will adopt a receding horizon method on Algorithm 5.1.

At each instant $t \in \mathscr{T}$, the coordination is regarded as a finite-horizon optimization over a window of T instants, denoted by $\mathscr{T}_{|t} \triangleq \{t, \ldots, t+T-1\}$. Under this scheme, at t, the system predicts the PEV population, denoted by $\mathscr{N}_{|t}$ and the net base demand trajectory $\boldsymbol{d}_{|t} - \boldsymbol{R}_{|t} \equiv \{d_{i|t} - R_{i|t}, i \in \mathscr{T}_{|t}\}$, and the RES trajectory $\boldsymbol{R}_{|t} \equiv \{R_{i|t}, i \in \mathscr{T}_{|t}\}$. Then based on $\mathscr{N}_{|t}$, $\boldsymbol{d}_{|t}$ and $\boldsymbol{R}_{|t}$ and by applying Algorithm 5.1, each PEV obtains a corresponding optimal coordination strategy, denoted by $\boldsymbol{u}^*_{n,\cdot|t}(\mathscr{N}_{|t}, \boldsymbol{d}_{|t} - \boldsymbol{R}_{|t}) \equiv \{u^*_{ni|t}, i \in \mathscr{T}_{|t}\}$.

By applying Algorithm 5.1, a strategy $\boldsymbol{u}^*_{n,\cdot|t}$ can be obtained for each PEV n under the parameters predicted at instant t. However, in keeping with receding horizon strategies, only the rate $u^*_{nt|t}$ is applied to instant t. Denote the solution considering forecast errors by $\boldsymbol{u}^{\mathrm{rh}}_n \equiv \{u^{\mathrm{rh}}_{nt}, t \in \mathscr{T}\}$, which can be implemented by Algorithm 5.2 specified below.

Algorithm 5.2 Receding horizon based algorithm.

Require:
  Initialize $t = 1$;
Ensure:
  Coordination behaviors of PEVs $\boldsymbol{u}^{\mathrm{rh}}$
1: **while** true **do**
2:   At t, update the PEV population $\mathscr{N}_{|t}$, and the net base demand trajectory $\boldsymbol{d}_{|t} - \boldsymbol{R}_{|t}$;
3:   Solve Problem 5.1 by Algorithm 5.1 and obtain the solution $\boldsymbol{u}^*_{n,\cdot|t}(\mathscr{N}_{|t}, \boldsymbol{d}_{|t} - \boldsymbol{R}_{|t})$;
4:   Set $u^{\mathrm{rh}}_{nt} = u^*_{nt|t}$;
5:   Update $\mathrm{soc}_{n,t+1}$ by (2.1);
6:   Update $t := t + 1$;
7: **end while**

Note: At each time instant t, only the rate $u^*_{nt|t}$ is applied for t, i.e. set $u^{\mathrm{rh}}_{nt} = u^*_{nt|t}$. At the next instant $t+1$, the system updates the forecast values, and the process repeats.

5.5.1 Results and Discussions

The simulation results demonstrate the developed performance via applying the proposed method to a system with $N = 600$ PEVs over a multi-time period from 20:00 on one day to 20:00 on the next day, and the duration of each instant is set to be 15 min.

In this section, the codes for the cases under study are executed by using MATLAB R2014 program and carried out on MacBook pro Intel core i5, 3.1 GHz personal computer with 16 GB RAM, for both centralized and the proposed decentralized method.

5.5.1.1 Convergence and Optimality

For the purpose of demonstration, here simply suppose that all the PEVs are homogeneous with the parameters: $\mathscr{T}_n = \mathscr{T}$, $\Gamma_n = 20\,\text{kWh}$, $\text{soc}_{n,\min} = 20\%$ and $\text{soc}_{n,\text{ref}} = \text{soc}_{n,\max} = 90\%$ for all $n \in \mathscr{N}$. Also suppose that all the PEVs share a common initial SOC value of 30%; then the total required energy of all the PEVs over the period of $\mathscr{T}$ is valued with 12 kWh. Suppose that the generation cost is $c(G_t) = 10^{-5}G_t^2 + 0.02G_t$, and the battery degradation cost is as formulated in [19], such that $f_n(u_{nt}) = 0.004u_{nt}^2 + 0.075u_{nt} + 0.003$. Consider a net base demand trajectory, the difference of the inelastic base demand and RES productions, as displayed in Fig. 5.13.

Figure 5.13 displays the optimal coordination by solving Problem 5.1 with weighting parameter η equal to 1.0 and 0.8 respectively. As illustrated, for each pair of adjacent instants t and s, (5.5) holds in case $d_t - R_t \le d_s - R_s$, which is consistence with Theorems 5.1 and 5.2. However as observed, this property may not hold for a pair of instants which are not adjacent to each other. For example, the net base demand $d_{21:00} - R_{21:00} > d_{16:00} - R_{16:00}$, but $G^*_{21:00} < G^*_{16:00}$. Moreover, (5.7) holds for the sub-intervals $\{20:00, 20:15, \ldots, 22:00\}$, $\{24:00, 24:15, \ldots, 7:15\}$ and $\{9:00, 9:15, \ldots, 19:15\}$, respectively, which is consistent with Lemma 5.1.

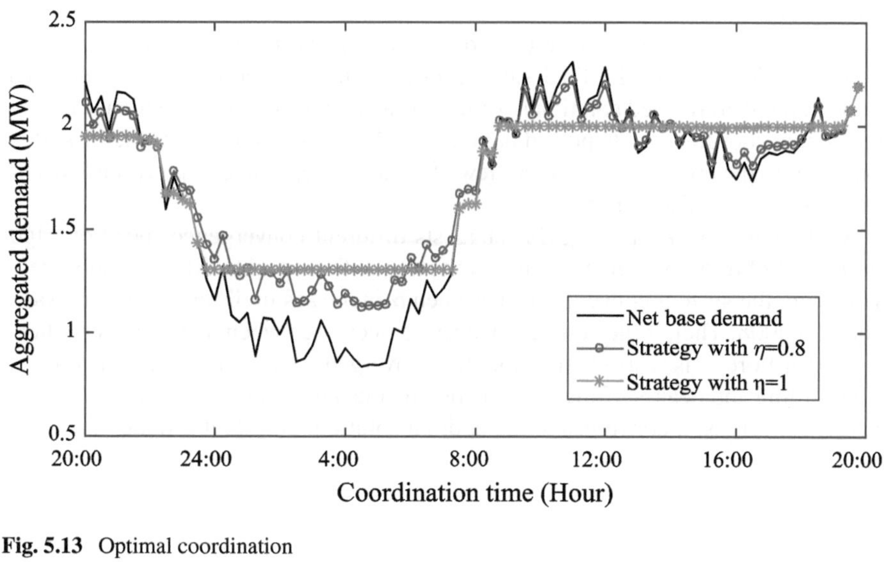

Fig. 5.13 Optimal coordination

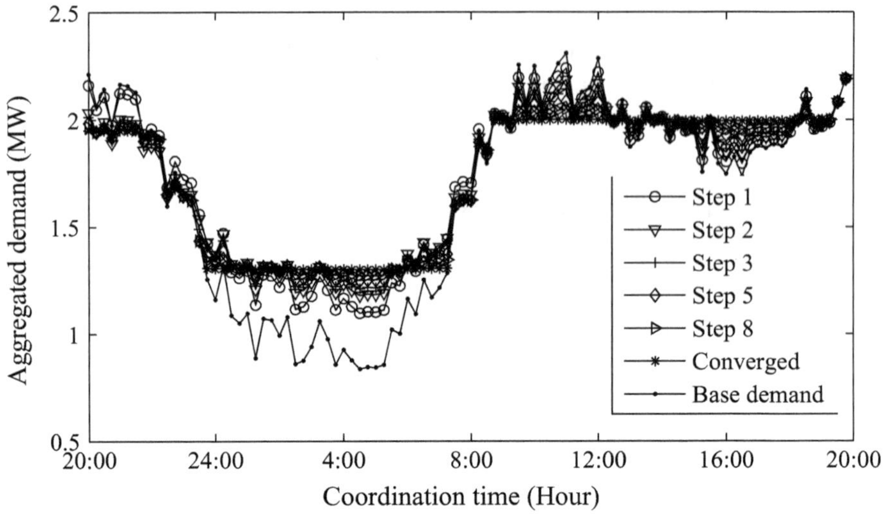

Fig. 5.14 Updates of the coordinated behaviors under Algorithm 5.1, with α satisfies the convergence condition

The decentralized coordination is implemented following Algorithm 5.1. As stated in Theorem 5.3, the condition of α is explicated for the convergence of the system, say $0 < \alpha < 0.00166$ in this case. Take $\alpha = 0.0014$ which satisfies the condition of the convergence. The system converges in 20 iterations, and the updates of the total demand in the system, which is the summation of the net base demand and the accumulated PEVs' charging/discharging power, are given in Fig. 5.14 with respect to the iteration steps.

For the purpose of demonstration, only some of the total steps are shown, say Steps 1, 2, 3, 5, 8 and the final step, to provide the change of the total demand. Moreover, the converged solution is valley-filling and peak-shifting which is exactly the optimal coordination displayed in Fig. 5.13. As illustrated in Fig. 5.14, the system converges to the optimal solution in a few of iteration steps under Algorithm 5.1. This is consistent with Theorem 5.3.

As the value of α varies, (5.19) suggests different convergence speed of Algorithm 5.1. In Fig. 5.15, it gives the iteration steps with respect to different step size α. As shown, the system converges faster when α increases until reaching some value around 0.0008. Then it the convergence speed decreases when α increases further.

Furthermore, it is worth to note that the converged solution is always the optimal valley-filling and peak-shifting coordination displayed in Fig. 5.13. That is, if the the system converges, it converges to the optimal solution, despite the value of α.

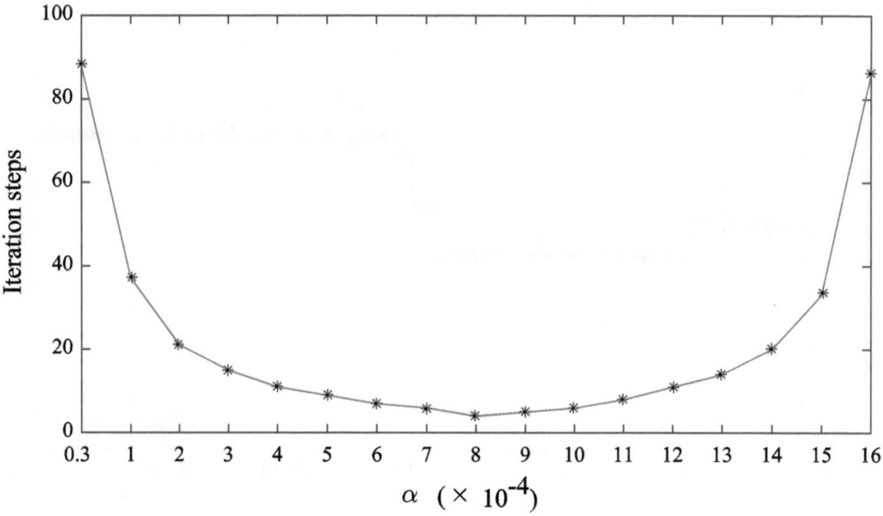

Fig. 5.15 Convergence rate with respect to α

Table 5.1 Amounts of PEVs in each PEV category

	T1 (20 kWh)	T2 (25 kWh)	T3 (30 kWh)	T4 (35 kWh)
M1 (70%)	135	135	120	0
M2 (80%)	20	10	15	120
M3 (90%)	0	15	15	15

5.5.1.2 Case Study Considering PEV Heterogeneity

For approaching realistic characteristics of PEV populations, consider 12 categories of PEVs with the parameters and amounts given in Table 5.1. The types of T1–T4 represent the PEVs with different battery energy capacities respectively, while M1–M3 represent those with different SOC levels when they plug out the grid respectively. For each PEV category, e.g. Type T1&M1 representing such PEVs whose battery capacity is 20 kWh and plug-out SOC level is 70%, suppose that the initial SOC values of PEVs, $\{soc_{n0}; n \in \mathcal{N}\}$, approximately satisfy a Gaussian distribution [26] whose expectation is 0.5 and standard deviation is 0.1.

The iteration updates of the aggregated coordination behaviors of PEVs are illustrated in Fig. 5.16a. It takes 7 iterations to converge and the converged solution is also valley-filling and peak-shifting. Note that this solution is different from those illustrated in Figs. 5.13 and 5.14, since the heterogeneity in the PEV population has various effects on the demand. Similarly, the system converges to the optimal solution in a few of iteration steps in this heterogeneous case. Figure 5.16b displays the evolutions of the SOC trajectories for PEVs of Type T4&M3 with different initial SOC values respectively. As can be seen it is inferred, the SOC trajectory remains

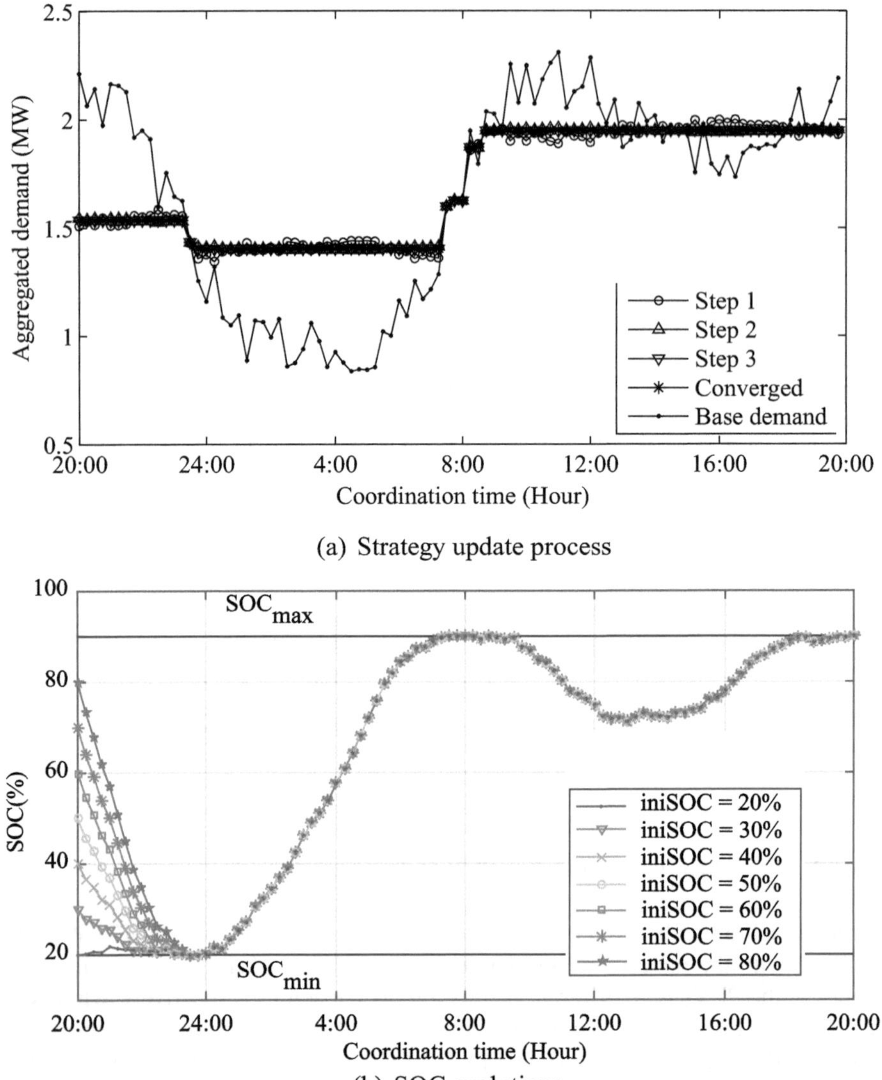

(a) Strategy update process

(b) SOC evolutions

Fig. 5.16 Updates of the coordinated behaviors under Algorithm 5.1 and corresponding SOC trajectory in the heterogeneous case

within the region of [20%, 90%], which implies the feasibility of the SOC over the whole operation period.

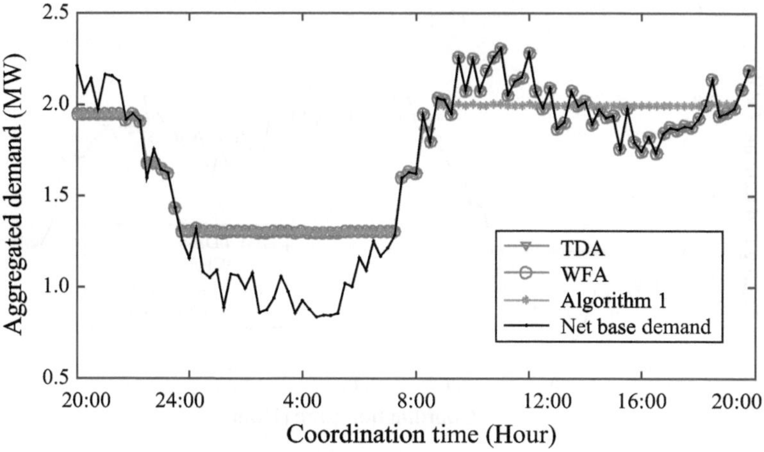

Fig. 5.17 Optimal solution versus the implemented solutions by applying TDA, WFA and Algorithm 5.1 under the first net base demand trajectory

5.5.1.3 Comparisons with Decentralized Methods in the Literature

To overcome the difficulties concerning the centralized method, some research work in the literature also has designed decentralized methods to implement the coordination behaviors. For example, in [4], a time division based algorithm (TDA) is proposed where the total operation period is divided into serval sub-intervals by comparing the net base demand of each instant with the average value of the net base demand trajectory over the whole period. In [5, 6], a water filling based algorithm (WFA) is proposed where the whole period is partitioned into two sub-intervals, such that all the PEVs are only allowed to be discharged during the first sub-interval, and to be charged during the second one.

The results for the performance evaluation of Algorithm 5.1 and the decentralized methods TDA and WFA are compared considering different shapes of the net base demand trajectory. In Figs. 5.17 and 5.18, they display the optimal solution and the implemented solutions by applying TDA, WFA and Algorithm 5.1 under two shapes of the net base demand trajectory respectively. It is shown that the performances of TDA and WFA are dependent upon the shape of the net base demand trajectory.

5.5.2 *Receding Horizon Algorithm*

In the following, the performance of the receding horizon based algorithm given in Algorithm 5.2 is demonstrated through simulation results.

The size of the receding window is 96 which corresponds to 24 h since each time instant is 15 min, and a total of 96 iterations ($t = 1, \ldots, 96$) are proceeded by

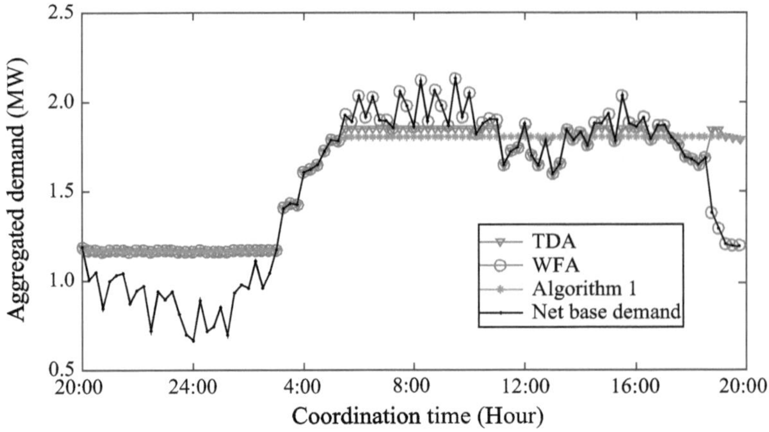

Fig. 5.18 Optimal solution versus the implemented solutions by applying TDA, WFA and Algorithm 5.1 under the second net base demand trajectory

applying Algorithm 5.2. The evolution of the receding process is given in Fig. 5.19, and for purpose of demonstration, only 4 iterations are displayed among the whole 96 iterations, say the iterations at time 20:00, 2:00, 8:00 and 14:00 respectively. From Fig. 5.19, it can seen that the strategy obtained by the receding horizon algorithm helps minimize the net base demand variance. The reduction of the net base demand variance can greatly improve the operation efficiency and security of the power grid.

At the beginning of the operation period, the net base demand is predicted as given in Fig. 5.19a and the corresponding optimal PEV strategies are valley-filling and peak-shift for this net base demand. this strategy is called off-line strategy. As time goes on, the net base demand changes and the actual net base demand trajectory in the end of the operation period could be obtained; then the off-line strategy under the actual net base demand is displayed as the green line in Fig. 5.20. The blue line in Fig. 5.20 gives the coordination solution obtained by Algorithm 5.2, while the red line is the optimal strategy implemented for the actual net base demand by Algorithm 5.1. The optimal strategy under actual net base demand can only be calculated in the end of the operation period, which cannot be implemented in practice. As observed, the strategy obtained by receding horizon based algorithm is calculated online and better than the off-line strategy.

5.6 Conclusions

The charging & discharging coordination problem has been studied in this chapter in which PEVs serve as ESSs and reduce the fluctuation of the RES. Firstly the underlying coordination problem is formulated as an optimization problem, such that the system deals with the tradeoff between the generation cost and the aggregated battery

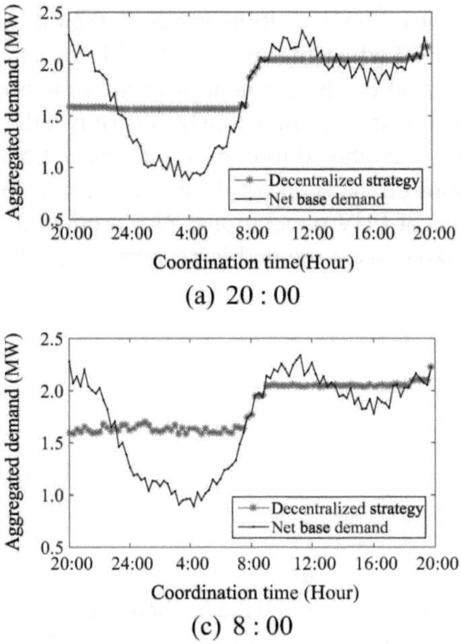

Fig. 5.19 Coordination of PEVs at different times considering both PEVs and net base demand forecast errors

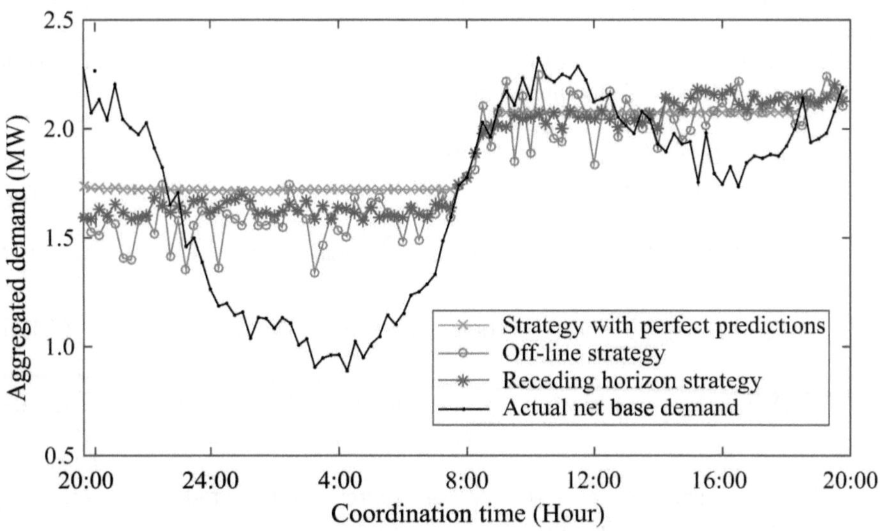

Fig. 5.20 PEVs' behaviors under Algorithms 5.1 and 5.2 respectively

degradation cost over a multi-time period. The optimal solution possesses the valley-filling and peak-shift properties on each pair of adjacent time instants. Moreover, a decentralized method has been presented based on the gradient projection method such that PEVs' strategies were updated locally and simultaneously. It is shown that, under certain mild conditions, the system is guaranteed to converge to the optimal solution without imposing any considerations on the shape of the demand curve. To address the forecast errors in the predictions of PEV populations, RES productions and inelastic load profiles, a receding horizon based method is designed such that the prediction proceeds at each iteration.

5.7 Appendices

5.7.1 *Proof of Theorem 5.1*

Firstly for notational simplicity, here it considers $\boldsymbol{u} \equiv \boldsymbol{u}^*$.

The optimal charging strategy is implemented by applying the method of Lagrange multiplier, see [29]. Denote by $\boldsymbol{\gamma}^- \equiv (\gamma_{nt}^-; n \in \mathscr{N}, t \in \mathscr{T})$ and $\boldsymbol{\gamma}^+ \equiv (\gamma_{nt}^+; n \in \mathscr{N}, t \in \mathscr{T})$, the Lagrange multipliers for the inequality constraints (2.2a), and $\boldsymbol{\lambda}^- \equiv (\lambda_{nt}^-; n \in \mathscr{N}, t \in \mathscr{T})$, $\boldsymbol{\lambda}^+ \equiv (\lambda_{nt}^+; n \in \mathscr{N}, t \in \mathscr{T})$ and $\boldsymbol{A} \equiv (A_n; n \in \mathscr{N})$, the Lagrange multipliers for the inequality constraints (5.1a), (5.1b), and (5.1c), respectively.

Thus the KKT conditions of Problem 5.1 are the constraints given in (2.2a) and (5.1) together with the following relations:

$$\gamma_{nt}^- \geq 0,\ \gamma_{nt}^-\big(- u_{nt} - \varUpsilon_n^-\big) = 0, \tag{5.10a}$$

$$\gamma_{nt}^+ \geq 0,\ \gamma_{nt}^+\big(u_{nt} - \varUpsilon_n^+\big) = 0, \tag{5.10b}$$

$$\lambda_{nt}^- \geq 0,\ \lambda_{nt}^-\Big(\varGamma_n(\text{soc}_{n,\min} - \text{soc}_{n0}) - \sum_{r\in[0,t]} u_{nr}\Big) = 0, \tag{5.10c}$$

$$\lambda_{nt}^+ \geq 0,\ \lambda_{nt}^+\Big(\sum_{r\in[0,t]} u_{nr} - \varGamma_n(\text{soc}_{n,\max} - \text{soc}_{n0})\Big) = 0, \tag{5.10d}$$

$$[\nabla J(\boldsymbol{u})]_{nt} - \gamma_{nt}^- + \gamma_{nt}^+ - \sum_{r\in[t,T]} \lambda_{nr}^- + \sum_{r\in[t,T]} \lambda_{nr}^+ - A_n = 0, \tag{5.10e}$$

for all $n \in \mathscr{N}$ and $t \in \mathscr{T}$, where $\nabla J(\boldsymbol{u})$ represents the gradient of $J(\boldsymbol{u})$ and $[\nabla J(\boldsymbol{u})]_{nt}$ is its entry at the nth row and tth column, such that

$$[\nabla J(\boldsymbol{u})]_{nt} = \eta c'(D_t) + (1-\eta) f_n'(u_{nt}).$$

Consider each pair of adjacent instants $t, s \in \mathscr{T}$. Without loss of generality, it considers $s = t + 1$. It firstly shows (5.5a) by proof of contradiction. Suppose that $u_{nt} < u_{ns}$; then by $D_{-n,t}(\boldsymbol{u}) \leq D_{-n,s}(\boldsymbol{u})$, $D_t = D_{-n,t} + u_{nt} < D_s = D_{-n,s} + u_{ns}$.

By Assumptions 3.2 and 3.6, it obtains that

$$\eta c'(D_t) + (1-\eta) f_n'(u_{nt}) < \eta C_s'(D_s) + (1-\eta) f_n'(u_{ns});$$

then by (5.10e), it can obtain that

$$-\gamma_{nt}^- + \gamma_{nt}^+ - \lambda_{nt}^- + \lambda_{nt}^+ > -\gamma_{ns}^- + \gamma_{ns}^+. \tag{5.11}$$

Also since $u_{nt} < u_{ns}$ and by (2.2a), $u_{nt} < \Upsilon_n^+$ and $u_{ns} > -\Upsilon_n^-$; then (5.10a) and (5.10b) give $\gamma_{nt}^+ = 0$ and $\gamma_{ns}^- = 0$, by which together with (5.11), it gets the following

$$\lambda_{nt}^+ > \lambda_{nt}^- + \gamma_{nt}^- + \gamma_{ns}^+ \geq 0. \tag{5.12}$$

This, together with (5.10d), gives that $\sum_{r \in [0,t]} u_{nr} = \Gamma_n(\text{soc}_{n,\max} - \text{soc}_{n0})$ which implies the battery of PEV n is full charged at t. Thus $u_{nt} \geq 0$ and $u_{ns} \leq 0$, which is contradicted with $u_{nt} < u_{ns}$. As a consequence, it obtains that (5.5a) holds.

The inequality (5.5b) can be verified as well by applying the same technique used above to show (5.5a). ∎

5.7.2 *Proof of Theorem 5.2*

For notational simplicity, it considers $\boldsymbol{u} \equiv \boldsymbol{u}^*$ in the proof below. By Theorem 5.1, the optimal solution $\boldsymbol{u}$ satisfies the KKT conditions specified in (5.10).

Consider each pair of adjacent instants $t, s \in \mathscr{T}$ with $\mathscr{N}_t = \mathscr{N}_s$ and let $s = t + 1$.

Firstly show (5.5a) in case $d_t \leq d_s$ by proof of contradiction in the following.

Suppose that there exists a PEV n such that $u_{nt} < u_{ns}$, and $f_n'(u_{nt}) < f_n'(u_{ns})$ by Assumptions 3.2 and 3.6. Then by (5.10e), it can obtain that

$$\begin{aligned} &\eta\big(c'(D_t) - c'(D_s)\big) \\ &> \gamma_{nt}^- - \gamma_{nt}^+ + \lambda_{nt}^- - \lambda_{nt}^+ - \gamma_{ns}^- + \gamma_{ns}^+. \end{aligned} \tag{5.13}$$

Since $u_{nt} < u_{ns}$ and by (2.2a), $u_{nt} < \Upsilon_n^+$ and $u_{ns} > -\Upsilon_n^-$. Then (5.10a) and (5.10b) imply that $\gamma_{nt}^+ = 0$ and $\gamma_{ns}^- = 0$. Since $\lambda_{nt}^+ \geq 0$, it considers the case $\lambda_{nt}^+ > 0$ firstly. By (5.10d), it gives that $\sum_{r \in [0,t]} u_{nr} = \Gamma_n(\text{soc}_{n,\max} - \text{soc}_{n0})$, i.e., the SOC value of the battery of PEV n reaches the maximum value at t; then $u_{nt} \geq 0$ and $u_{ns} \leq 0$, which is contradicted with $u_{nt} < u_{ns}$. Thus $\lambda_{nt}^+ = 0$.

Substitute $\gamma_{nt}^+ = \gamma_{ns}^- = \lambda_{nt}^+ = 0$ into (5.13), it gets

$$\eta\big(c'(D_t) - c'(D_s)\big) > \gamma_{nt}^- + \lambda_{nt}^- + \gamma_{ns}^+ \geq 0, \tag{5.14}$$

by which together with Assumptions 3.2 and 3.6, it obtains $D_t > D_s$. Thus by $d_t \leq d_s$, $u_{nt} < u_{ns}$ and $\mathscr{N}_t = \mathscr{N}_s$, there must exist a PEV m such that $u_{mt} > u_{ms}$.

Meanwhile, the KKT conditions of (5.10) hold for PEV m. As $u_{mt} > u_{ms}$ and (2.2a), it obtains that $u_{mt} > -\Upsilon_n^-$ and $u_{ms} < \Upsilon_n^+$, both of which together imply that $\gamma_{mt}^- = 0$ and $\gamma_{ms}^+ = 0$ by (5.10a) and (5.10b). By (5.10e), (5.14) and $u_{mt} > u_{ms}$, the following holds:

$$\lambda_{mt}^- > \gamma_{mt}^+ + \lambda_{mt}^+ + \gamma_{ms}^- \geq 0; \tag{5.15}$$

then by (5.10c), $\sum_{r \in [0,t]} u_{mr} = \Gamma_n(\mathrm{soc}_{n,\min} - \mathrm{soc}_{n0})$, that is, the battery state of PEV m is at its minimum value at t, and it gives $u_{mt} \leq 0$ and $u_{ms} \geq 0$, which is contradicted with $u_{mt} > u_{ms}$. Hence, there does not exist a PEV n satisfies $u_{nt} < u_{ns}$, i.e. (5.5a) holds for all PEVs.

Similarly, (5.5b) can be verified by applying the same technique used above to show (5.5a). ■

5.7.3 *Proof of Theorem 5.3*

Consider a pair of distinct strategies $\boldsymbol{u}, \boldsymbol{v} \in \mathscr{U}$ and define $U_t = \sum_{n \in \mathscr{N}} u_{nt}$ and $V_t = \sum_{n \in \mathscr{N}} v_{nt}$; then

$$\begin{aligned} &\|\nabla J(\boldsymbol{u}) - \nabla J(\boldsymbol{v})\|_2^2 \\ &= \sum_{t \in \mathscr{T}} \sum_{n \in \mathscr{N}} \Big\{ \eta \big(c'(D_t + U_t) - c'(D_t + V_t) \big)^2 \\ &\quad + (1 - \eta) \big(f_n'(u_{nt}) - f_n'(v_{nt}) \big)^2 \Big\}. \end{aligned} \tag{5.16}$$

As the first and second derivatives of $c(\cdot)$ and $f_n(\cdot)$ are continuous on $\mathbb{R}$, then by the mean value theorem, there exists some W_t between $D_t + U_t$ and $D_t + V_t$, and w_{nt} between u_{nt} and v_{nt} such that

$$C''(W_t) = \frac{c'(D_t + U_t) - c'(D_t + V_t)}{U_t - V_t}, \tag{5.17a}$$

$$f_n''(w_{nt}) = \frac{f_n'(u_{nt}) - f_n'(v_{nt})}{u_{nt} - v_{nt}}, \tag{5.17b}$$

by which together with (5.16), it can be obtained that

$$\begin{aligned}
&\|\nabla J(\boldsymbol{u}) - \nabla J(\boldsymbol{v})\|_2^2 \\
&= \sum_{t\in\mathscr{T}} \Big\{\eta N[C''(W_t)]^2(U_t - V_t)^2 \\
&\quad + (1-\eta)\sum_{n\in\mathscr{N}} [f_n''(w_{nt})]^2(u_{nt} - v_{nt})^2\Big\} \\
&\leq \sum_{t\in\mathscr{T}} \Big\{\eta N\kappa^2(U_t - V_t)^2 + (1-\eta)\sum_{n\in\mathscr{N}} \rho^2(u_{nt} - v_{nt})^2\Big\} \\
&\leq \Big(\eta N^2\kappa^2 + (1-\eta)\rho^2\Big)\|\boldsymbol{u} - \boldsymbol{v}\|_2^2,
\end{aligned}$$

where the last inequality holds by the Jensen's inequality below,

$$\Big(\sum_{n\in\mathscr{N}} u_{nt} - \sum_{n\in\mathscr{N}} v_{nt}\Big)^2 \leq N\sum_{n\in\mathscr{N}} (u_{nt} - v_{nt})^2,$$

and $\|\boldsymbol{u} - \boldsymbol{v}\|_2^2 \triangleq \sum_{t\in\mathscr{T}}\sum_{n\in\mathscr{N}}(u_{nt} - v_{nt})^2$. It implies that

$$\begin{aligned}
&\|\nabla J(\boldsymbol{u}) - \nabla J(\boldsymbol{v})\|_2 \\
&\leq \Big(\eta N^2\kappa^2 + (1-\eta)\rho^2\Big)^{1/2}||\boldsymbol{u} - \boldsymbol{v}||_2;
\end{aligned} \tag{5.18}$$

then, the system cost $J(\cdot)$ monotonically decreases on $\boldsymbol{u}^k$ under Algorithm 5.1 in case $0 < \alpha < 2\big(\eta N^2\kappa^2 + (1-\eta)\rho^2\big)^{-1/2}$ [25], and

$$\begin{aligned}
&J(\boldsymbol{u}^{k+1}) - J(\boldsymbol{u}^k) \\
&\leq \Big(\frac{1}{2}\big(\eta N^2\kappa^2 + (1-\eta)\rho^2\big)^{1/2} - \frac{1}{\alpha}\Big)\|\boldsymbol{u}^{k+1} - \boldsymbol{u}^k\|_2^2.
\end{aligned} \tag{5.19}$$

Hence Algorithm 5.1 converges to an equilibrium, denoted by $\boldsymbol{u}^\dagger$, since $J(\boldsymbol{u})$ is bounded from below.

Based on Proposition 2.3.2 in [25] and (5.18), it obtains that $\boldsymbol{u}^\dagger$ is stationary, i.e., it satisfies

$$\nabla J(\boldsymbol{u}^\dagger)^{\mathrm{T}}(\boldsymbol{u} - \boldsymbol{u}^\dagger) \geq 0, \quad \forall\, \boldsymbol{u} \in \mathscr{U}, \tag{5.20}$$

where $\boldsymbol{x}^{\mathrm{T}}$ represents the transpose of the vector $\boldsymbol{x}$. By Assumptions 3.2 and 3.6, $c(\cdot)$ and $f_n(\cdot)$ are all convex, for all $n \in \mathscr{N}$ and $t \in \mathscr{T}$; so the system cost $J(\boldsymbol{u})$ is convex. Hence, by Proposition 2.1.2 in [25], $\boldsymbol{u}^\dagger = \boldsymbol{u}^*$, that is, Algorithm 5.1 converges to the optimal solution $\boldsymbol{u}^*$ that minimizes the system cost $J(\boldsymbol{u})$. ■

References

1. Y. Ma, T. Houghton, A. Cruden, D. Infield, Modeling the benefits of vehicle-to-grid technology to a power system. IEEE Trans. Power Syst. **27**(2), 1012–1020 (2012)
2. W. Kempton, J. Tomic, Vehicle-to-grid power fundamentals: calculating capacity and net revenue. J. Power Sources **144**(1), 268–279 (2005)
3. Y. He, B. Venkatesh, L. Guan, Optimal scheduling for charging and discharging of electric vehicles. IEEE Trans. Smart Grid **3**(3), 1095–1105 (2012)
4. Z. Chen, Z. Hu, Optimal coordination of charging and discharging control of large-scale plug-in electric vehicles, in *Transportation Electrification Asia-Pacific* (2014), pp. 1–6
5. H. Xing, M. Fu, Z. Lin, Y. Mou, Decentralized optimal scheduling for charging and discharging of plug-in electric vehicles in smart grids. IEEE Trans. Power Syst. (2016)
6. H. Li, Z. Ma, A distributed charging and discharging coordination for large-population plug-in electric vehicles, in *Proceedings of the 2015 Chinese Intelligent Systems Conference*, pp. 415–424. Accessed 8 Nov 2015
7. M. Zeng, S. Leng, Y. Zhang, Power charging and discharging scheduling for V2G networks in the smart grid, in *IEEE International Conference on Communications Workshops*, Budapest, pp. 1052–1056. Accessed 9–13 June 2013
8. S. Stüdli, E. Crisostomi, Optimal real-time distributed V2G and G2V management of electric vehicles. Int. J. Control **87**(6), 1153–1162 (2014)
9. M. Esmaili, M. Rajabi, Optimal charging of plug-in electric vehicles observing power grid constraints. IET Gener. Transm. Distrib. **8**(4), 583–590 (2014)
10. S. Falahati, S.A. Taher, M. Shahidehpour, Grid frequency control with electric vehicles by using of an optimized fuzzy controller. Appl. Energy **178**, 918–928 (2016)
11. T.N. Pham, H. Trinh, L.V. Hien, Load frequency control of power systems with electric vehicles and diverse transmission links using distributed functional observers. IEEE Trans. Smart Grid **7**(1), 238–252 (2016)
12. D.Q. Xu, G. Joós, M. Lévesque, M. Maier, Integrated V2G, G2V, and renewable energy sources coordination over a converged fiber-wireless broadband access network. IEEE Trans. Smart Grid **4**(3), 1381–1390 (2013)
13. L. Igualada, C. Corchero, M. Cruz-Zambrano, F.-J. Heredia, Optimal energy management for a residential microgrid including a vehicle-to-grid system. IEEE Trans. Smart Grid **5**(4), 2163–2172 (2014)
14. C. Chen, S. Duan, T. Cai, B. Liu, G. Hu, Optimal allocation and economic analysis of energy storage system in microgrids. IEEE Trans. Power Electron. **26**(10), 2762–2773 (2011)
15. K.N. Kumar, B. Sivaneasan, P.H. Cheah, P.L. So, V2G capacity estimation using dynamic EV scheduling. IEEE Trans. Smart Grid **5**(2), 1051–1060 (2014)
16. L. Jian, Y. Zheng, X. Xiao, C.C. Chan, Optimal scheduling for vehicle-to-grid operation with stochastic connection of plug-in electric vehicles to smart grid. Appl. Energy **146**, 150–161 (2015)
17. H. Zhang, Z. Hu, Z. Xu, Y. Song, Evaluation of achievable vehicle-to-grid capacity using aggregate PEV model. IEEE Trans. Power Syst. **32**(1), 784–794 (2017)
18. Z. Ma, S. Zou, X. Liu, A distributed charging coordination for large-scale plug-in electric vehicles considering battery degradation cost. IEEE Trans. Control Syst. Technol. **23**(5), 2044–2052 (2015)
19. Z. Ma, S. Zou, L. Ran, X. Shi, I. Hiskens, Efficient decentralized coordination of large-scale plug-in electric vehicle charging. Automatica **69**, 35–47 (2016)
20. C. Wu, H. Mohsenian-Rad, J. Huang, Vehicle-to-aggregator interaction game. IEEE Trans. Smart Grid **3**(1), 434–442 (2012)
21. S. Zou, Z. Ma, X. Liu, I. Hiskens, An efficient game for coordinating electric vehicle charging. IEEE Trans. Autom. Control **62**(5), 2374–2389 (2017)
22. M.A. Ortega-Vazquez, Optimal scheduling of electric vehicle charging and vehicle-to-grid services at household level including battery degradation and price uncertainty. IET Gener. Transm. Distrib. **8**(6), 1007–1016 (2014)

23. D.T. Hoang, P. Wang, D. Niyato, E. Hossain, Charging and discharging of plug-in electric vehicles (PEVs) in vehicle-to-grid (V2G) systems: a cyber insurance-based model. IEEE Access **5**, 732–754 (2017)
24. Y.M. Atwa, E.F. El-Saadany, Optimal allocation of ESS in distribution systems with a high penetration of wind energy. IEEE Trans. Power Syst. **25**(4), 1815–1822 (2010)
25. D.P. Bertsekas, *Nonlinear Programming*, 2nd edn. (Athena Scientific, Belmont, 1999)
26. Z. Luo, Z. Hu, Y. Song, Z. Xu, H. Lu, Optimal coordination of plug-in electric vehicles in power grids with cost-benefit analysis—Part I: enabling techniques. IEEE Trans. Power Syst. **28**(4), 3546–3555 (2013)
27. Z. Luo, Z. Hu, Y. Song, Z. Xu, H. Lu, Optimal coordination of plug-in electric vehicles in power grids with cost-benefit analysis—Part II: a case study in China. IEEE Trans. Power Syst. **28**(4), 3556–3565 (2013)
28. J. Zheng, X. Wang, K. Men, C. Zhu, S. Zhu, Aggregation model-based optimization for electric vehicle charging strategy. IEEE Trans. Smart Grid **4**(2), 1058–1066 (2013)
29. S. Boyd, L. Vandenberghe, *Convex Optimization* (Cambridge University Press, Cambridge, 2004)

Chapter 6
Dynamic Price-Based Coordination Methods

Abstract This work is concerned with large-scale incorporation of PEVs, which are typically inter-temporal shifting loads, into an electricity markets over multi time intervals. In the context of deregulated market structures with deep concerns for the social efficiency, demand response (DR) is promising to enable interaction of PEVs with the grid. It designs a decentralized dynamic pricing mechanism for DR based on auction rules, to induce PEVs to behave efficiently and obtain system wide benefit. The system implements an optimal allocation with respect to the submitted bid profile, and broadcasts an associated system price to all the PEVs. The expected system price dictates PEVs' best response; on the other hand, the resulting PEV pattern will alter this price. Moreover, it proposes a dynamic algorithm to implement the efficient results such that all the PEVs simultaneously update their best responses with respect to the threshold price set in last step. It specifies the sufficient conditions for the proposed method, under which the system converges to the efficient solution.

6.1 Introduction

Pressures of rapid consumption of exhaustible energy, increasing market competition and reliability have led to an increasing push for demand-side participation, especially in the form of increased price responsiveness of electricity demand, in competitive power markets [1, 2]. Meanwhile, consumers are getting exposed to a more volatile electricity price and they may decide to modify the profile of their demand to reduce their electricity costs [3]. Demand Response (DR) appears as a very promising opportunity for consumers and brings several advantages for the whole system [4–6]. DR can be defined as the changes in electricity usage by end-use customers in response to changes in the price of electricity over time, or the incentive payments designed to induce lower electricity use at times of high electricity prices or when system reliability is jeopardized [7, 8].

Certain customer load units, like PEVs, may migrate from one time period to another dependent on the price prevailing at other time periods, and shiftable loads are said to be lossless if the total energy use over the full time horizon is fixed [1, 3]. This brings attention to inter-temporal factors of price elasticity or customer utility.

Z. Ma, *Decentralized Charging Coordination of Large-scale Plug-in Electric Vehicles in Power Systems*, https://doi.org/10.1007/978-981-13-7652-8_6

In many researches, the demand-price interaction is characterized by price elasticity matrices [9, 10], which represent the sensitivity of electricity demand with respect to the change of price. Moreover, the concept of "cross-elasticity" is developed in [11], to characterize the inter-temporal factors of DR systems, such as [5, 12–16]. However, there may be approximation in obtaining the values of the elasticity, e.g. a simplified linearized price demand curve in [17] and a stepwise linear price demand curve in [18]. In addition, the elasticity values are fixed in the above researches, whereas in actual they are uncertain due to a variety of reasons including latency in communication and change in consumption behavior. Therefore, consumer loads have different response patterns to the electricity prices under different scenarios. In [19], the authors try to consider this uncertainty and inexact DR information by allowing the price-elastic demand curve to vary within a certain range, and propose an efficient robust unit commitment approach for the solution. To fix the related issues of elasticity values, it originally applies an equality energy constraint of total electricity use in this chapter to reflect the inter-temporal load characteristic. Further, this constraint is transformed as a deviation cost that is added into the valuation function of end-users.

In this work, it considers large-scale incorporation of PEVs, which are kinds of inter-temporal shifting loads, into an electricity markets. Centrally administered demand management methods was an important feature of many regulated utilities, like Lagrangian/Newton based methods [3], interior point method [13], linear programming [20] and mixed integer programming [19, 21]. However, the valuation function and the constraint sets of PEVs are private, and it will be burden for the market operator to solve an optimization problem directly coordinating a significantly large number of customers. Seasoned to cope with these issues, hierarchical DR management framework is proposed via DR aggregators between the market operators and the end-users in [2, 22, 23], such that the resultant large optimization problem can be decoupled into separate subproblems. In [24], an efficient pricing method based on VCG mechanism is proposed to seek social benefit. In [1, 16], DR is incorporated with auction based mechanisms where loads simply respond to real-time or forecasted prices. Since auctions in electricity markets considering demand response may have greater potential to increase efficiency than relying solely on a more passive demand response, it designs a new auction based pricing mechanism to induce PEVs behave efficiently and obtain the maximum social welfare.

In the auction based pricing mechanism, each PEV submits a multi dimension bid related to the inter-temporal factor to the system. The system implements an optimal allocation with respect to the submitted bid profile, and broadcasts an associated system price to all the PEVs. Due to the cross-elastic correlation among the intervals, the best updated response of PEVs, at an instant, is determined by both the demand at this instant and the total demand over the whole horizon. In addition, it develops a dynamic dispatch algorithm for distributing the energy where all the individual PEVs update their best bidding responses simultaneously with respect to the system price which is determined by the previous collection of bid profile. The expected system price dictates PEVs' best response; on the other hand, the resulting load pattern will alter this price. As a result, the algorithm continues until the system price is fixed.

It can be shown that, following the update procedure of the individual best response behaviors, the system converges to the efficient coordination solution under certain mild sufficient conditions.

The rest of this work is organized as follows. In Sect. 6.2, it models PEVs as a class of the price responsive loads and formulate a class of PEV coordination programs to maximize the social welfare in power markets. Next, in Sect. 6.3, it introduces a novel decentralized dynamic pricing method based upon auctions for the underlying PEV coordination problems, under which the system converges to the efficient solution under certain mild conditions. Simulations are studied in Sect. 6.4 to demonstrate the results developed in this chapter. Section 6.5 presents conclusions of this part.

6.2 System Model and Problem Formulation

Consider a smart power system with a single energy provider and a collection of loads, say PEVs, as part of the general wholesale electricity market.

Load units in the underlying system are specified as two parts, price taking and price responsive loads, where price taking loads are unwilling to adjust their demand even with sufficient price incentives. In this chapter, it formulates the fixed demand of price taking loads as inelastic base demand, which is denoted by $\boldsymbol{d} \equiv (d_t, t \in \mathscr{T})$, and each price responsive load, say a individual PEV here, is assumed to be an independent end-user belonging to $\mathscr{N}$.

Like formulated in (2.2) in Chap. 2, PEVs' operating restrictions can be included in the mathematical model through the lower and upper limits of demand in each interval and energy constraints. For analytical simplicity, it is supposed that the lower limit is zero, say $\Upsilon_n^- = 0$, for all $n \in \mathscr{N}$ and the upper limit Υ_n^+ is sufficiently large. To indicate the inter-temporal characteristics of the price responsive loads, it introduces an equality constraint to model a fixed, specified amount of energy consumption during the time horizon, since it relates variables that are associated to different intervals. Due to the above statements, together with (2.2) and (2.3), a charging coordination behavior is called admissible if the following holds:

$$u_{nt} \geq 0, \forall t \in \mathscr{T} \text{ and } \Sigma(\boldsymbol{u}_n) = \frac{\Gamma_n}{\varsigma_n}(\mathrm{soc}_{n,\max} - soc_{n0}).$$

The feasible set of PEV n is defined as

$$\mathscr{U}_n \triangleq \{\boldsymbol{u}_n | u_{nt} \geq 0, \forall t \in \mathscr{T}, \Sigma(\boldsymbol{u}_n) = \Gamma_n\},$$

and $\mathscr{U} = \prod_{n\in\mathscr{N}} \mathscr{U}_n$ is the set of consumptions of the PEV population.

Model the behavior of different PEVs through their different valuation functions, denoted by $v_n(\boldsymbol{u}_n)$. For dealing with the inter-temporal energy constraint, it is formulated as a deviation cost represented in the valuation function, which is assumed to have the following form,

$$v_n(\boldsymbol{u}_n) = -\sum_{t\in\mathscr{T}} f_n(u_{nt}) - \delta_n(\Sigma(\boldsymbol{u}_n) - \Gamma_n)^2, \tag{6.1}$$

where $f_n(\cdot)$ is a function to reflect the essential cost of PEV n, and δ_n is a proportional parameter related to the intertemporal demand constraint, which reflects the relative desire to reach the total energy consumption over the whole time slots [25]. Moreover, as specified in previous chapters, here it is still supposed that $f_n(\cdot)$ satisfies Assumption 3.1 given in Chap. 3 which indicates that $v_n(\boldsymbol{u}_n)$ is monotonically increasing, concave and differentiable.

6.2.1 Optimal Demand Response Program

Suppose that the system deals the tradeoff between the *total cost* and the *benefit* to supply the energy for all the loads over the whole time horizon. Hence, subject to a collection of admissible controls $\boldsymbol{u} \equiv (\boldsymbol{u}_n; n \in \mathscr{N})$, the system cost, denoted by J, is specified as the following:

$$J(\boldsymbol{u}) \triangleq \sum_{t\in\mathscr{T}} c(D_t) - \sum_{n\in\mathscr{N}} v_n(\boldsymbol{u}_n), \tag{6.2}$$

where $c(D_t)$ denotes the generation cost with respect to the total demand D_t such that $D_t = d_t + \Sigma(\boldsymbol{u}_t)$ with $\Sigma(\boldsymbol{u}_t) \equiv \sum_{n\in\mathscr{N}} u_{nt}$, and d_t is the aggregated inelastic base demand in the grid at interval t.

The optimal objective of the system is to determine an energy consumption $\boldsymbol{u} \in \mathscr{U}$ to minimize the system cost specified in (6.2). Denote the socially optimal (*efficient*) energy consumption by $\boldsymbol{u}^{**}$ that can be implemented with the Lagrange multiplier methods below, see [26].

There exist Lagrange multipliers $\boldsymbol{\mu} \equiv (\mu_{nt}, n \in \mathscr{N}, t \in \mathscr{T})$ such that $\boldsymbol{u}^{**}$ satisfies the following,

$$\begin{aligned} c'(D_t) - \tfrac{\partial}{\partial u_{nt}} v_n(\boldsymbol{u}_n) - \mu_{nt} &= 0 \\ \mu_{nt} \geq 0, \quad \mu_{nt}u_{nt} &= 0 \end{aligned}$$

where for notational simplicity, $c'(\cdot)$ represents the partial derivative of function $c(\cdot)$ on u_{nt}. Then

$$c'(D_t^{**}) \begin{cases} = \tfrac{\partial}{\partial u_{nt}^{**}} v_n(\boldsymbol{u}_n^{**}), & \text{in case } u_{nt}^{**} > 0 \\ \geq \tfrac{\partial}{\partial u_{nt}^{**}} v_n(\boldsymbol{u}_n^{**}), & \text{in case } u_{nt}^{**} = 0 \end{cases}, \tag{6.3}$$

where $D_t^{**} = d_t + \Sigma(\boldsymbol{u}_t^{**})$.

Concerning $c(\cdot)$, same as previous chapter, it satisfies Assumption 3.2 in this chapter, see [27, 28] as well.

Hence by the concavity and convexity of $v_n(\cdot)$ and $c(\cdot)$, see [26], it can obtain that the efficient energy consumption variable $\boldsymbol{u}^{**}$ can be characterized by (6.3) uniquely.

Consider $p_t \equiv c'(D_t)$, for all $t \in \mathscr{T}$, i.e., $\boldsymbol{p}$ represents the generation marginal cost over the time horizon with respect to $\boldsymbol{u}$, and

$$p_t^{**} = c'(D_t^{**}) = \frac{\partial}{\partial u_{nt}^{**}} v_n(\boldsymbol{u}_n^{**}), \tag{6.4}$$

for each of PEVs $n \in \mathscr{N}$ with $\boldsymbol{u}^{**} > 0$.

The centralized efficient coordination behavior can be effectively implemented in the case that the system has complete information and can directly schedule all the loads. However, in practice the individuals are not willing to share their private information with others and the centralized control method might be computationally infeasible. Thus as detailed in the ensuing section, it proposes a decentralized coordination method, such that each of PEVs updates its energy consumption with respect to a common electricity price.

6.3 Decentralized Dynamic Pricing Mechanism for Demand Response

In Sect. 6.3.1, it firstly designs an auction mechanism and introduce a system price for the underlying demand response program; then under this auction mechanism, it proposes a novel dynamic method in Algorithm 6.1, to implement the efficient coordination solution.

6.3.1 Auction Mechanism for Demand Response Problems

In the auction, each PEV $n \in \mathscr{N}$ reports a $2T$-dimensional bid, denoted by $\boldsymbol{b}_n$, such that

$$\boldsymbol{b}_n \equiv ((\beta_{nt}, r_{nt}); t \in \mathscr{T}),$$

which specifies the maximum per unit price β_{nt} that PEV n is willing to pay at instant t and demands up to r_{nt} units of the electricity resources. Denote by $\mathscr{B}_n$ the set of (admissible) bids of PEV n specified above.

A bid profile of all the PEVs is denoted by $\boldsymbol{b} \equiv (\boldsymbol{b}_n, n \in \mathscr{N})$, and the bid profile of n's opponents is defined as $\boldsymbol{b}_{-n} \equiv (\boldsymbol{b}_1, \ldots, \boldsymbol{b}_{n-1}, \boldsymbol{b}_{n+1}, \ldots, \boldsymbol{b}_N)$.

The system assigns an allocation profile to the consumers to minimize the total monetary cost with respect to a bid profile $\boldsymbol{b}$; that is, to determine an allocation, denoted by $\boldsymbol{u}^*(\boldsymbol{b})$, such that

$$\boldsymbol{u}^*(\boldsymbol{b}) = \underset{\boldsymbol{u}\in[\boldsymbol{0},\boldsymbol{r}]}{\operatorname{argmin}} \widehat{J}(\boldsymbol{u}; \boldsymbol{b}), \tag{6.5}$$

where it calls $\widehat{J}(\boldsymbol{u}; \boldsymbol{b})$ the revealed system cost that is specified below:

$$\widehat{J}(\boldsymbol{u}; \boldsymbol{b}) \triangleq \sum_{t\in\mathscr{T}} c(D_t) - \sum_{n\in\mathscr{N}} \boldsymbol{\beta}_n^\tau \boldsymbol{u}_n,$$

where $\boldsymbol{\beta}_n^\tau$ denotes the transposition of the column vector $\boldsymbol{\beta}_n$; and $\boldsymbol{u}^*(\boldsymbol{b})$ can be implemented by applying the Lagrange multiplier method.

Besides, it defines a system price, denoted by $p(\boldsymbol{b})$, with respect to a bid profile $\boldsymbol{b}$, as below:

$$p_t(\boldsymbol{b}_t) \triangleq \beta_t^{\text{sys}}(\boldsymbol{b}_t) + \eta\big(c'(D_t^*) - \beta_t^{\text{sys}}(\boldsymbol{b}_t)\big), \tag{6.6}$$

for all $t \in \mathscr{T}$ where $\eta \in (0, 1]$; $D_t^* = d_t + \Sigma(\boldsymbol{u}_t^*)$ with $\boldsymbol{u}^* \equiv \boldsymbol{u}^*(\boldsymbol{b})$ representing the allocation of n by (6.5); and

$$\beta_t^{\text{sys}}(\boldsymbol{b}_t) = \begin{cases} c'(d_t), & \text{in case } \max_{n\in\mathscr{N}}\{\beta_{nt}\} \le c'(d_t) \\ \sup\{\max_{n\in\mathscr{N}}\{\beta_{nt}\}, c'(d_t)\}, & \text{in case } \boldsymbol{r}_t = \boldsymbol{0} \\ \inf_{n\in\mathscr{N},\beta_{nt}>c'(D_t^*)}\{\beta_{nt}\}, & \text{otherwise} \end{cases}. \tag{6.7}$$

Remark: (6.6) implies that

$$p_t(\boldsymbol{b}_t) \begin{cases} = c'(d_t), & \text{in case } \beta_{nt}(\boldsymbol{b}_t) < c'(d_t), \\ = \beta_t^{\text{sys}}, & \text{in case } c'(D_t^*) = \beta_t^{\text{sys}} \\ \in \left(c'(D_t^*), \beta_t^{\text{sys}}\right), & \text{in case } c'(D_t^*) < \beta_t^{\text{sys}} \end{cases}, \tag{6.8}$$

with $\beta_t^{\text{sys}} \equiv \beta_t^{\text{sys}}(\boldsymbol{b}_t)$ defined in (6.7).

6.3.2 Individual Best Response w.r.t. Given System Prices

It considers another allocation rule with respect to a bid profile $\widehat{\boldsymbol{b}} \equiv ((\widehat{\beta}_{nt}, \widehat{r}_{nt}); t \in \mathscr{T}, n \in \mathscr{N})$ and the system price $\boldsymbol{p}$, to determine the allocation to PEV n, denoted by $\widehat{\boldsymbol{u}}_n^* \equiv \widehat{\boldsymbol{u}}_n^*(\widehat{\boldsymbol{b}}_n; \boldsymbol{p})$, such that:

$$\widehat{u}_{nt}^* = \begin{cases} \widehat{r}_{nt}, & \text{in case } \widehat{\beta}_{nt} \ge p_t \\ 0, & \text{otherwise} \end{cases}. \tag{6.9}$$

That is to say, given a system price $\boldsymbol{p}$, PEV n will

- win the auction at instance t and be allocated with $\widehat{r}_{nt}$ in case its bid price $\widehat{\beta}_{nt} \geq p_t$;
- lose the auction at t and be not allocated in case its bid price $\widehat{\beta}_{nt} < p_t$.

The payoff of PEV n, with respect to $(\widehat{\boldsymbol{b}}_n; \boldsymbol{p})$, denoted by $f_n(\widehat{\boldsymbol{b}}_n; \boldsymbol{p})$, is defined as below:

$$f_n(\widehat{\boldsymbol{b}}_n; \boldsymbol{p}) = v_n(\widehat{\boldsymbol{u}}_n^*) - \widehat{\boldsymbol{\beta}}_n^T \widehat{\boldsymbol{u}}_n^*; \tag{6.10}$$

and PEV n implements a best response with respect to $\boldsymbol{p}$, denoted by $\widehat{\boldsymbol{b}}_n^*(\boldsymbol{p})$, such that

$$\widehat{\boldsymbol{b}}_n^*(\boldsymbol{p}) = \max_{\widehat{\boldsymbol{b}}_n \in \mathscr{B}_n} f_n(\widehat{\boldsymbol{b}}_n; \boldsymbol{p}). \tag{6.11}$$

Denote by $\widehat{\mathscr{B}}_n^*(\boldsymbol{p})$ the set of best responses of load n with respect to $\boldsymbol{p}$.

Lemma 6.1 (Individual best response w.r.t. a given system price) *The best response of load n with respect to a system price* $\boldsymbol{p}$, $\widehat{\boldsymbol{b}}_n^*(\boldsymbol{p})$, *satisfies the following property:*

$$\frac{\partial}{\partial \widehat{r}_{nt}^*} v_n(\widehat{\boldsymbol{r}}_n^*) \begin{cases} = \widehat{\beta}_{nt}^* = p_t, & \text{in case } \widehat{r}_{nt}^* > 0 \\ \leq \widehat{\beta}_{nt}^* < p_t, & \text{in case } \widehat{r}_{nt}^* = 0 \end{cases}; \tag{6.12}$$

and furthermore $\boldsymbol{u}_n^*(\widehat{\boldsymbol{b}}^*) = \widehat{\boldsymbol{r}}_n^*$, *for all* $n \in \mathscr{N}$.

Lemma 6.1 can be obtained by the Lagrange solving of (6.11). ■

By Lemma 6.1, the best response of PEV n with respect to a system price $\boldsymbol{p}$ is unique in case its bid demand is larger than zero; and is not unique in case its bid demand equals to zero, but in this case its bid price is less than the system price.

Lemma 6.2 (Efficient bid profile) *Define a bid profile* $\boldsymbol{b}^{**}$ *such that* $r_{nt}^{**} = u_{nt}^{**}$ *for all* $t \in \mathscr{T}$ *and* $n \in \mathscr{N}$, *where* $\boldsymbol{u}^{**}$ *represents the efficient allocation specified in (6.3). Hence,* $\boldsymbol{u}^*(\boldsymbol{b}^{**}) = \boldsymbol{u}^{**}$ *holds.*

It is straightforward to verify Lemma 6.2 by (6.3) and Lemma 6.1. ■

6.3.3 *Updates of System Price Under Individual Best Responses*

It defines a notion $\widehat{\boldsymbol{p}}^*(\boldsymbol{p})$ with respect to a system price $\boldsymbol{p}$, such that $\widehat{p}_t^*(\boldsymbol{p}) \triangleq \widehat{p}_t\left(\widehat{\boldsymbol{b}}^*(\boldsymbol{p})\right)$ for all $t \in \mathscr{T}$. That is to say, $\widehat{\boldsymbol{p}}^*(\boldsymbol{p})$ represents the updated system price under the collection of best responses $\left\{\widehat{\boldsymbol{b}}_n^*(\boldsymbol{p}); n \in \mathscr{N}\right\}$ with respect to a given system price $\boldsymbol{p}$.

In Theorem 6.1, it will analyze the property of $\widehat{\boldsymbol{p}}^*(\boldsymbol{p})$ with respect to a certain specified system price $\boldsymbol{p}$. Before that, it firstly introduces a bunch of notions below:

$$\psi \triangleq \sup_{n\in\mathcal{N}} \sup_{\boldsymbol{b}_n\in\mathcal{B}_n} \left\{ \frac{\partial^2}{\partial r_{nt}^2} v_n(\boldsymbol{r}_n) \right\}; \tag{6.13a}$$

$$\kappa \triangleq \sup_{y\in[0,\infty)} \left\{ c''(D_t) \right\}; \tag{6.13b}$$

then, by (6.1), it can verify that $\psi = 2\delta_n - \min_{n\in\mathcal{N}} \left\{ \min_{d\in[0,\Gamma_n]} \{f_n''(r)\} \right\}$.

Consider a system price $\boldsymbol{p}$ such that

$$p_t \begin{cases} \geq p_t^{**}, & \text{in case } t \in \mathcal{T}_0 \\ > p_t^{**}, & \text{otherwise} \end{cases}; \tag{6.14}$$

where $\mathcal{T}_0 \subset \mathcal{T}$ such that $\mathcal{T}_0 \triangleq \left\{ t \in \mathcal{T};\ \text{s.t. } p_t^{**} = c'(D_t) \right\}$, with $\boldsymbol{p}^{**}$ specified in (6.4). That is to say, $\mathcal{T}_0$ represents the set of instances at which there are no energy consumptions with respect to the efficient allocation.

Also define a notion γ, such that

$$\gamma \triangleq \max_{t\in\mathcal{T}^+} \left\{ \frac{\Sigma(\boldsymbol{r}_t^{**})}{N_t^+ \min\limits_{n\in\mathcal{N}} \{r_{nt}^{**}\}} \right\}, \tag{6.15}$$

where $\mathcal{T}^+ \triangleq \mathcal{T}/\mathcal{T}^0$, and N_t^+ represents the number of those loads $n \in \mathcal{N}$ satisfying $r_{nt}^{**} > 0$.

Theorem 6.1 *Consider a system price* $\mathbf{p}$ *satisfying (6.14), and suppose that* $\eta \in \left(0, \frac{|\psi|}{N\gamma\kappa+|\psi|}\right)$; *then it can show that the updated system price curve* $\widehat{\mathbf{p}}^*(\mathbf{p})$ *satisfies the following:*

$$\mathbf{p}^{**} < \widehat{\mathbf{p}}^*(\mathbf{p}) \leq \mathbf{p},\ \text{and}\ \widehat{p}_t^*(\mathbf{p}) < p_t \text{ for some } t \in \mathcal{T}. \tag{6.16}$$

It will be shown that $\widehat{\mathbf{p}}^*(\mathbf{p}) \leq \mathbf{p}$ *and* $\widehat{p}_t^*(\mathbf{p}) < p_t$, *for some* $t \in \mathcal{T}$ *in Sect. 6.6.1, and show* $\mathbf{p}^{**} < \widehat{\mathbf{p}}^*(\mathbf{p})$ *in case* $\eta \in \left(0, \frac{|\psi|}{N\gamma\kappa+|\psi|}\right)$ *in Sect. 6.6.2.* ■

Remark: Considering the auction problems for the homogeneous load populations, $\gamma = 1$; then Theorem 6.1 holds in case $\eta \in \left(0, \frac{|\psi|}{N\kappa+|\psi|}\right)$.

6.3.4 Decentralized Dynamic Algorithm

Here it specifies a decentralized iteration algorithm to update bid profiles in Algorithm 6.1 below. Essentially each PEV updates its best response maximizing its individual payoff specified in (6.10) with respect to the system price implemented in last iteration step. In case the update procedure converges the behavior is the implemented bid profile equilibrium.

Algorithm 6.1 (*Dynamic implementation algorithm*)

- Initialize an aggregated inelastic demand $\boldsymbol{d}$, an initial system price $\boldsymbol{p}^{(0)}$ and a termination parameter $\varepsilon_{\text{stop}}$;
- Set $k = 0$ and consider a positive ε;
- While $\varepsilon > \varepsilon_{\text{stop}}$
  - Implement a bid $\boldsymbol{b}_n^{(k+1)}$ for all n simultaneously w.r.t. $\boldsymbol{p}^{(k)}$, such that

$$\boldsymbol{b}_n^{(k+1)}(\boldsymbol{p}^{(k)}) \triangleq \underset{\boldsymbol{b}_n \in \mathscr{B}_n}{\operatorname{argmax}} \left\{ v_n(\boldsymbol{u}_n^*) - \boldsymbol{\beta}_n^{(k),\tau} \boldsymbol{u}_n^* \right\},$$

  with $\boldsymbol{u}_n^*$ obtained in (6.9) w.r.t. b_n and $\boldsymbol{p}^{(k)}$;
  - Implement $\boldsymbol{p}^{(k+1)}$ w.r.t. $\boldsymbol{b}^{(k+1)}$ by solving (6.6);
  - Update $\varepsilon := \|\boldsymbol{p}^{(k+1)} - \boldsymbol{p}^{(k)}\|_1$;
  - Update $k := k + 1$. ■

Following the analysis in Theorem 6.1, it shows in Corollary 6.1 below that the auction system can converge to the efficient solution by applying the proposed method.

Corollary 6.1 (Convergence of Algorithm) *Suppose* $\eta \in \left(0, \frac{|\psi|}{N\gamma\kappa + |\psi|}\right)$ *and consider an initial system price* $\mathbf{p}^{(0)}$ *such that* $\mathbf{p}^{(0)}$ *satisfies (6.14); then by applying Algorithm 6.1, the following asymptotic property holds:*

$$\lim_{k\to\infty} \boldsymbol{p}^{(k)} = \boldsymbol{p}^{**}, \tag{6.17}$$

where $\mathbf{p}^{(k)}$ *represents the updated system price at iteration step k.*

Proof: *Given an initial system price such that* $\mathbf{p}^{(0)}$ *satisfying (6.14), and in case* $\eta \in \left(0, \frac{|\psi|}{N\gamma\kappa + |\psi|}\right)$, *by Theorem 6.1, it obtains that*

$$\boldsymbol{p}^{**} < \boldsymbol{p}^{(k+1)} \leq \boldsymbol{p}^{(k)}, \text{ and } p_t^{(k+1)} < p_t^{(k)}, \text{ for some } t \in \mathscr{T}, \tag{6.18}$$

which implies that $\{\mathbf{p}^{(k)}; k = 0, 1, \ldots\}$ *is a monotonic decreasing sequence with a bound from low; then* $\mathbf{p}^{(k)}$ *converges to a limit, denoted by* $\underline{\mathbf{p}}$, *as k goes to infinity, and*

$$\boldsymbol{p}^{(k)} > \underline{\boldsymbol{p}} = \lim_{l\to\infty} \boldsymbol{p}^{(l)}, \forall k. \tag{6.19}$$

Thus $\underline{\boldsymbol{p}} \geq \boldsymbol{p}^{**}$.

Considering $\underline{\boldsymbol{p}} > \boldsymbol{p}^{**}$, *by Theorem 6.1, in case* $\eta \in (0, |\psi|/(N\gamma\kappa + |\psi|))$, *it can obtain that the updated system price* $\widehat{\boldsymbol{p}}^{*}(\underline{\boldsymbol{p}})$ *w.r.t.* $\underline{\boldsymbol{p}}$ *satisfies* $\widehat{\boldsymbol{p}}^{*} > \boldsymbol{p}^{**}$, *which is a contradiction of (6.19); then it obtains that* $\underline{\boldsymbol{p}} = \boldsymbol{p}^{**}$ *by (6.18) and (6.19), which indicates (6.17).* ■

Remark: By Corollary 6.1, it claims that there always exists a parameter η for any demand response problems such that the designed auction system can converge to the efficient solution.

In summary, by Lemma 6.2 and Corollary 6.1, in case $\eta \in \left(0, \frac{|\psi|}{N\gamma\kappa+|\psi|}\right)$ and $\boldsymbol{p}^{(0)}$ satisfying (6.14), the auction system converges to the efficient solution $\boldsymbol{u}^{**}$ with the decentralized update algorithm.

6.4 Numerical Examples

Consider a population of loads with population size of $N = 7,000$ over a common time interval from 12:00 a.m. on a day to 12:00 a.m. on next day. In the literature, e.g. [27–29], the electricity generation cost $c(\cdot)$, has been widely considered in a quadratic form on the supply, say $c(D_t) = \frac{1}{2}\mathsf{a}D_t^2 + \mathsf{b}D_t + \mathsf{c}$. Here it may be supposed that the generation cost function $c(D_t) = 5 \times 10^{-7}D_t^2 + 0.02D_t$. with a typical 24 h inelastic demand $\boldsymbol{d}$ in a hot summer season in the region of the MISO of US, as illustrated in Fig. 6.1. Suppose that the essential cost of PEV n is approximately specified as below:

$$f_n(u_{nt}) = 0.0027u_{nt}^2 + 0.05u_{nt}. \tag{6.20}$$

It considers that all of the loads share an identical maximum energy consumption $\Gamma_n = 33.75$ KWh, and the parameter δ_n related to the inter-temporal energy constraint is set as $\delta_n = 0.02$. The efficient load consumptions $\boldsymbol{u}^{**}$ is shown in Fig. 6.1.

With the generation cost specified above, $\kappa = 10^{-6}$. Also by (6.20) and $\delta_n = 0.02$ for all n, $|\psi| \doteq 0.0347$; then it obtains that

$$\frac{|\psi|}{N\gamma\kappa + |\psi|} = \frac{0.0347}{7000 \times 10^{-6} + 0.0347} \doteq 0.83,$$

with $\gamma = 1$ due to the homogeneity of load populations. In this simulation, it considers $\eta = 0.6 \in (0, 0.83)$ which satisfies $\eta \in \left(0, \frac{|\psi|}{N\gamma\kappa+|\psi|}\right)$ specified in Corollary 6.1.

It considers an initial system price $\boldsymbol{p}^{(0)}$, with $p_t^{(0)} = 0.35$\$/kWh for all t. It is direct to verify that

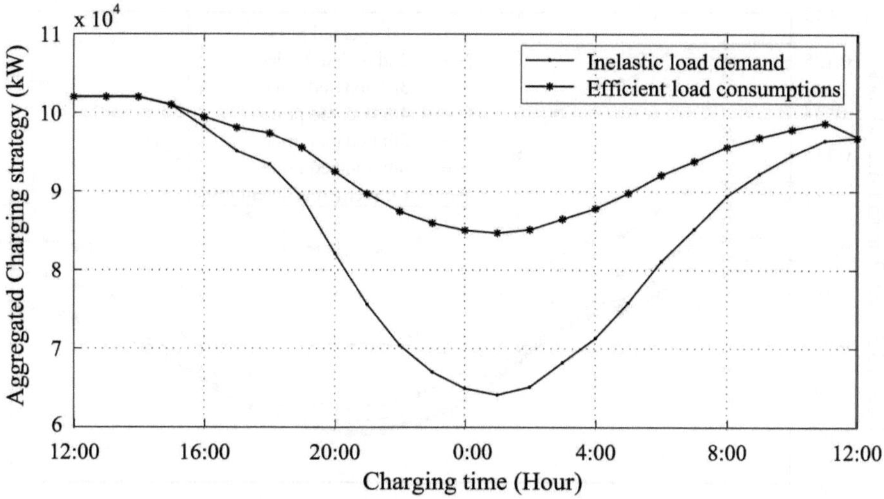

Fig. 6.1 The efficient load consumptions and inelastic load demand

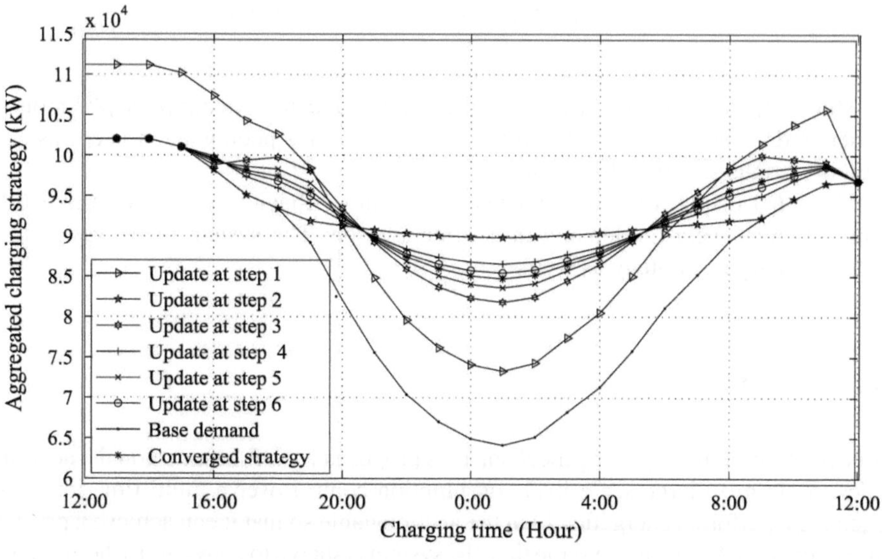

Fig. 6.2 Update procedure of load consumptions following Algorithm 6.1

$$p_t^{(0)} > p_t^{**} = 10^{-6}\,(d_t + Nu_t^{**}) + 0.02, \text{ for all } t.$$

Figure 6.2 displays the updates of the load consumptions following Algorithm 6.1, such that the system converges to the efficient coordination solution in a few of steps. This is consistent with Corollary 6.1.

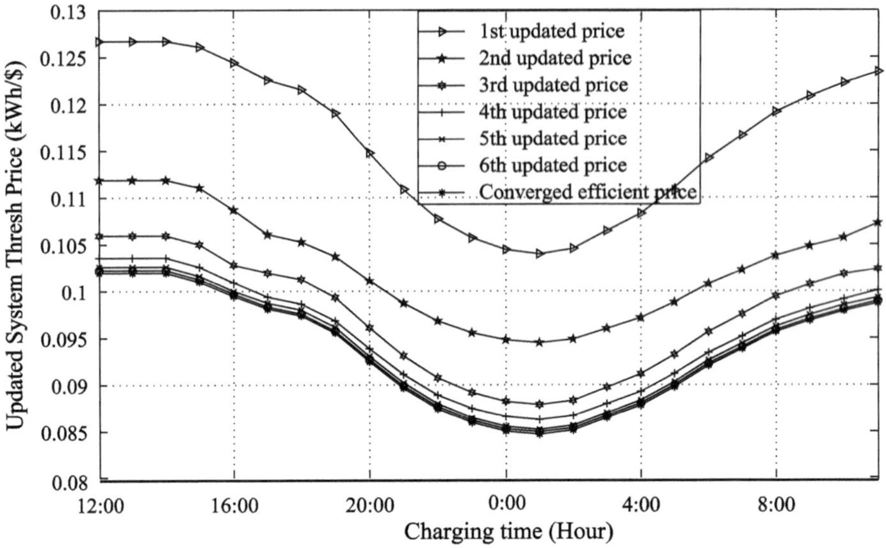

Fig. 6.3 Iteration updates of the system price by applying Algorithm 6.1

Moreover from Fig. 6.3, it can observe that the updated system price $\boldsymbol{p}^{(k)}$ monotonically decreases with respect to iteration step k. This phenomenon is consistent with the results stated in Theorem 6.1.

Figure 6.4 displays the the updates of consume behaviors under Algorithm 6.1 with a typical inelastic demand in a winter season. The system converges to the efficient charging strategy as well.

6.5 Conclusions

A novel decentralized pricing mechanism is proposed for DR based upon the auction rules to optimize large scale load consumption control over a multi-time interval. Each PEV performs energetically in the auction game so that it can achieve a private maximum benefit, and at the same time the system is shown to converge to the efficient solution under the specified sufficient condition. Through simulation examples, it can observe that the proposed dynamic pricing mechanism improves the performance of the system by encouraging PEVs to reduce their power consumption and shift their loads to off-peak hours.

The pricing mechanism developed in this chapter can be extended in several directions. For example, a system with multiple energy providers can be considered, as well as a system with on-site distributed generation and storage, and hierarchical system structure with microgrids as the medium.

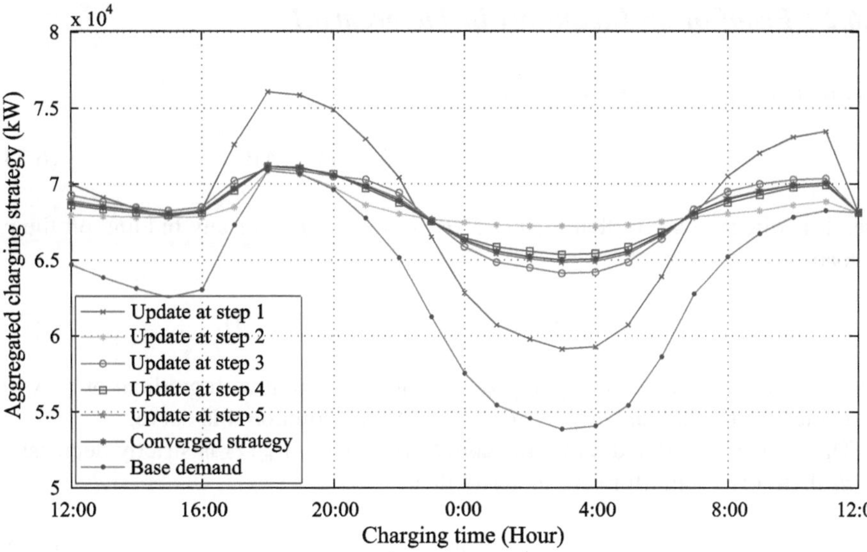

Fig. 6.4 Updates of load consumptions by Algorithm 6.1 in a winter season

6.6 Appendices

6.6.1 Proof of Some Inequalities in Theorem 6.1

By (6.7) and (6.12),

$$\widehat{\boldsymbol{\beta}}^{\text{sys}} \leq \boldsymbol{p}; \tag{6.21}$$

which, together with (6.6) and $c'(\widehat{D}_t^*) \leq \beta_t^{\text{sys}}(\boldsymbol{b}_t)$ by (6.8), implies $\widehat{\boldsymbol{p}}^*(\boldsymbol{p}) \leq \boldsymbol{p}$.

Next it show $\widehat{p}_t^*(\boldsymbol{p}) < p_t$, *for some* $t \in \mathscr{T}$.

In case $\boldsymbol{p}$ satisfying (6.14), $\Sigma(\widehat{\boldsymbol{r}}_n^*) < \Sigma(\boldsymbol{r}_n^{**})$ for all n; then there must exist an instant t such that $\Sigma(\widehat{\boldsymbol{r}}_t^*) < \Sigma(\boldsymbol{r}_t^{**})$ which, under Assumption 3.2, implies that

$$c'(d_t + \Sigma(\widehat{\boldsymbol{r}}_t^*)) < c'(d_t + \Sigma(\boldsymbol{r}_t^{**})) = p_t^{**}, \tag{6.22}$$

where the equality holds by (6.3). Hence

$$\begin{aligned}
\widehat{p}_t^*(\boldsymbol{p}) &= \widehat{\beta}_t^{\text{sys}} + \eta\left(c'(\widehat{D}_t^*) - \widehat{\beta}_t^{\text{sys}}\right), \quad \text{by (6.6)} \\
&\leq (1-\eta)p_t + \eta c'(\widehat{D}_t^*), \quad \text{by (6.21)} \\
&< (1-\eta)p_t + \eta p_t^{**}, \quad \text{by (6.22)} \\
&< p_t,
\end{aligned}$$

where the last inequality holds since $p_t > p_t^{**}$ in case $t \in \mathscr{T}/\mathscr{T}_0$. ■

6.6.2 Proof of an Inequality in Theorem 6.1

Firstly define $w_n(r, A)$ and $\widehat{w}_{nt}(r)$ below:

$$w_n(r, A) \triangleq -f_n'(r) + 2\delta_n \cdot (\Gamma_n - A); \tag{6.23}$$

then it is direct to verify that $w_n(r_{nt}, \Sigma(\boldsymbol{r}_n)) = \frac{\partial}{\partial r_{nt}} v_n(\boldsymbol{r}_n)$ for any bidding profile $\boldsymbol{r}_n$ of load n.

$$\widehat{w}_{nt}(r) \triangleq \psi \cdot (r - r_{nt}^{**}) + c'(d_t + \Sigma(\boldsymbol{r}_t^{**})), \tag{6.24}$$

with $d \in [0, r_{nt}^{**}]$, that is to say, $\widehat{w}_{nt}(r)$ is a linear function with gradient of ψ, with ψ defined in (6.13a), and $(r_{nt}^{**}, c'(d_t + \Sigma(\boldsymbol{r}_t^{**}))$ is a point on this curve.

Thus since $\psi < 0$ and under the convexity of $f_n(\cdot)$, $\widehat{w}_{nt}(r)$ is strictly decreasing on d; then it is straightforward to verify that

$$\widehat{w}_{nt}(r) \leq w_n(r, \Sigma(\boldsymbol{r}_n^{**})), \tag{6.25}$$

with $d \in [0, r_{nt}^{**}]$, where $w_n(r, \Sigma(\boldsymbol{r}_n^{**}))$ is specified in (6.23).

In case $\boldsymbol{p}$ satisfying (6.14), $\Sigma(\widehat{\boldsymbol{r}}_n^*) < \Sigma(\boldsymbol{r}_n^{**})$ which implies that

$$w_n(r, \Sigma(\boldsymbol{r}_n^{**})) < w_n(r, \Sigma(\widehat{\boldsymbol{r}}_n^*)), \text{ for all feasible } d. \tag{6.26}$$

Hence by (6.25) and (6.26), the following holds

$$\widehat{w}_{nt}(r) < w_n(r, \Sigma(\widehat{\boldsymbol{r}}_n^*)), \text{ for all } d \in [0, r_{nt}^{**}]. \tag{6.27}$$

By the auction mechanism (6.6), the updated system price w.r.t. $\boldsymbol{p}$ is specified below:

$$\widehat{p}_t^*(\boldsymbol{p}) = \widehat{\beta}_t^{\text{sys}} + \eta \cdot \left(c'(\widehat{D}_t^*) - \widehat{\beta}_t^{\text{sys}}\right), \tag{6.28}$$

where $\widehat{D}_t^* = d_t + \Sigma(\widehat{\boldsymbol{r}}_t^*)$ and $\widehat{\beta}_t^{\text{sys}}$ is specified in (6.7).

For notational simplicity, it considers

$$e_t^{**} \equiv c'(d_t + \Sigma(\boldsymbol{r}_t^{**})), \text{ and } \widehat{e}_t^* \equiv c'(d_t + \Sigma(\widehat{\boldsymbol{r}}_t^*)).$$

It will consider the following two cases concerning the relationship between $\Sigma(\widehat{\boldsymbol{r}}_t^*)$ and $\Sigma(\boldsymbol{r}_t^{**})$ in (I) – (III) respectively below:

(I) *In case* $\Sigma(\widehat{\boldsymbol{r}}_t^*) < \Sigma(\boldsymbol{r}_t^{**})$:

In this case and under Assumption 3.2, $\widehat{e}_t^* < e_t^{**}$.

Further consider (I.a) and (I.b) below:

(I.a) *In case* $\Sigma(\widehat{\boldsymbol{r}}_t^*) < \Sigma(\boldsymbol{r}_t^{**})$ *and* $\widehat{\beta}_t^{\text{sys}} < p_t$:

In this case, by (6.12), $\widehat{\boldsymbol{r}}_t^* = \mathbf{0}$. Also by (6.7), $\widehat{\beta}_t^{\text{sys}} \geq \widehat{\beta}_{nt}^*$ for $n \in \mathcal{N}$; then together with (6.25), the following holds:

$$\widehat{\beta}_{nt}^* \geq w_n(\widehat{r}_{nt}^*, \Sigma(\widehat{\boldsymbol{r}}_n^*)) > \widehat{w}_{nt}(\widehat{r}_{nt}^*), \tag{6.29}$$

for $n \in \mathcal{N}$, where the first inequality holds by the specification of w_n and (6.12); and the second inequality holds by (6.27).

Suppose that $r_{nt}^{**} > 0$ for some n; then by the specification of constant valued γ in (6.15), the following holds

$$r_{nt}^{**} - \widehat{r}_{nt}^* = r_{nt}^{**} \geq \frac{1}{N\gamma}\Sigma(\boldsymbol{r}_t^{**}) \geq \frac{1}{N\gamma}\left(\Sigma(\boldsymbol{r}_t^{**}) - \Sigma(\widehat{\boldsymbol{r}}_t^*)\right);$$

then, by the specification of $\widehat{w}_{nt}$ given in (6.24),

$$\begin{aligned} \widehat{w}_{nt}(\widehat{r}_{nt}^*) &= \psi \cdot (\widehat{r}_{nt}^* - r_{nt}^{**}) + e_t^{**} \\ &\geq \frac{|\psi|}{N\gamma}\left(\Sigma(\boldsymbol{r}_t^{**}) - \Sigma(\widehat{\boldsymbol{r}}_t^*)\right) + e_t^{**} \\ &\geq \frac{|\psi|}{N\gamma\kappa}\left(p_t^{**} - \widehat{e}_t^*\right) + p_t^{**}, \end{aligned} \tag{6.30}$$

where the last equality holds by the specification of κ given in (6.13b) and $e_t^{**} \equiv c'(d_t + \Sigma(\boldsymbol{r}_t^{**})) = p_t^{**}$ by (6.3).

Hence by (6.28), the following holds

$$\begin{aligned} \widehat{p}_t^* &= (1-\eta)\cdot\widehat{\beta}_t^{\text{sys}} + \eta\widehat{e}_t^*, \qquad \text{since } \widehat{e}_t^* = c'(\widehat{D}_t^*) \\ &> (1-\eta)\cdot\widehat{w}_{nt}(\widehat{r}_{nt}^*) + \eta\widehat{e}_t^*, \qquad \text{by (6.29)} \\ &\geq (1-\eta)\cdot\left(\frac{|\psi|}{\gamma\kappa}(p_t^{**} - \widehat{e}_t^*) + p_t^{**}\right) + \eta\widehat{e}_t^*, \qquad \text{by (6.30)} \\ &= p_t^{**} + \frac{|\psi|}{N\gamma\kappa}\left(1 - \frac{N\gamma\kappa + |\psi|}{|\psi|}\eta\right)(p_t^{**} - \widehat{e}_t^*) \\ &\geq p_t^{**}, \end{aligned} \tag{6.31}$$

where the last equality holds in case $\eta < \frac{|\psi|}{N\gamma\kappa+|\psi|}$ since $p_t^{**} > \widehat{e}_t^*$.

(I.b) *In case $\Sigma(\widehat{\boldsymbol{r}}_t^*) < \Sigma(\boldsymbol{r}_t^{**})$ and $\widehat{\beta}_t^{\text{sys}} = p_t$:*

In this case, by (6.12), it obtains that $\widehat{\beta}_{mt}^* = p_t$, for all $m \in \mathcal{N}$, and there must exist an $n \in \mathcal{N}$ such that

$$r_{nt}^{**} - \widehat{r}_{nt}^* \geq \frac{1}{N}\left(\Sigma(\boldsymbol{r}_t^{**}) - \Sigma(\widehat{\boldsymbol{r}}_t^*)\right);$$

then, parallel to the analysis of (6.30), by the specification of $\widehat{w}_{nt}$ given in (6.24),

$$\widehat{w}_{nt}(\widehat{r}_{nt}^{*}) \geq \frac{|\psi|}{N\kappa}\left(p_t^{**} - \widehat{e}_t^{*}\right) + p_t^{**}. \tag{6.32}$$

Hence parallel with the analysis of (6.31),

$$\widehat{p}_t^{*} > p_t^{**} + \frac{|\psi|}{N\kappa}\left(1 - \frac{N\kappa + |\psi|}{|\psi|}\eta\right)\left(p_t^{**} - \widehat{e}_t^{*}\right) \geq p_t^{**}, \tag{6.33}$$

where the last equality holds in case $\eta < |\psi|/(N\gamma\kappa + |\psi|)$, which is larger than or equal to $|\psi|/(N\gamma\kappa + |\psi|)$ since $\gamma \geq 1$ by the specification of (6.15).

In summary, in (I.a) and (I.b) above, it has shown that, considering $\Sigma(\widehat{\boldsymbol{r}}_t^{*}) < \Sigma(\boldsymbol{r}_t^{**}), \widehat{p}_t^{*} > p_t^{**}$ in case $\eta \in \left(0, \dfrac{|\psi|}{N\gamma\kappa + |\psi|}\right)$.

Next in (II) below, it will show that, considering $\Sigma(\widehat{\boldsymbol{r}}_t^{*}) < \Sigma(\boldsymbol{r}_t^{**}), \widehat{p}_t^{*} > p_t^{**}$, for all $\eta \in [0, 1]$.

(II) *In case* $\Sigma(\widehat{\boldsymbol{r}}_t^{*}) > \Sigma(\boldsymbol{r}_t^{**})$

Firstly, considering $\Sigma(\widehat{\boldsymbol{r}}_t^{*}) > \Sigma(\boldsymbol{r}_t^{**})$ and under Assumption 3.2,

$$\widehat{e}_t^{*} \equiv c'(d_t + \Sigma(\widehat{\boldsymbol{r}}_t^{*})) > e_t^{**} \equiv c'(d_t + \Sigma(\boldsymbol{r}_t^{**})) = p_t^{**}, \tag{6.34}$$

where the last equality holds by (6.3); then by (6.28),

$$\begin{aligned} \widehat{p}_t^{*} &= (1-\eta)\widehat{\beta}_t^{\text{sys}} + \eta\widehat{e}_t^{*}, \qquad \text{since } \widehat{e}_t^{*} = c'(\widehat{D}_t^{*}) \\ &> (1-\eta)\widehat{e}_t^{*} + \eta p_t^{**} \\ &> (1-\eta)p_t^{**} + \eta p_t^{**} = p_t^{**}, \end{aligned}$$

where the 2nd inequality from last holds, for all $\eta \in [0, 1]$, by (6.12) and (6.34).

(III) *In case* $\Sigma(\widehat{\boldsymbol{r}}_t^{*}) = \Sigma(\boldsymbol{r}_t^{**})$:

In this case and under Assumption 3.2, $\widehat{e}_t^{*} < e_t^{**}$.

Furthermore consider (III.a) and (III.b) below:

(III.a) *In case* $\Sigma(\widehat{\boldsymbol{r}}_t^{*}) = \Sigma(\boldsymbol{r}_t^{**}) = 0$: $\widehat{p}_t^{*} \geq e_t^{**} = p_t^{**}$.

(III.b) *In case* $\Sigma(\widehat{\boldsymbol{r}}_t^{*}) = \Sigma(\boldsymbol{r}_t^{**}) > 0$:

Following the same technique adapted in (I), it can show that, considering $\Sigma(\widehat{\boldsymbol{r}}_t^{*}) = \Sigma(\boldsymbol{r}_t^{**}) > 0, \widehat{p}_t^{*} > p_t^{**}$ in case $\eta \in \left(0, \frac{|\psi|}{N\gamma\kappa+|\psi|}\right)$.

In summary, following the analysis in (I) – (III) above, the conclusion holds. ■

References

1. J. Wang, S. Kennedy, J. Kirtley, A new wholesale bidding mechanism for enhanced demand response in smart grids, in *Innovative Smart Grid Technologies (ISGT)*, Gaithersburg, MD, pp. 1–8. Accessed 19–21 Jan 2010
2. M. Parvania, M. Fotuhi-Firuzabad, M. Shahidehpour, Optimal demand response aggregation inwholesale electricity markets. IEEE Trans. Smart Grid **4**(4), 1957–1965 (2013)
3. L. Goel, Q. Wu, P. Wang, Y. Ding, Reliability enhancement of deregulated power systems considering demand-price elasticity, in *7th IEEE International Power Engineering Conference (IPEC)* (2005), pp. 934–938
4. R. Walawalkar, S. Fernands, N. Thakur, K.R. Chevva, Evolution and current status of demand response (DR) in electricity markets: insights from PJM and NYISO. Energy **35**(4), 1553–1560 (2010)
5. H.A. Aalami, M.P. Moghaddam, G.R. Yousefi, Demand response modeling considering interruptible/curtailable loads and capacity market programs. Appl. Energy **87**(1), 243–250 (2010)
6. G. Liu, K. Tomsovic, A full demand response model in co-optimized energy and reservemarket. Electr. Power Syst. Res. **111**, 62–70 (2014)
7. M.H. Albadi, E.F. El-Saadany, A summary of demand response in electricity markets. Electr. Power Syst. Res. **78**(11), 1989–1996 (2008)
8. J. Aghaei, M.I. Alizadeh, Demand response in smart electricity grids equipped with renewable energy sources: a review. Renew. Sustain. Energy Rev. **18**, 64–72 (2013)
9. R. Rajaraman, J.V. Sarlashkar, F.L. Alvarado, The effect of demand elasticity on security prices for the poolco and multi-lateral contract models. IEEE Trans. Power Syst. **12**(3), 1177–1184 (1997)
10. M.G. Lijesen, The real-time price elasticity of electricity. Energy Econ. **29**(2), 249–258 (2007)
11. D.S. Kirschen, G. Strbac, P. Cumperayot, D. de Paiva Mendes, Factoring the elasticity of demand in electricity prices. IEEE Trans. Power Syst. **15**(2), 612–617 (2000)
12. A.K. David, Y.Z. Li, Effect of inter-temporal factors on the real time pricing of electricity. IEEE Trans. Power Syst. (Inst. Electr. Electron. Eng.) **8**(1), 44–52 (1993)
13. W. Uturbey, A.S. Costa, Optimal power flow with inter-temporal constraints as an aiding tool for demand-side management. IEE Proc. Gener. Transm. Distrib. **149**(1), 37–43 (2002)
14. P. Faria, Z. Vale, Demand response in electrical energy supply: an optimal real time pricing approach. Energy **36**(8), 5374–5384 (2011)
15. J. Aghaei, M.I. Alizadeh, Multi-objective self-scheduling of CHP (combined heat and power)-based microgrids considering demand response programs and ESSs (energy storage systems). Energy **55**, 1044–1054 (2013)
16. L. Goel, Q. Wu, P. Wang, Nodal price volatility reduction and reliability enhancement of restructured power systems considering demand-price elasticity. Electr. Power Syst. Res. **78**(10), 1655–1663 (2008)
17. C.L. Su, D. Kirschen, Quantifying the effect of demand response on electricity markets. IEEE Trans. Power Syst. **24**(3), 1199–1207 (2009)
18. A. Khodaei, M. Shahidehpour, S. Bahramirad, SCUC with hourly demand response considering intertemporal load characteristics. IEEE Trans. Smart Grid **2**(3), 564–571 (2011)
19. C. Zhao, J. Wang, J.P. Watson, Y. Guan, Multi-stage robust unit commitment considering wind and demand response uncertainties. IEEE Trans. Power Syst. **28**(3), 2708–2717 (2013)
20. M. Ali, J. Jokisalo, K. Siren, M. Lehtonen, Combining the demand response of direct electric space heating and partial thermal storage using LP optimization. Electr. Power Syst. Res. **106**, 160–167 (2014)
21. D. Setlhaolo, X. Xia, J. Zhang, Optimal scheduling of household appliances for demand response. Electr. Power Syst. Res. **116**, 24–28 (2014)
22. N. Gatsis, G.B. Giannakis, Decomposition algorithms for market clearing with large-scale demand response. IEEE Trans. Smart Grid **4**(4), 1976–1987 (2013)
23. J. Medina, N. Muller, I. Roytelman, Demand response and distribution grid operations: opportunities and challenges. IEEE Trans. Smart Grid **1**(2), 193–198 (2010)

24. P. Samadi, H. Mohsenian-Rad, R. Schober, V.W.S. Wong, Advanced demand side management for the future smart grid using mechanism design. IEEE Trans. Smart Grid **3**(3), 1170–1180 (2012)
25. S. Han, S. Han, K. Sezaki, Development of an optimal vehicle-to-grid aggregator for frequency regulation. IEEE Trans. Smart Grid **1**(1), 65–72 (2010)
26. S. Boyd, L. Vandenberghe, *Convex Optimization* (Cambridge University Press, Cambridge, 2004)
27. E. Bompard, Y. Ma, R. Napoli, G. Abrate, The demand elasticity impacts on the strategic bidding behavior of the electricity producers. IEEE Trans. Power Syst. **22**(1), 188–197 (2007)
28. V.P. Gountis, A.G. Bakirtzis, Bidding strategies for electricity producers in a competitive electricity marketplace. IEEE Trans. Power Syst. **19**(1), 356–365 (2004)
29. F.S. Wen, A.K. David, Strategic bidding for electricity supply in a day-ahead energy market. Electr. Power Syst. Res. **59**, 197–206 (2001)

Chapter 7
PEV Charging Coordination in Constrained Distribution Networks

Abstract It studies in this chapter the charging coordination of PEVs in distribution networks with capacity-constrained feeder lines. It is usually challenging to design an effective decentralized method to implement the optimal solution for the formulated constrained optimization problems due to the coupling relationship of charging behaviors on the constrained distribution network system. Alternatively, in this chapter, in order to avoid possible overloading on the feeder lines, it proposes a gradient-projection based decentralized method such that the step size is properly adjusted. It is shown that by applying the proposed method, the system converges and the implemented solution satisfies the capacity constraints of feeder lines. Simulation results verify that the proposed method can efficiently avoid possible overloading on the feeder lines in different distribution networks.

7.1 Introduction

The uncoordinated charging behaviors of PEV populations in the distribution networks may induce localized overloading, large power losses and voltage problems [1–6]. How to coordinate the PEV charging behaviors in distribution networks has been studied in the literature, to minimize the power losses [3], to maximize the profits of the utility considering voltage constraints [7], to minimize the battery degradation cost and deviation cost from the referred SOC value under the constraints of transformers [5], and to shave the peak demand and minimize the power losses considering voltage regulation [8].

Besides the above listed coordination methods, some research work devotes to avoiding overloading in distribution networks. In [9], the PEV charging is coordinated to minimize individual costs in response to the time-varying prices with additional incentives provided to mitigate potential overloads in the network. In [10], the energy costs of the distribution networks are minimized with overload limitation by adopting Lyapunov optimization and Lagrange dual decomposition techniques. while in [11], the variance of the aggregate load is minimized with feeder overload constraints relaxed by adding an overload cost function to the objective function. A decentralized PEV coordinating method is proposed in [12] by solving Lagrangian dual problem

Z. Ma, *Decentralized Charging Coordination of Large-scale Plug-in Electric Vehicles in Power Systems*, https://doi.org/10.1007/978-981-13-7652-8_7

to avoid the feeder overloading, while it doesn't consider the total amount of the energy charged by each PEV and the optimal solution was implemented on a single time slot. The work decouples the overload constraints by implementing Lagrangian multipliers with slack parameters, which may violate the feeder capacity. In addition, the application of slack parameters may result in suboptimal solutions.

Due to the coupling relationship among the charging behaviors of individual PEVs introduced by feeder overload constraints, it is usually challenging to design a method to coordinate the charging behaviors of PEV populations in decentralized ways. In this chapter, it formulates a class of charging coordination problems of PEV populations in a distribution network by considering the hard constraint of the feeder capacity, and develop a gradient-projection based decentralized method for the underlying constrained optimization problems.

In case the capacity constraints of feeder lines are not considered, then by the decoupling relationship among the admissible sets of coordination behaviors of all the PEVs, the gradient projection on an individual admissible (charging) set is independent upon the admissible sets of all the others. Following this modeling characteristic, the coordination behaviors of all the PEVs can be updated locally and simultaneously. However, as demonstrated via numerical simulations, the updated behaviors in the decentralized form may not belong to the admissible space under capacity constraints. Consequently, in order to avoid the overloading on the feeder lines during the dynamic updates, it properly adjusts the step size for each update step to satisfy the capacity constraints. More specifically, if the updated behaviors are not satisfied with the capacity constraints at some iteration step; it re-implements the decentralized gradient projection with an adjusted smaller valued step-size parameters, and this process continues until the updated behaviors satisfy the constraints. Furthermore, suppose that the updated behaviors reach the exact capacity constraints on some feeder line at some instant; then it considers that the charging behaviors of all the PEVs on these feeder lines shall be fixed and won't be updated any longer.

It is shown that by applying the proposed method, the system converges and the implemented behavior satisfies the hard capacity constraints of distribution feeders. The developed results are demonstrated with some numerical examples.

The rest of the chapter is organized as follows. In Sect. 7.2, it formulates a class of charging coordination problems in distribution networks under feeder capacity constraints. In Sect. 7.3, it proposes a decentralized gradient projection method with variable step sizes for the underlying PEV charging problems, and show the convergence and the optimality performance of the implemented strategy. In Sect. 7.4, numerical examples are studied to illustrate the results developed in the chapter. Finally, conclusions and future research directions are drawn in Sect. 7.5.

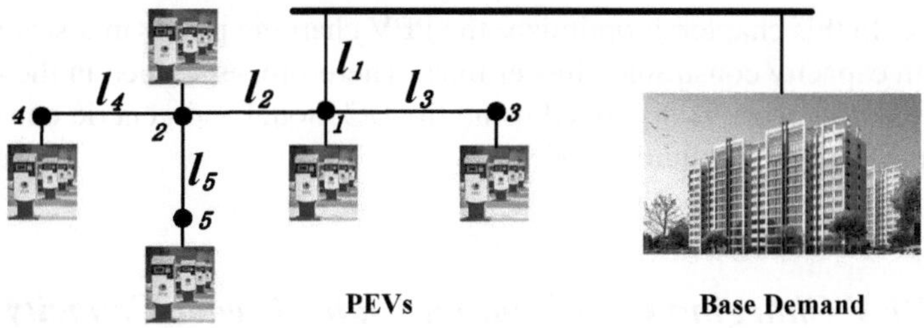

Fig. 7.1 A distribution network with 5 feeder lines

7.2 PEV Charging Coordination in Distribution Networks

This section introduces an optimal charging coordination of a set of PEVs $\mathscr{N}$ in a distribution network with feeder capacity constraints. It firstly specifies the characteristics of the distribution network.

Denote by $\mathscr{L}$ and $\mathscr{M}$ the set of feeders and nodes (excluding the root node) respectively in the distribution network. Denote by $\mathscr{N}_m$, with $m \in \mathscr{M}$, a collection of PEVs each of which is charged at node m, and denote by $\mathscr{N}_l$, with $l \in \mathscr{L}$, a collection of PEVs each of which is supplied with the electricity via feeder line l.

Use a matrix $\mathbb{A}$, $\mathbb{A} \equiv [a_{lm}]_{L\times M}$, with $L \equiv |\mathscr{L}|$ and $M \equiv |\mathscr{M}|$, to represent the topology of the distribution network, such that $a_{lm} = 1$ in case $\mathscr{N}_m \subset \mathscr{N}_l$ and $a_{lm} = 0$ otherwise. For example, concerning the distribution network displayed in Fig. 7.1, $\mathbb{A} = \begin{bmatrix} 1\,1\,1\,1\,1 \\ 0\,1\,0\,1\,1 \\ 0\,0\,1\,0\,0 \\ 0\,0\,0\,1\,0 \\ 0\,0\,0\,0\,1 \end{bmatrix}$.

Suppose that PEVs are charged over a multi-time interval $\mathscr{T}$, and denote by $\mathscr{T}_n$, with $\mathscr{T}_n \subset \mathscr{T}$, the charging interval of PEV n. Denote by $\boldsymbol{d} = (d_t, t \in \mathscr{T})$ the total inelastic base demand trajectory. Suppose that both of the base demand and the PEVs are supplied by the same generation source, see an illustration of the system in Fig. 7.1. As usual, still let $\boldsymbol{u}_n \equiv (u_{nt}; t \in \mathscr{T})$ denote the charging strategy of PEV n, and say $\boldsymbol{u}_n$ is admissible, if (3.1) holds. Also the admissible set of charging strategies of PEV n is denoted by $\mathscr{U}_n$.

Let $\boldsymbol{u} \equiv (\boldsymbol{u}_n; n \in \mathscr{N}) \in \mathbb{R}_+^{NT}$ denote the charging profile of all the PEVs in the distribution network, and the admissible set of charging strategies for PEV populations is denoted by $\mathscr{U}$, such that

$$\mathscr{U} \triangleq \mathscr{U}_1 \times \cdots \times \mathscr{U}_N. \tag{7.1}$$

For notational simplicity, it considers $U_t(\boldsymbol{u}) \equiv \sum_{n\in\mathscr{N}} u_{nt}$, and $U_{lt}(\boldsymbol{u}) \equiv \sum_{n\in\mathscr{N}_l} u_{nt}$, which represents the aggregated charging power supplied via the feeder line l at time t under the charging profile $\boldsymbol{u}$. Define a matrix $\mathbb{U}(\boldsymbol{u}) \equiv [U_{lt}]_{L\times T}$. It specifies another matrix, denoted by $\mathbb{V}(\boldsymbol{u}) \equiv [V_{mt}]_{M\times T}$, such that $V_{mt} \equiv \sum_{n\in\mathscr{N}_m} u_{nt}$, i.e., V_{mt} represents the total charging power of all the PEVs plugged in node m at time t. Thus, it can verify that $\mathbb{U} = \mathbb{A}\mathbb{V}$.

Remark: In this chapter, it optimizes the PEV charging power in distribution networks with capacity constrained feeder lines. Therefore, as studied in the literature, e.g. [10, 13], this work is discussed using the DC model and generic constraints in AC networks are not explicitly modeled.

7.2.1 PEV Charging Coordination Under Feeder Capacity Constraints

Denote by $\boldsymbol{\Pi} \equiv [\Pi_l]_{L\times 1}$ the power capacity of the feeders in the distribution network. Let $\mathbb{B} \equiv [\boldsymbol{\Pi} \cdots \boldsymbol{\Pi}]$ which is an $L \times T$ matrix. Also define a set of charging profiles satisfying feeder capacity constraints, such that

$$\mathscr{C}_\varepsilon \triangleq \{\boldsymbol{u}, \text{ s.t. } U_{lt}(\boldsymbol{u}) \le \Pi_l + \varepsilon, \forall (l,t) \in \mathscr{L} \times \mathscr{T}\}, \tag{7.2}$$

where ε is a given positive valued parameter.

To demonstrate whether the implemented charging profile $\boldsymbol{u}$ satisfies the feeder capacity constraints, it uses a notion $\boldsymbol{\xi}(\boldsymbol{u})$ defined as below:

$$\xi_{lt}(\boldsymbol{u}) \triangleq \frac{1}{\Pi_l} U_{lt}(\boldsymbol{u}), \quad \forall (l,t) \in \mathscr{L} \times \mathscr{T}, \tag{7.3}$$

i.e., $\boldsymbol{\xi}_l(\boldsymbol{u})$ denotes the normalized charging trajectory on feeder line l.

Note: A collection of charging profiles $\boldsymbol{u}$ satisfies the feeder capacity constraints over the whole charging interval $\mathscr{T}$, in case $U_{lt}(\boldsymbol{u}) \le \Pi_l$, or equivalently $\xi_{lt}(\boldsymbol{u}) \le 1$, for all $(l,t) \in \mathscr{L} \times \mathscr{T}$.

Problem 7.1 (PEV charging coordination under feeder capacity constraints)

$$\min_{\boldsymbol{u} \in \mathscr{U} \cap \mathscr{C}_\varepsilon} \{J(\boldsymbol{u})\}, \quad \text{with } J(\boldsymbol{u}) \triangleq \sum_{t \in \mathscr{T}} c(d_t + U_t(\boldsymbol{u})). \tag{7.4}$$

■

Note: As earlier chapters, in this part, it still considers the specification of generation cost curve $c(\cdot)$ as given in Assumption 3.2, say $c(\cdot)$ could also be considered as power generation costs that is convex. Under Assumption 3.2, as analyzed in previous chapters and in the literature, like [14–17], the objective of the system is to fill the valley of the base demand curve.

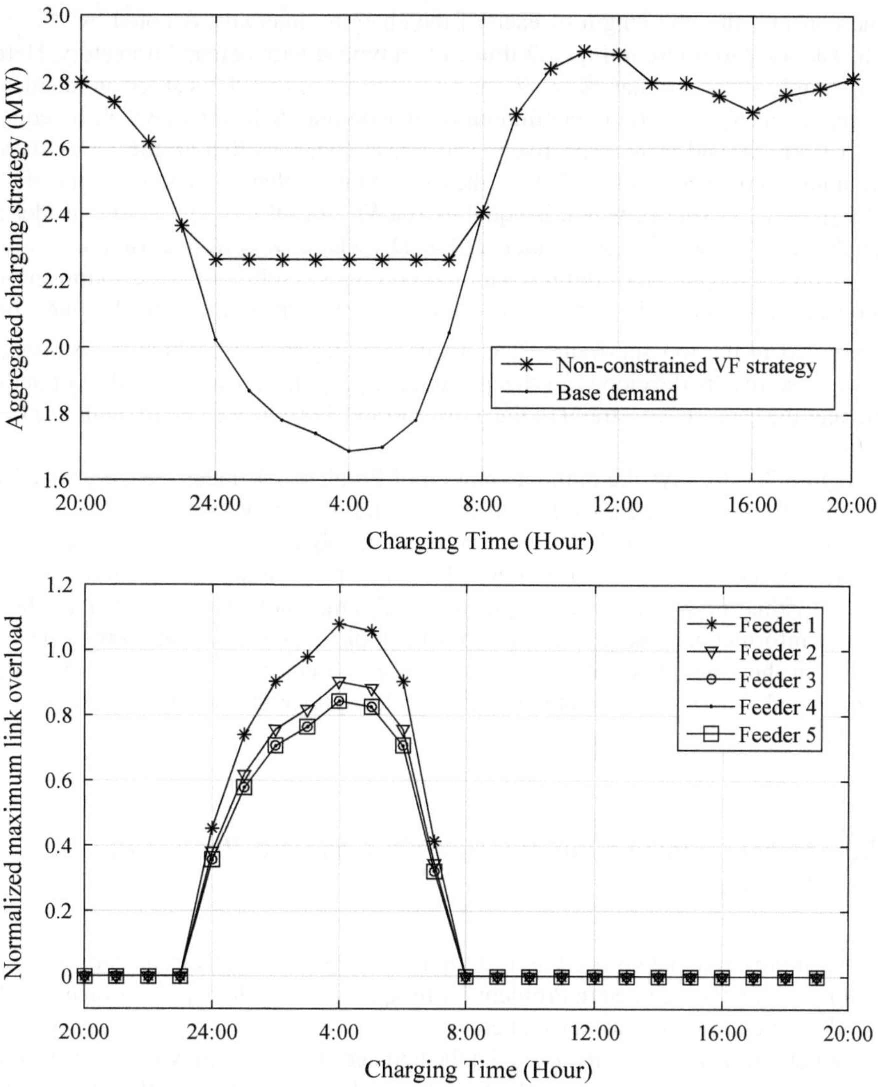

Fig. 7.2 Valley-filling behaviors which don't satisfy the capacity constraints

7.2.2 Case Studies

It specifies a simple PEV charging coordination problem in a distribution network as illustrated in Fig. 7.1, and suppose that the capacities of the feeder lines are set as $\boldsymbol{\Pi} = [530, 385, 140, 140, 140]$ (kW).

It specifies the PEV charging coordination with a population size of $N = 350$ over a multi-time interval from 8:00 p.m. on one day to 8:00 p.m. on the next day, and consider that the length of each of the charging intervals, denoted by ΔT, is 1 h. The top sub-figure in Fig. 7.2 illustrates a typical base demand trajectory. Here it is simply supposed that $\mathscr{T}_n = \mathscr{T}$ for all n, and all of the PEVs share an identical battery capacity size, common minimum and maximum SOC limits which are equal to 20 kWh, 20 and 90% respectively, and it also considers that all the PEVs share a common initial SOC of 40%; then the total required charging energy is specified as $\Gamma_n(soc_{max} - soc_{n0})$, which is equal to 10kWh, for all $n \in \mathscr{N}$. Also consider a quadratic form for $c(D_t)$ such that $c(D_t) = D_t^2$, where $D_t = d_t + U_t(\boldsymbol{u})$.

Figure 7.2 displays the optimal charging coordination solution for the optimization problem $\min_{\boldsymbol{u} \in \mathscr{U}} J(\boldsymbol{u})$, which does not consider capacity constraints of feeder lines. As illustrated in the top sub-figure, it is a valley-filling strategy, which is consistent with those results developed in the literature, e.g., [14, 15], however this strategy violates the capacity constraint of line l_1 at 4:00 and 5:00 since $\xi_{lt} > 1.0$ with $(l, t) = (l_1, 4{:}00)$, $(l_1, 5{:}00)$.

Figure 7.3 displays the optimal solution of Problem 7.1 under feeders' capacity constraints. As illustrated, all the capacity constraints are satisfied.

The centralized optimal solution can be effectively implemented in the case that the system has complete information and has a permission to directly coordinate all the individual PEVs. However, in practice, PEVs may not be willing to share their private information, and the transmission of complete information may create heavy traffic in the communication networks. Hence, the centralized method is usually impractical. Alternatively, in this chapter, it will propose a decentralized coordination method.

7.3 Decentralized Coordination via Gradient Projection Method

A decentralized update procedure is designed in Algorithm 7.1 for the coordination problem of PEVs specified in Problem 7.1 by applying a gradient projection method. Firstly some terms are introduced below.

Denote by $\widehat{\boldsymbol{u}}(\boldsymbol{u}) \equiv (\widehat{u}_n(\boldsymbol{u}), n \in \mathscr{N})$ the updated strategy of PEVs with respect to a given charging strategy $\boldsymbol{u}$ by adopting the gradient projection method [18]; then $\widehat{\boldsymbol{u}}(\boldsymbol{u})$ shall be given as below:

$$\widehat{\boldsymbol{u}}(\boldsymbol{u}) = [\boldsymbol{u} - \alpha \nabla J(\boldsymbol{u})]^+_{\mathscr{U} \cap \mathscr{C}_\varepsilon} \tag{7.5}$$

where $\nabla J(\boldsymbol{u})$ is the gradient of the cost function $J(\boldsymbol{u})$, α is the step size, and $[\boldsymbol{x}]^+_{\mathscr{U} \cap \mathscr{C}_\varepsilon}$ represents the projection of $\boldsymbol{x}$ on the hyperplane $\mathscr{U} \cap \mathscr{C}_\varepsilon$. By the specification of $\mathscr{C}_\varepsilon$ given in (7.2), $\widehat{u}_n(\boldsymbol{u})$ are coupled with each other by $\mathscr{C}_\varepsilon$.

Suppose that an updated strategy $\widehat{\boldsymbol{u}}(\boldsymbol{u})$ specified in (7.5) is equal to $\widetilde{\boldsymbol{u}}(\boldsymbol{u})$ which is given as below:

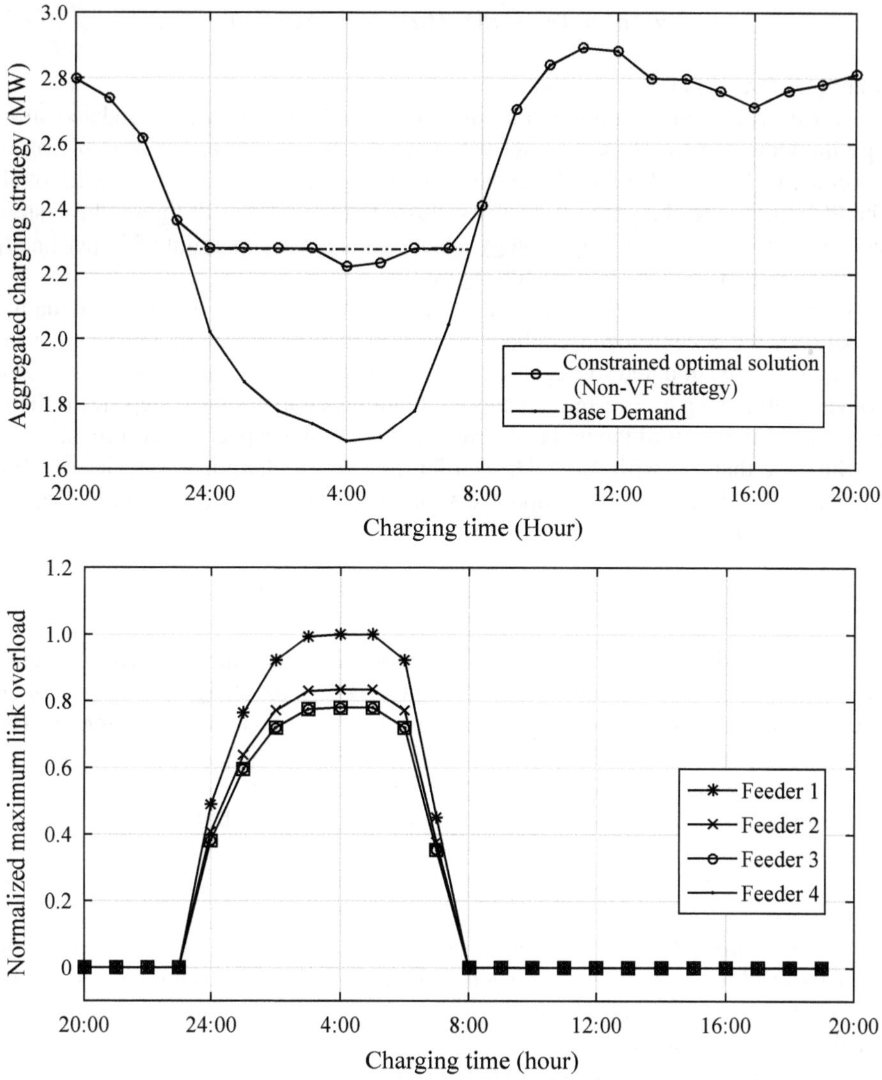

Fig. 7.3 Optimal behaviors satisfying the feeders' capacity constraints

$$\widetilde{\boldsymbol{u}}(\boldsymbol{u}) = [\boldsymbol{u} - \alpha \nabla J(\boldsymbol{u})]^+_{\mathscr{U}}, \tag{7.6}$$

that is to say, the projected strategy $\widetilde{\boldsymbol{u}}(\boldsymbol{u})$ on $\mathscr{U}$ satisfies the capacity constraints of feeder lines defined in $\mathscr{C}_\varepsilon$. Thus, due to the decoupling property of the admissible set defined in (7.1), say $\mathscr{U} = \mathscr{U}_1 \times \cdots \mathscr{U}_{N-1} \times \mathscr{U}_N$, it can be verified that the updated strategy given in (7.5) can be implemented by the following decentralized form:

$$\widehat{\boldsymbol{u}}_n(\boldsymbol{u}) = [\boldsymbol{u}_n - \alpha[\nabla J(\boldsymbol{u})]_n]^+_{\mathscr{U}_n}, \quad \forall n \in \mathscr{N}, \tag{7.7}$$

where $[\nabla J(\boldsymbol{u})]_n$ is the nth row of $\nabla J(\boldsymbol{u})$.

However, as demonstrated via a numerical simulation below, the updated strategy implemented by (7.7), may not belong to the admissible space $\mathscr{U} \cap \mathscr{C}_\varepsilon$. More specifically, Fig. 7.4 displays the updates of the aggregated charging strategies of all the PEVs over the charging interval. As illustrated in the top sub-figure, the system converges to the valley-filling strategy which violates the constraint of feeder line l_1 at 4:00 and 5:00, see the bottom sub-figure.

Consequently, it should find a way to avoid the violations during the dynamic updates of the decentralized gradient projection method presented in (7.7). Essentially, in case the updated strategies are not satisfied with the capacity constraint(s) at some iteration step k, it re-implements (7.7) with a smaller valued step size α. This process continues until the updated strategies satisfy the capacity constraints.

Suppose that $\mathscr{V}_n$, with $\mathscr{V}_n \subset \mathscr{U}_n$, is a feasible set of charging strategies of PEV n. Denote by $\widehat{\boldsymbol{u}}_n(\boldsymbol{u}; \mathscr{V}_n, \alpha)$ the updated strategies by applying (7.7) with respect to $(\boldsymbol{u}_n, \mathscr{V}_n, \alpha)$, such that

$$\widehat{\boldsymbol{u}}_n(\boldsymbol{u}; \mathscr{V}_n, \alpha) = [\boldsymbol{u}_n - \alpha[\nabla J(\boldsymbol{u})]_n]^+_{\mathscr{V}_n}. \tag{7.8}$$

Next, consider the case that the updated strategies reach the exact capacity constraint on some feeder line at some instant; then it keeps the charging strategies of all the PEVs on this feeder line at this instant fixed, and not be updated any longer. The modeling issues related to this consideration will be specified in (7.9)–(7.11) later.

Consider $\mathscr{S}$ as a set such that $\mathscr{S} \subset \mathscr{L} \times \mathscr{T}$; then it defines a subset $\mathscr{W} \subset \mathscr{S}$ with respect to a collection of charging strategies $\boldsymbol{v} \in \mathscr{U}$, such that

$$\mathscr{W}(\mathscr{S}, \boldsymbol{v}; \varepsilon) \triangleq \big\{(l, t) \in \mathscr{S}, \text{ s.t. } |U_{lt}(\boldsymbol{v}) - \Pi_l| \leq \varepsilon\big\}. \tag{7.9}$$

Interpretation of $\mathscr{W}(\mathscr{S}, \boldsymbol{v}; \varepsilon)$ *given in* (7.9): $\mathscr{W}(\mathscr{S}, \boldsymbol{v}; \varepsilon)$ represents a collection of pairs of $(l, t) \in \mathscr{S} \subset \mathscr{L} \times \mathscr{T}$ such that, subject to the charging strategy $\boldsymbol{v}$, the aggregated charging behavior on line l at instant t is nearly fully occupied with a small valued variation ε.

Define $\widehat{\mathscr{S}}(\mathscr{S}, \boldsymbol{v})$ with respect to $(\mathscr{S}, \boldsymbol{v})$ such that

$$\widehat{\mathscr{S}}(\mathscr{S}, \boldsymbol{v}) \triangleq \mathscr{S} \setminus \mathscr{W}, \text{ with } \mathscr{W} \equiv \mathscr{W}(\mathscr{S}, \boldsymbol{v}; \varepsilon). \tag{7.10}$$

Interpretation of (7.10): By (7.9), $\widehat{\mathscr{S}}$ represents a subset of $\mathscr{S}$, subject to a charging strategy $\boldsymbol{v}$, such that all the fully occupied pairs of $(l, t) \in \mathscr{W}$ are excluded from $\mathscr{S}$. Moreover by the dynamic process proposed in Algorithm 7.1, $\mathscr{S}^k$ evolves following (7.10) with respect to iteration step k; then, by the initialization of $\mathscr{S}^0 = \mathscr{L} \times \mathscr{T}$, $\mathscr{S}^k$ is a subset of $\mathscr{L} \times \mathscr{T}$, such that all the fully occupied pairs of (l, t) defined in (7.9) up to iteration step k are excluded from $\mathscr{L} \times \mathscr{T}$.

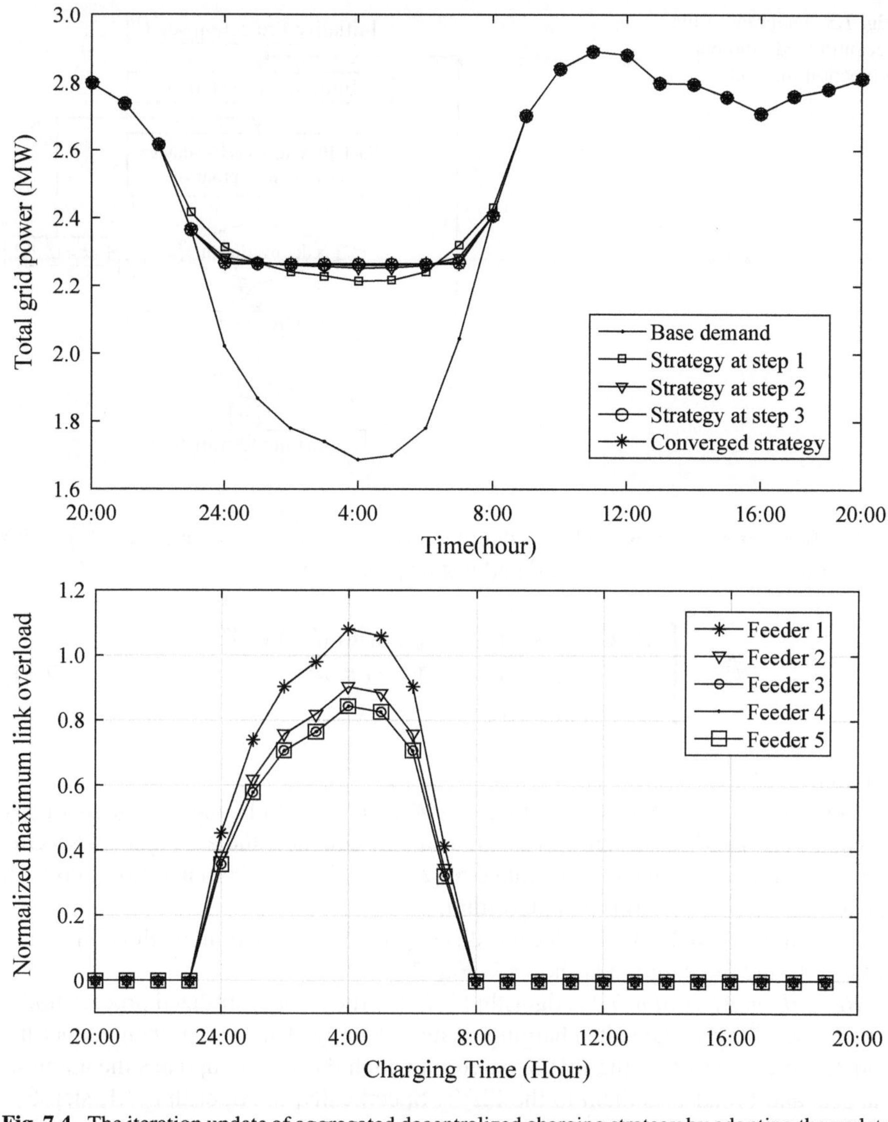

Fig. 7.4 The iteration update of aggregated decentralized charging strategy by adopting the update (7.7)

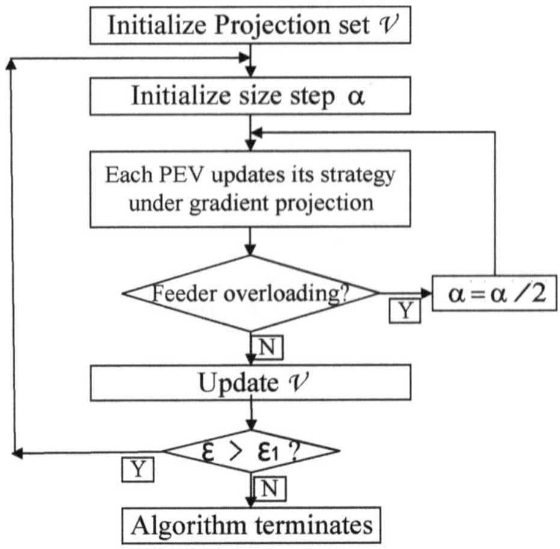

Fig. 7.5 Diagram of the decentralized gradient projection method

Further define a subset of $\mathscr{U}$, denoted by $\widehat{\mathscr{V}} \equiv \left(\widehat{\mathscr{V}}_n(\mathscr{V}, \mathscr{S}, \boldsymbol{v}); n \in \mathscr{N}\right)$, with respect to $\mathscr{V} \subset \mathscr{U}$, $\mathscr{S} \subset \mathscr{L} \times \mathscr{T}$ and $\boldsymbol{v} \in \mathscr{U}$, such that

$$\widehat{\mathscr{V}_n} \triangleq \begin{cases} \left\{\boldsymbol{u}_n \in \mathscr{V}_n;\ \text{s.t. } u_{nt} = v_{nt},\ \text{with } (l, t) \in \mathscr{W}\right\} \\ \qquad \text{if } n \in \mathscr{N}_l,\ \text{s.t. } \exists (l, t) \in \mathscr{W} \\ \mathscr{V}_n, \quad \text{otherwise} \end{cases}, \tag{7.11}$$

for all $n \in \mathscr{N}$.

Remark on (7.11): Suppose that $(l, t) \in \mathscr{W}$; then by (7.11), the charging strategy of each PEV in $\mathscr{N}_l$ at instant t is fixed; as a consequence, the aggregated charging power on line l at t is fixed in the range of $[\Pi_l - \varepsilon, \Pi_l + \varepsilon]$ from iteration step k till the termination of the proposed algorithm.

Now it can establish the decentralized gradient projection method in Algorithm 7.1, and the diagram is shown in Fig. 7.5.

Remark on Algorithm 7.1: Algorithm 7.1 specifies a decentralized process under which each PEV updates its charging strategy locally. The system then collect the updated strategies of all the PEVs, based on which the system updates the admissible sets and broadcasts them to the PEVs. Specifically, in Algorithm 7.1, step 6 is implemented in each PEV, while steps 4, 9, 10, 11 are implemented in the system.

Theorem 7.1 *Consider Assumption 3.2 and apply Algorithm 7.1 with* $\alpha \in (0, \dfrac{2}{N\rho})$*; then the system converges to an admissible solution, denoted by* $(\boldsymbol{u}^\dagger, \mathscr{V}^\dagger, \mathscr{S}^\dagger)$*, such that*

(I) $U_{lt}(\boldsymbol{u}^\dagger) \le \Pi_l + \varepsilon_1$*, for all* $(l, t) \in \mathscr{L} \times \mathscr{T}$*;*
(II) $\boldsymbol{u}^\dagger = \underset{\boldsymbol{u} \in \mathscr{V}^\dagger}{argmin}\{J(\boldsymbol{u})\}$.

Proof Please see Appendix 7.6.1. ■

Algorithm 7.1 Implementation of decentralized charging coordination under capacity constraints.

Require:
Initialize base demand $\boldsymbol{d}$ and feeder capacity $\boldsymbol{\Pi}$;
Give step size $\alpha > 0$ and termination criteria $\varepsilon_1, \varepsilon_1 > 0$;
Set $\boldsymbol{u}^0 = \boldsymbol{u}_0$, $\mathscr{S}^0 = \mathscr{L} \times \mathscr{T}$ and $\mathscr{V}^0 = \mathscr{U}$;
Set $k = 0$ and $\varepsilon > \varepsilon_1$;

Ensure:
Charging strategies of PEVs;

while $\varepsilon > \varepsilon_1$ **do**
  Implement $\widehat{u}_n(\boldsymbol{u}^k; \mathscr{V}^k, \alpha)$ by (7.8), $\forall n \in \mathscr{N}$;
  Set $j = 0$;
  while $\exists (l, t) \in \mathscr{S}^k$ with $U_{lt}(\widehat{\boldsymbol{u}}(\boldsymbol{u}^k; \mathscr{V}^k, \frac{\alpha}{2^j})) > \Pi_l + \varepsilon_1$ **do**
    $j := j + 1$;
    Implement $\widehat{u}_n(\boldsymbol{u}^k; \mathscr{V}^k, \frac{\alpha}{2^j})$ by (7.8), $\forall n \in \mathscr{N}$;
  end while
  Update $\boldsymbol{u}_n^{k+1} := \widehat{u}_n(\boldsymbol{u}^k, \mathscr{V}^k; \frac{\alpha}{2^j})$, $\forall n \in \mathscr{N}$;
  Update $\mathscr{S}^{k+1}$ w.r.t. $\mathscr{S}^k$, $\boldsymbol{u}^{k+1}$ by (7.10);
  Update $\mathscr{V}^{k+1}$ w.r.t. $\mathscr{V}^k$, $\mathscr{S}^k$, $\boldsymbol{u}^{k+1}$ by (7.11);
  Set $\varepsilon := \|\boldsymbol{u}^{k+1} - \boldsymbol{u}^k\|_1$;
  Set $k := k + 1$;
end while

7.4 Numerical Results

The designed algorithms is evaluated using Matlab on a 5-feeder test system first in Case Study I, and on a 5-feeder system considering PEVs' diversity in Case Study II. In addition, it considers a 12-feeder test system in Case Study III.

7.4.1 Case Study I

Algorithm 7.1 will be applied to implement the coordination behaviors under constrained capacity constraints of feeder lines. Here, unless specified, it uses the parameters considered in the example in Sect. 7.2.

By applying Algorithm 7.1, it firstly initializes $(\boldsymbol{u}^0, \mathscr{S}^0, \mathscr{V}^0)$ as $\boldsymbol{u}^0 = \boldsymbol{u}_0$, $\mathscr{S}^0 = \mathscr{L} \times \mathscr{T}$ and $\mathscr{V}^0 = \mathscr{U}$.

At the initial iteration step, say $k = 0$, it implements $\widehat{u}_n(\boldsymbol{u}^0; \mathscr{V}^0, \alpha)$ by (7.8), for all $n \in \mathscr{N}$. The top sub-figure in Fig. 7.6 displays the aggregated charging behaviors on each line l, say $U_{lt}(\widehat{\boldsymbol{u}}(\boldsymbol{u}^0; \mathscr{V}^0, \alpha))$, for all $t \in \mathscr{T}$. As illustrated, the resulting charging strategies violate the capacity constraint of feeder line l_1 at 4:00 and 5:00; then by applying Algorithm 7.1, the system implements $\widehat{u}_n(\boldsymbol{u}^0; \mathscr{V}^0, \frac{\alpha}{2^j})$, by (7.8), for all $n \in \mathscr{N}$ successively, with $j \in \{1, 2, \cdots\}$. The procedure continues until $U_{lt}(\widehat{\boldsymbol{u}}(\boldsymbol{u}^0; \mathscr{V}^0, \frac{\alpha}{2^{\widehat{j}}})) \leq \Pi_l + \varepsilon_1$, for all $(l, t) \in \mathscr{S}^0$, for some $\widehat{j}$, that is to say,

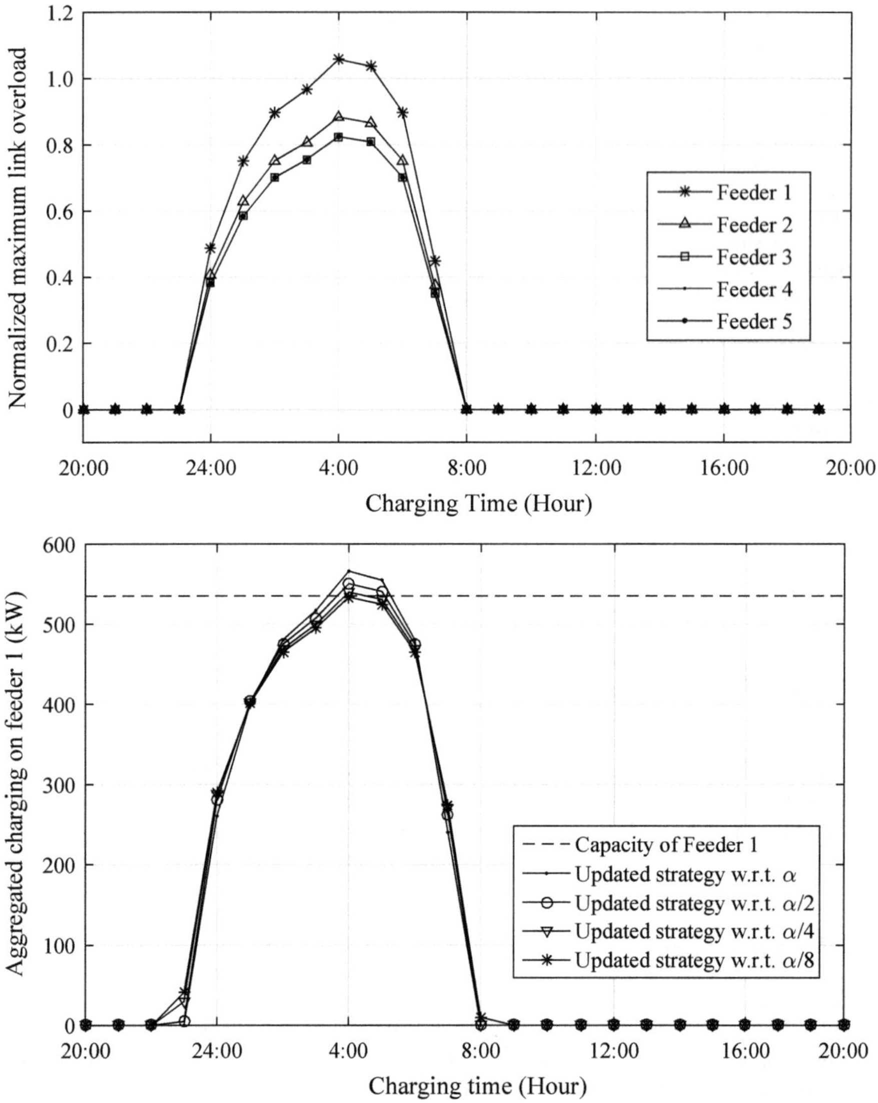

Fig. 7.6 Updates of the aggregated behaviors on feeder l_1 with respect to a sequence of parameters $\frac{\alpha}{2^j}$

the charging strategy $\widehat{\boldsymbol{u}}_n\big(\boldsymbol{u}^0;\, \mathscr{V}^0,\, \frac{\alpha}{2^{\widehat{j}}}\big)$ satisfies capacity constraints of all the feeder lines. As illustrated in the bottom sub-figure in Fig. 7.6, the charging strategies satisfy the capacity constraints with $\widehat{j} = 3$.

Note: For the purpose of demonstration, in the bottom sub-figure in Fig. 7.6, it only displays the updates of the aggregated charging behaviors on feeder l_1, with respect to a sequence of parameters $\frac{\alpha}{2^j}$, where the capacity constraint was violated subject to $\widehat{\boldsymbol{u}}_n(\boldsymbol{u}^0;\, \mathscr{V}^0, \alpha)$.

Hence by applying Algorithm 7.1, the charging strategy is updated with $\boldsymbol{u}_n^1 = \widehat{\boldsymbol{u}}_n(\boldsymbol{u}^0, \mathscr{V}^0;\, \frac{\alpha}{2^{\widehat{j}}})$, for all $n \in \mathscr{N}$.

As observed, the updated behavior $\boldsymbol{u}^1$ satisfies $|U_{lt}(\boldsymbol{u}^1) - \Pi_l| \le \varepsilon$, with $(l, t) = (l_1, 4{:}00)$; then, by Algorithm 7.1, $\mathscr{W}(\mathscr{S}^0, \boldsymbol{u}^1, \varepsilon) = \{(l_1, 4{:}00)\}$ by (7.9), and

$$\mathscr{S}^1 = \mathscr{S}^0 / \mathscr{W}(\mathscr{S}^0, \boldsymbol{u}^1, \varepsilon) = \mathscr{S}^0 / \{(l_1, 4{:}00)\}, \quad \text{by (7.10)},$$

$$\mathscr{V}_n^1 = \begin{cases} \left\{ u_n \in \mathscr{V}_n^0;\ \text{s.t. } u_{n,4:00} = u^1_{n,4:00} \right\}, & \forall n \in \mathscr{N}_{l_1} \\ \mathscr{V}_n^0, & \text{otherwise} \end{cases},$$

by (7.11). i.e., after the 1st iteration step, say $k = 1$, the system is updated with $\big(\boldsymbol{u}^1, \mathscr{S}^1, \mathscr{V}^1\big)$.

In summary, after the 1st iteration step, the implemented charging strategy $\boldsymbol{u}^1$ satisfies the capacity constraints of all the feeder lines and the aggregated charging behaviors reach the capacity constraint of feeder line l_1 at 4:00. Moreover, the charging behaviors of all the PEVs of $\mathscr{N}_{l_1}$ at 4:00 are fixed and aren't updated any longer.

The process continues until it terminates at iteration step k with $\|\boldsymbol{u}^{k+1} - \boldsymbol{u}^k\|_1 \le \varepsilon_1$. Figure 7.7 displays the iteration updates of the decentralized behaviors following Algorithm 7.1 such that the system converges in a few of iteration steps. Also as compared with Fig. 7.3, the implemented solution is globally optimal.

7.4.2 Case Study II

For approaching realistic characteristics of PEV populations, it is supposed that the initial SOC values of PEV populations, $\{\text{soc}_{n0};\, n \in \mathscr{N}\}$, approximately satisfy a Gaussian distribution, denoted by $\text{N}(\hat{\mu}, \hat{\gamma})$, see [19, 20]. It considers all of the PEVs share an identical battery capacity size of 30 kWh. For the purpose of demonstration, it considers $\hat{\mu} = 0.5$ and $\hat{\gamma} = 0.1$. Figures 7.8, 7.9, 7.10 illustrate the converged solution following Algorithm 7.1.

More specifically, Fig. 7.8 illustrates the converged solution which is optimal such that the system reaches the upper bound of the power that the system can supply over the period of 10:00–13:00; while Figs. 7.9 and 7.10 display the normalized aggregated

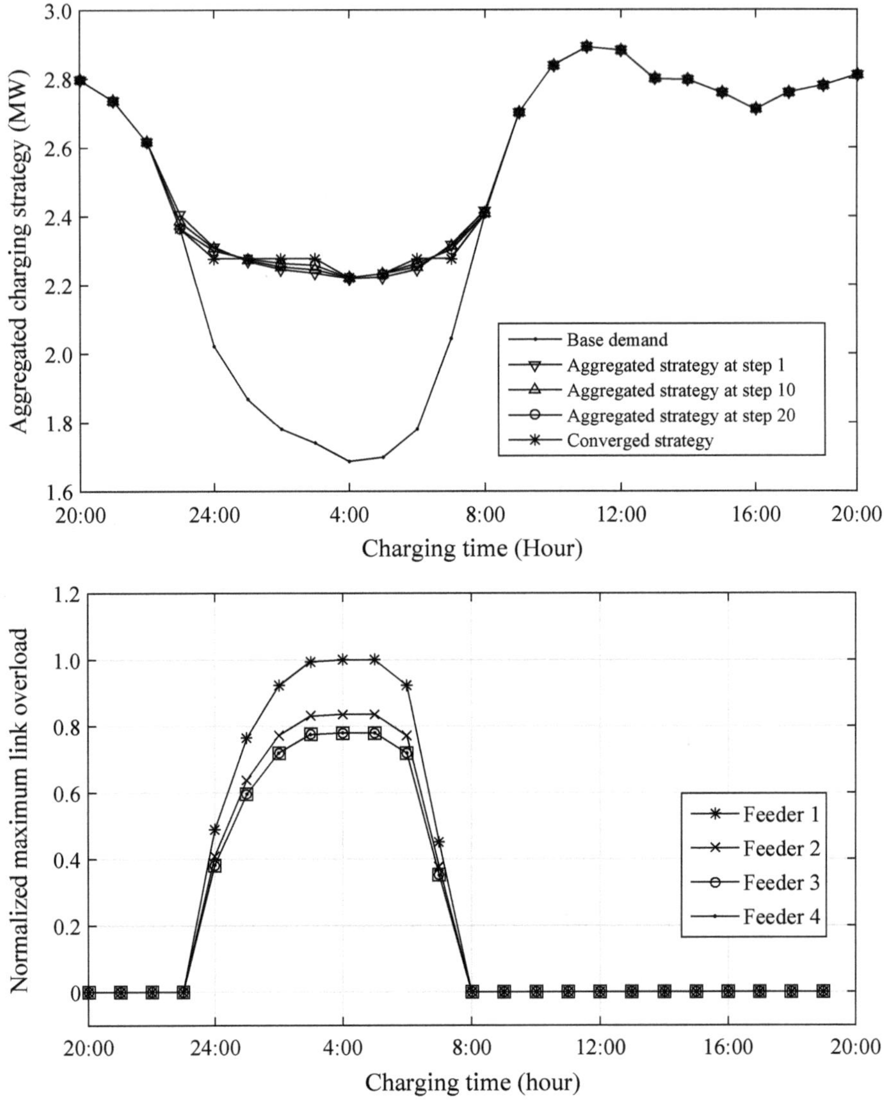

Fig. 7.7 Iteration updates of aggregated behaviors by applying Algorithm 7.1

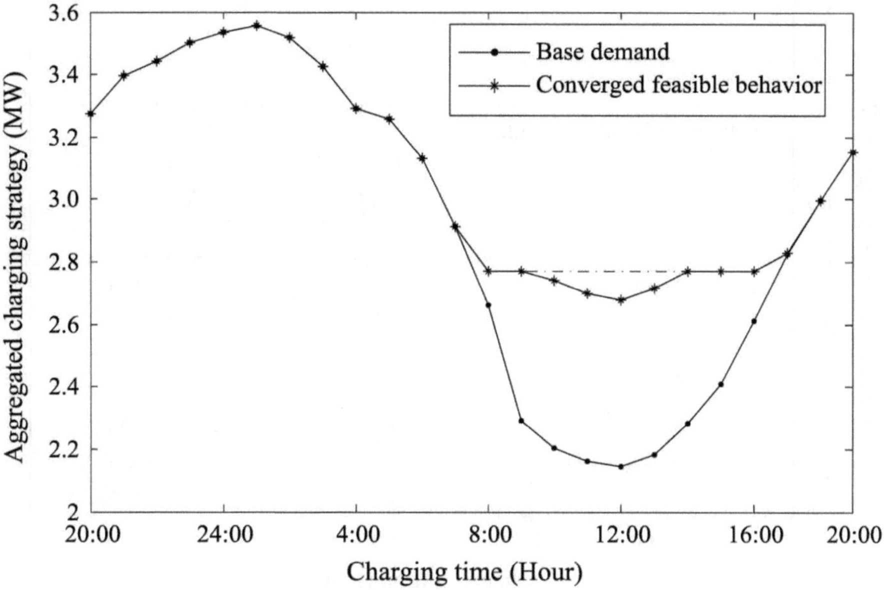

Fig. 7.8 Implemented feasible coordination solution with heterogeneous PEV populations by applying Algorithm 7.1 for all feeders

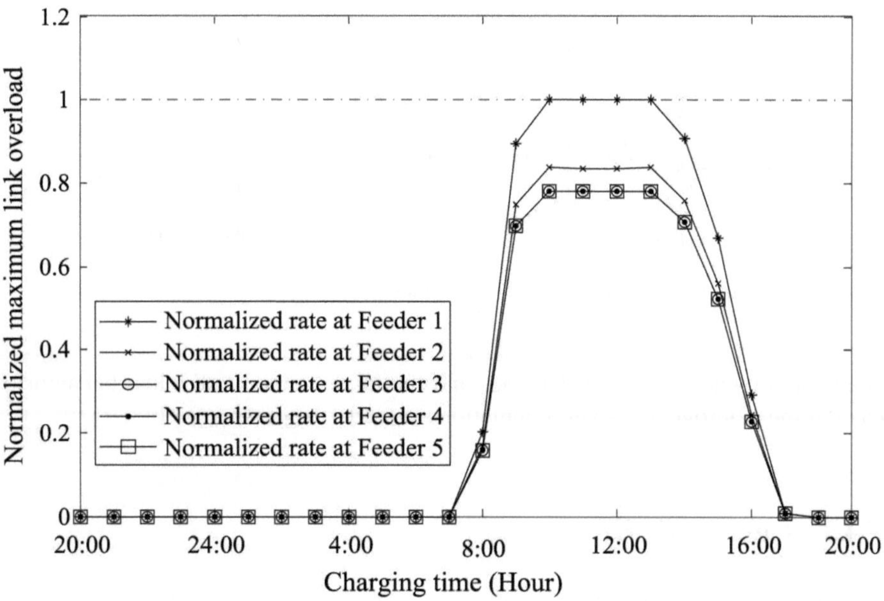

Fig. 7.9 Implemented coordination solution with heterogeneous PEV populations by applying Algorithm 7.1

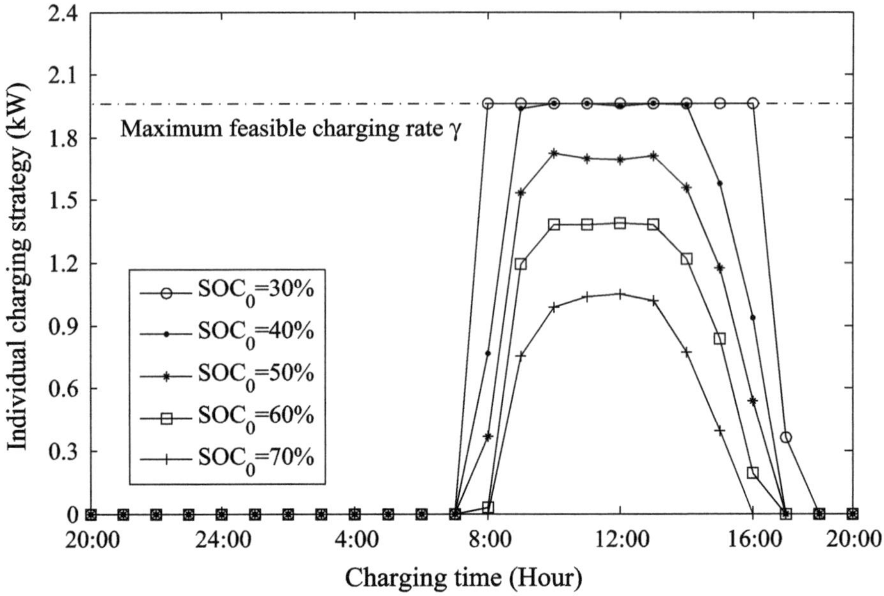

Fig. 7.10 Implemented coordination solution with heterogeneous PEV populations by applying Algorithm 7.1 with respect to different initial SOCs

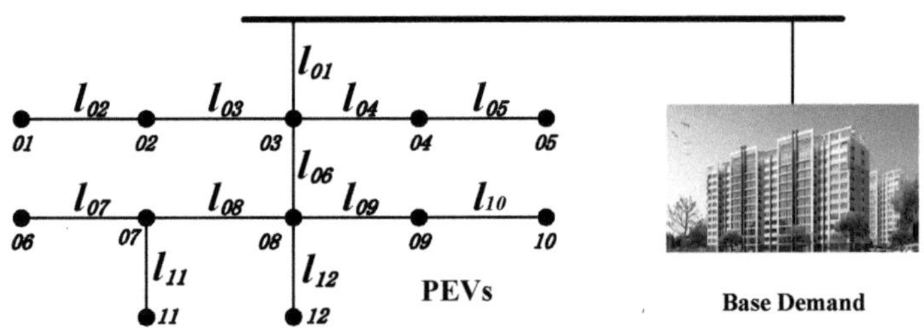

Fig. 7.11 A distribution network with 12 feeder lines

behaviors on each of the feeder lines and the implemented individual behavior of each of the heterogeneous sub-populations at feeder m_{01} respectively.

7.4.3 Case Study III

It further studies the charging coordination problems in a distribution network with 12 feeder lines, see an illustration in Fig. 7.11.

The capacities of all the feeder lines are listed in Table 7.1 respectively. Suppose that there are 50 PEVs at each of the feeders and these PEVs possess the characteristics given in Sect. 7.2.2 respectively.

Table 7.1 Capacities of 12 feeder lines

Feeder line	l_{01}	l_{02}	l_{03}	l_{04}	l_{05}	l_{06}
Capacity (kW)	920	100	180	130	100	590
Feeder line	l_{07}	l_{08}	l_{09}	l_{10}	l_{11}	l_{12}
Capacity (kW)	100	200	180	100	100	100

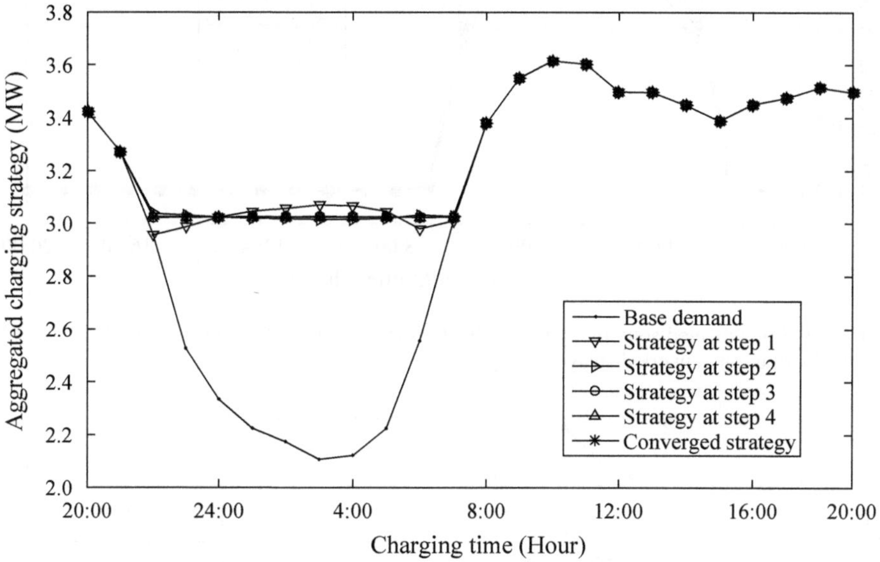

Fig. 7.12 Update procedure by applying the gradient projection method presented in (7.7) for the 12-feeder distribution network displayed in Fig. 7.11

Figures 7.12, 7.13 and 7.14 display the updates of aggregated charging behaviors by adopting the decentralized gradient projection method presented in (7.7).

As illustrated, the system converges to the optimal valley-filling strategy within a few of iteration steps. However as illustrated in the bottom sub-figure, the implemented behavior violates the capacity constraints of line l_4 and line l_8 over the period of 01:00–05:00 on the next day.

Figure 7.15 displays the updates of aggregated charging behaviors by adopting Algorithm 7.1. As illustrated in the top sub-figure, at several iteration steps, some feeder lines are occupied at some instants respectively.

For example, at iteration step 2, feeder line l_4 and line l_8 are fully occupied at instant 4:00 and 5:00 respectively. Nevertheless, the implemented aggregated behavior is same as that displayed in Fig. 7.12. For the purpose of demonstration, in Fig. 7.15, it only displays the aggregated demand curve at the first 7 iteration steps and the converged coordination behavior, and skips the other updated curves.

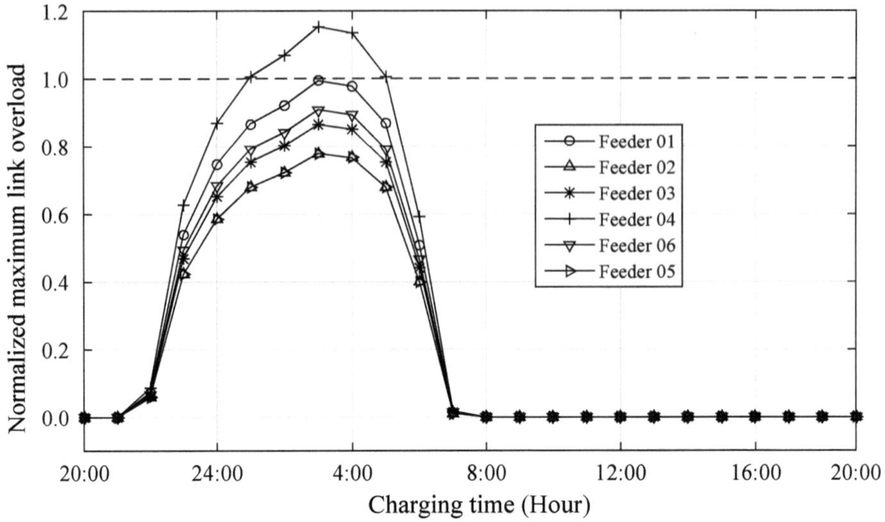

Fig. 7.13 Implemented solution by the gradient projection method presented in (7.7) for the 12-feeder distribution network displayed in Fig. 7.11

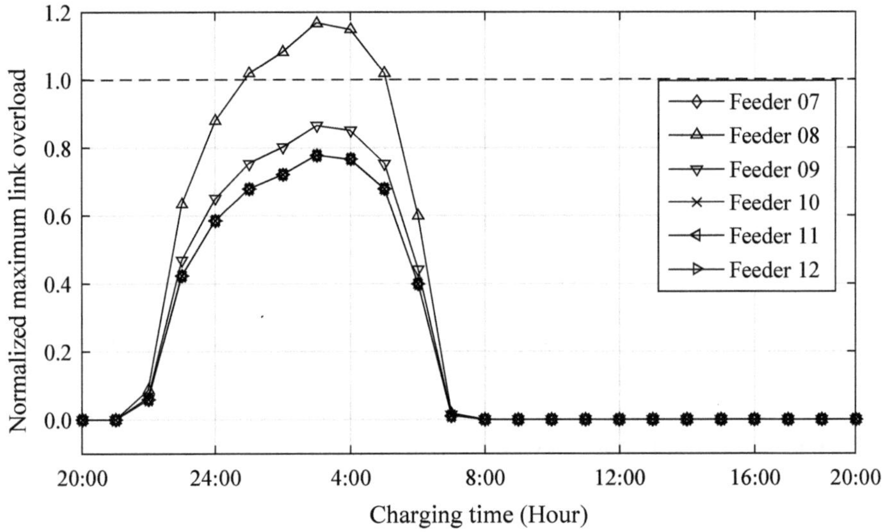

Fig. 7.14 Implemented solution by the gradient projection method presented in (7.7) for the 12-feeder distribution network displayed in Fig. 7.11

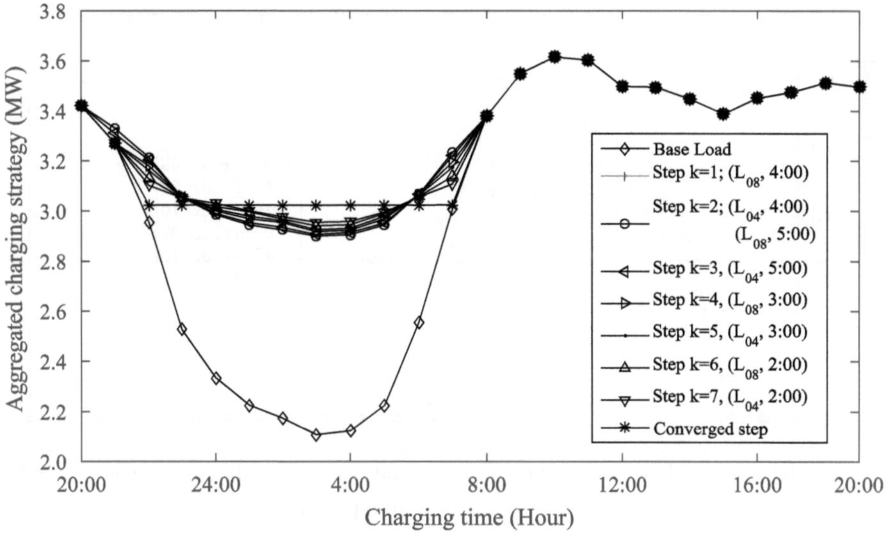

Fig. 7.15 Implemented solution by Algorithm 7.1 for 12-feeder distribution networks displayed in Fig. 7.11

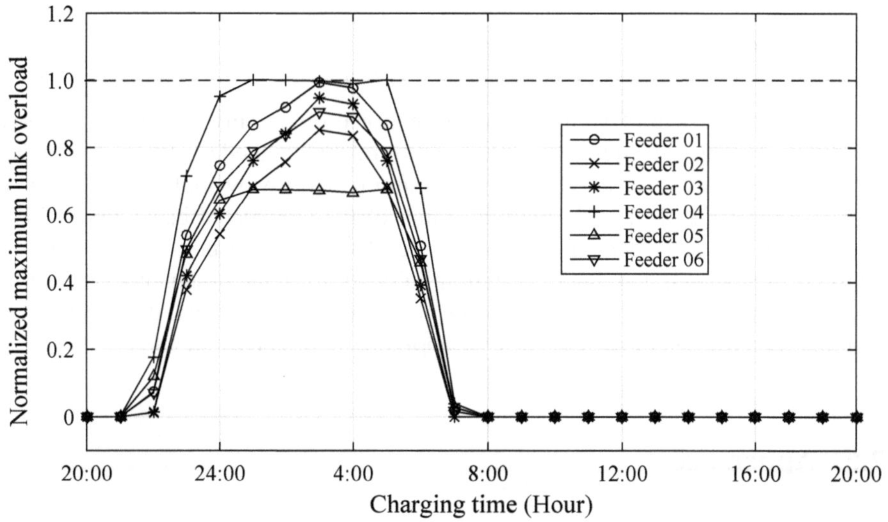

Fig. 7.16 Implemented feasible solution by Algorithm 7.1 at Feeders 1-6 for 12-feeder distribution networks displayed in Fig. 7.11

Also as illustrated in Figs. 7.16 and 7.17, the charging behaviors satisfy the capacity constraints of all the feeder lines.

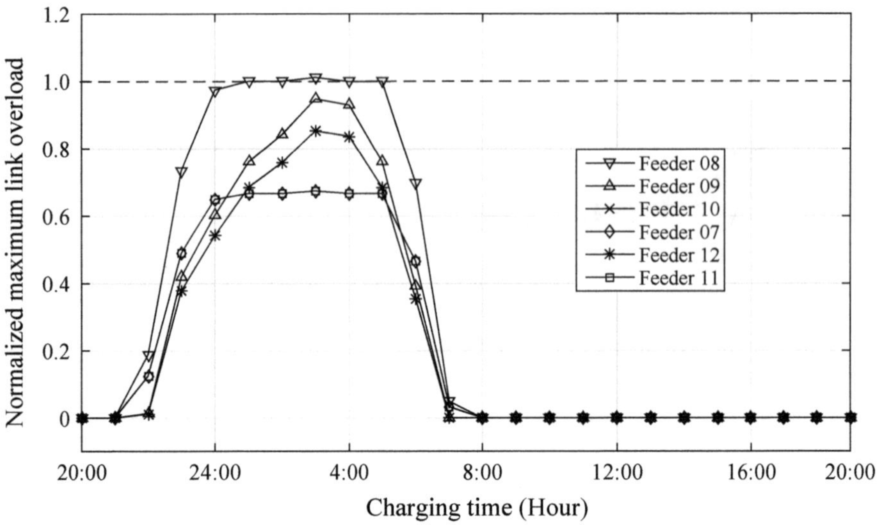

Fig. 7.17 Implemented feasible solution by Algorithm 7.1 at Feeders 7-12 for 12-feeder distribution networks displayed in Fig. 7.11

7.5 Conclusions

It formulates a class of PEV charging coordination problems in distribution networks with capacity constrained feeder lines. A decentralized method is proposed such that the strategies of all of the PEVs are updated simultaneously by applying the gradient projection method. To avoid the overloading of the aggregated behaviors of PEV populations on the feeder lines, the step size of the gradient projection method is adjusted. It is shown that, under certain mild conditions, the system converges to a coordination behavior which satisfies the capacity constraints, and the system cost is minimized over a converged space by applying the proposed method.

7.6 Appendices

7.6.1 Proof of Theorem 7.1

Suppose, at iteration step k, there exists some (l, t) such that $U_{lt}\Big(\widehat{\boldsymbol{u}}(\boldsymbol{u}^k; \mathscr{V}^k, \frac{\alpha}{2^j})\Big) > \Pi_l + \varepsilon_1$, i.e., the capacity constraint of feeder line l is violated at instant t; then by Algorithm 7.1, (7.8) is implemented repeatedly with the step size $\frac{\alpha}{2^j}$, until $U_{lt}\Big(\widehat{\boldsymbol{u}}(\boldsymbol{u}^k; \mathscr{V}^k, \frac{\alpha}{2^j})\Big) \leq \Pi_l + \varepsilon_1$, for all (l, t).

$$\widehat{\boldsymbol{u}}_n\left(\boldsymbol{u}^k; \mathscr{V}^k, \frac{\alpha}{2^j}\right) = \left[\boldsymbol{u}_n^k - \frac{\alpha}{2^j}\nabla_{\boldsymbol{u}_n} J(\boldsymbol{u}^k)\right]^+_{\mathscr{V}_n^k}, \quad \forall n \in \mathscr{N}. \tag{7.12}$$

Due to $\boldsymbol{u}_n^k \in \mathscr{V}_n^k$ and (7.12), it can obtain that there exists some $\widehat{j}$, such that $U_{lt}\left(\widehat{\boldsymbol{u}}(\boldsymbol{u}^k; \mathscr{V}^k, \frac{\alpha}{2^{\widehat{j}}})\right) \leq \Pi_l + \varepsilon_1$ holds for all $(l, t) \in \mathscr{S}^k$; then by Algorithm 7.1,

$$U_{lt}(\boldsymbol{u}^{k+1}) \leq \Pi_l + \varepsilon, \quad \text{with } \boldsymbol{u}_n^{k+1} = \widehat{\boldsymbol{u}}_n\left(\boldsymbol{u}^k, \mathscr{V}^k; \frac{\alpha}{2^{\widehat{j}}}\right), \tag{7.13}$$

for all $n \in \mathscr{N}$. Also by (7.9)–(7.11),

$$\begin{aligned}
\mathscr{S}^{k+1} &= \mathscr{S}^k / \mathscr{W}, \\
\mathscr{V}^{k+1} &= \widehat{\mathscr{V}}(\mathscr{V}^k, \mathscr{S}^k, \boldsymbol{u}^{k+1}),
\end{aligned}$$

with $\mathscr{W} = \left\{(l, t) \in \mathscr{S}^k, \text{ s.t. } |U_{lt}(\boldsymbol{u}^{k+1}) - \Pi_l| \leq \varepsilon\right\}$.

Note: By the above analyses together with $\mathscr{S}^0 = \mathscr{L} \times \mathscr{T}$ and $\mathscr{V}^0 = \mathscr{U}$, it obtains that $\mathscr{S}^k$ is a subset of $\mathscr{L} \times \mathscr{T}$, such that $U_{lt}(\boldsymbol{u}) < \Pi_l - \varepsilon$, for all $(l, t) \in \mathscr{S}^k$, say feeder line l is not fully occupied at instant t after the first k iteration steps.

Suppose that the sequences of $\{\mathscr{S}^k\}$ and $\{\mathscr{V}^k\}$ converge to $\mathscr{S}^\dagger$ and $\mathscr{V}^\dagger$ respectively after K iteration steps.

In the following it will show that, after $\widehat{K}$ iteration steps, with $\widehat{K} > K$, the system converges to an equilibrium solution, denoted by $\boldsymbol{u}^\dagger$, which minimizes the system cost $J(\boldsymbol{u})$ over the strategy space $\mathscr{V}^\dagger$.

For each strategy $\boldsymbol{u} \in \mathscr{V}^\dagger$, it defines $\nabla_{u_{nt}} J(\boldsymbol{u}) \triangleq \dfrac{\partial}{\partial u_{nt}} J(\boldsymbol{u})$; then by the system cost function defined in (7.4), $\nabla_{u_{nt}} J(\boldsymbol{u}) = c'(d_t + U_n(\boldsymbol{u}))$.

Consider a pair of distinct charging behaviors $\boldsymbol{u}, \boldsymbol{v} \in \mathscr{V}^\dagger$, and consider $U_t \equiv U_t(\boldsymbol{u})$ and $V_t \equiv U_t(\boldsymbol{v})$; then

$$\begin{aligned}
&||\nabla J(\boldsymbol{u}) - \nabla J(\boldsymbol{v})||_2^2 \\
&= \sum_{t \in \mathscr{T}} \sum_{n \in \mathscr{N}} (c'(d_t + U_t) - c'(d_t + V_t))^2
\end{aligned} \tag{7.14}$$

As the first and second derivatives of $c(x)$ are continuous on $\mathbb{R}$ under Assumption 3.2; then by the mean value theorem, there exists some D_t between $d_t + U_t$ and $d_t + V_t$, such that

$$c''(D_t) = \frac{c'(d_t + U_t) - c'(d_t + V_t)}{U_t - V_t}$$

by which together with (7.14), it can obtain that

$$\begin{aligned}||\nabla J(\boldsymbol{u}) - \nabla J(\boldsymbol{v})||_2^2 =& N \sum_{t \in \mathscr{T}} [c''(D_t)]^2 (U_t - V_t)^2 \\ \le& N\rho^2 \sum_{t \in \mathscr{T}} (U_t - V_t)^2\end{aligned}$$

where the inequality holds by $c''(D) \le \rho$ under Assumption 3.2. Hence

$$||\nabla J(\boldsymbol{u}) - \nabla J(\boldsymbol{v})||_2^2 \le N\rho^2 \sum_{t \in \mathscr{T}} (U_t - V_t)^2 \le N^2\rho^2 ||\boldsymbol{u} - \boldsymbol{v}||_2^2$$

where the last inequality holds by $(U_t - V_t)^2 \le N \sum_{n \in \mathscr{N}} (u_{nt} - v_{nt})^2$ due to the Jensen's inequality.

It implies that

$$||\nabla J(\boldsymbol{u}) - \nabla J(\boldsymbol{v})||_2 \le N\rho ||\boldsymbol{u} - \boldsymbol{v}||_2; \tag{7.15}$$

then, following the analysis in Proposition 2.3.2 in [18], $J(\cdot)$ monotonically decreases on $\boldsymbol{u}^k$ by applying the update of behaviors in a decentralized gradient projection method presented in (7.7) in case $0 < \alpha < \dfrac{2}{N\rho}$, such that

$$J(\boldsymbol{u}^{k+1}) - J(\boldsymbol{u}^k) \le \left(\frac{1}{2N\rho} - \frac{1}{\alpha}\right) ||\boldsymbol{u}^{k+1} - \boldsymbol{u}^k||_2^2. \tag{7.16}$$

Hence by (7.16), the sequence $\{J(\boldsymbol{u}^k); k = 1, 2, \ldots\}$ converges to a limit value, denoted by $J^\dagger$, since $J(\boldsymbol{u})$ is bounded from the below on $\mathscr{V}^\dagger$. It implies that the system converges to a stationary solution after certain K iteration steps, denoted by $\boldsymbol{u}^\dagger$, say

$$[\boldsymbol{u}_n^\dagger - \alpha \nabla_{\boldsymbol{u}_n} J(\boldsymbol{u}^\dagger)]^+_{\mathscr{V}_n^\dagger} = \boldsymbol{u}_n^\dagger, \quad \forall n \in \mathscr{N}.$$

By the *projection theorem* stated in Proposition 2.1.3.b, [18],

$$\left((\boldsymbol{u}^\dagger - \alpha \nabla_{\boldsymbol{u}} J(\boldsymbol{u}^\dagger)) - \boldsymbol{u}^\dagger\right)^{\mathrm{T}} \left(\boldsymbol{u} - \boldsymbol{u}^\dagger\right) \le 0, \quad \forall \boldsymbol{u} \in \mathscr{V}^\dagger,$$

where $\boldsymbol{x}^{\mathrm{T}}$ represents the transpose of vector $\boldsymbol{x}$, which implies

$$\nabla J(\boldsymbol{u}^\dagger)^{\mathrm{T}} (\boldsymbol{u} - \boldsymbol{u}^\dagger) \ge 0, \quad \forall \boldsymbol{u} \in \mathscr{U}. \tag{7.17}$$

Under Assumption 3.2, $c(\cdot)$ is convex, so $J(\boldsymbol{u})$ is convex over $\mathscr{V}^\dagger$. Hence by Proposition 2.1.2, [18], the stationary solution $\boldsymbol{u}^\dagger$ is globally optimal over $\mathscr{V}^\dagger$. ■

References

1. L. Kelly, A. Rowe, P. Wild, Analyzing the impacts of plug-in electric vehicles on distribution networks in British Columbia, in *IEEE Electrical Power and Energy Conference*, Montreal, Canada (2009)
2. L. Fernández, T. Román, R. Cossent, C. Domingo, P. Frías, Assessment of the impact of plug-in electric vehicles on distribution networks. IEEE Trans. Power Syst. **26**(1), 206–213 (2011)
3. K. Clement-Nyns, E. Haesen, J. Driesen, The impact of charging plug-in hybrid electric vehicles on a residential distribution grid. IEEE Trans. Power Syst. **25**(1), 371–380 (2010)
4. M.D. Galus, G. Andersson, Demand management of grid connected plug-in hybrid electric vehicles (PHEV), in *IEEE Energy 2030*, Atlanta, Georgia, pp. 1–8. Accessed 17–18 Nov 2008
5. R. Hermans, M. Almassalkhi, I.A. Hiskens, Incentive-based coordinated charging control of plug-in electric vehicles at the distribution-transformer level, in *American Control Conference (ACC)*, Montreal, Canada (2012), pp. 264–269
6. P. Richardson, D. Flynn, A. Keane, Impact assessment of varying penetrations of electric vehicles on low voltage distribution systems, in *IEEE Power and Energy Society General Meeting* (2010), pp. 1–6
7. K. Zhan, Z. Hu, Y. Song, Z. Luo, Z. Xu, L. Jia, Coordinated electric vehicle charging strategy for optimal operation of distribution network, in *3rd IEEE PES International Conference and Exhibition on Innovative Smart Grid Technologies (ISGT Europe)* (2012), pp. 1–6
8. A.S. Masoum, S. Deilami, P.S. Moses, M.A. Masoum, A. Abu-Siada, Smart load management of plug-in electric vehicles in distribution and residential networks with charging stations for peak shaving and loss minimization considering voltage regulation. IET Gener. Transm. Distrib. **5**(8), 877–888 (2011)
9. Mushfiqur R. Sarker, Miguel Ortega-Vazquez, Daniel S. Kirschen, Optimal coordination and scheduling of demand response via monetary incentives. IEEE Trans. Smart Grid **6**(3), 1341–1352 (2015)
10. B. Yang, J. Li, Q. Han, T. He, Distributed control for charging multiple electric vehicles with overload limitation. IEEE Trans. Parallel Distrib. Syst. **27**(12), 3441–3454 (2016)
11. Abouzar Ghavami, Koushik Kar, Aparna Gupta, Decentralized charging of plug-in electric vehicles with distribution feeder overload control. IEEE Trans. Autom. Control **61**(11), 3527–3532 (2016)
12. O. Ardakanian, C. Rosenberg, S. Keshav, Distributed control of electric vehicle charging, in *International Conference on Future Energy Systems* (2013), pp. 101–112
13. D. Wang, X. Guan, J. Wu, P. Li, P. Zan, H. Xu, Integrated energy exchange scheduling for multimicrogrid system with electric vehicles. IEEE Trans. Smart Grid **7**(4), 1762–1774 (2016)
14. Z. Ma, D. Callaway, I. Hiskens, Decentralized charging control of large populations of plug-in electric vehicles. IEEE Trans. Control Syst. Technol. **21**(1), 67–78 (2013)
15. Lingwen Gan, Ufuk Topcu, Steven H. Low, Optimal decentralized protocol for electric vehicle charging. IEEE Trans. Power Syst. **28**(2), 940–951 (2013)
16. Lu Xia, I. Mareels, T. Alpcan, M. Brazil, A distributed electric vehicle charging management algorithm using only local measurements, in *IEEE Power and Energy Society Innovative Smart Grid Technologies Conference* (2014), pp. 1–5
17. Zhongjing Ma, Suli Zou, Xiangdong Liu, A distributed charging coordination for large-scale plug-in electric vehicles considering battery degradation cost. IEEE Trans. Control Syst. Technol. **23**(5), 2044–2052 (2015)
18. D.P. Bertsekas, *Nonlinear Programming*, 2nd edn. (Athena Scientific, Belmont, 1999)
19. Z. Luo, Z. Hu, Y. Song, Z. Xu, H. Lu, Optimal coordination of plug-in electric vehicles in power grids with cost-benefit analysis - part I: enabling techniques. IEEE Trans.Power Syst. **28**(4), 3546–3555 (2013)
20. Z. Luo, Z. Hu, Y. Song, Z. Xu, H. Lu, Optimal coordination of plug-in electric vehicles in power grids with cost-benefit analysis - part II: a case study in china. IEEE Trans. Power Syst. **28**(4), 3556–3565 (2013)

Chapter 8
Efficient Game for Vehicle-to-Grid Coordination Problems

Abstract PEVs, as distributed energy sources, are promising to provide vehicle-to-grid (V2G) services for power grid, like frequency and voltage regulations, by coordinating their active and reactive power rates. However due to the autonomy of PEVs, it is challenging how to efficiently schedule the coordination behaviors among these units in a decentralized way. In this chapter it formulates the underlying coordination problems as a novel class of VCG-style auction games where players, power grid and PEVs, do not report a full cost or valuation function but only a multi-dimensional bid signal: the maximum active and reactive power quantities that power grid wants and the maximum per-unit prices it is willing to pay, the maximum active and reactive power quantities that a PEV can provide and the minimum per-unit prices it asks. We show the existence of the efficient Nash equilibrium for the underlying auction games, though there may exist other inefficient Nash equilibria. In order to deal with large-scale PEVs, in this chapter, it designs games with aggregator players each of which submits bid profiles representing the overall utility for a collection of PEVs, and extends the so-called quantized-PSP mechanism to the underlying auction games to implement the efficient Nash equilibrium.

8.1 Introduction

To operate safely, power grid has to be ensured with the amplitude of frequency and voltage close to their nominal values by balancing the production and consumption of both active and reactive power, see [1], (balance supply and demand of active power for frequency regulation and balance supply and demand of reactive power for voltage regulation). PEVs can play as distributed electricity storages for power grid, see [2]. Since PEVs can deliver electricity to power grid and consume electricity from power grid. Nowadays quite a few researches have been dedicated to studying how to apply vehicle-to-grid (V2G) structure for frequency regulations in smart grid, e.g. [3–7]. In practice, most of reactive power loads are inductive, so power compensation for voltage regulation mainly involves adjusting capacitors, e.g. [8–11]. More recently [12] proposed a V2G regulation model such that PEVs are applied to provide ancillary services of frequency and voltage regulations for power grid by coordinating their

Z. Ma, *Decentralized Charging Coordination of Large-scale Plug-in Electric Vehicles in Power Systems*, https://doi.org/10.1007/978-981-13-7652-8_8

active and reactive power rates, and further formulated the underlying regulation problems as two joint optimization problems under different pricing and contract scenarios.

In this chapter it proposes an auction-based decentralized method for V2G coordination problems initialized by [12] under the progressive second price (PSP) auction mechanism which was firstly proposed by [13, 14] to efficiently allocate a single network divisible resource.

Under the PSP auction mechanism, the incentive compatibility holds, i.e. the truth-telling bid profile is the best response of agent. Hence the PSP auction can be considered as the extension of the second price auction proposed in [15–17] to allocate indivisible items with truth-telling bid profiles in an efficient way. It is worth to note that the incentive compatibility does not hold in general for other sealed auction mechanisms, like uniform market clearing price and pay as bid auction mechanisms etc., see [18, 19], which have been widely adapted to economically dispatch generations in day-ahead deregulated electricity markets, see [20]. In the allocation problems of a single divisible resource under the PSP auction mechanism, each agent only reports a two-dimension bid profile which is composed of a maximum amount of demand and an associated buying price, and is used to replace his complete (private) utility function. The transfer money (or payment) of an agent is that, the utility of the whole system, when this agent does not join the auction process, minus the overall utility of the remaining system except this agent when he joins the auction process, that is to say, the payment of each agent is exactly the externality he imposes on the system through his participation. In [14], the authors verified that the efficient incentive compatible bid profile is a Nash equilibrium for PSP network auction games. More recently many progresses have been developed in PSP auction games. For example, [21] extended the results from single-side auction games to double-side ones in the network resource allocations; [22–24] proposed a decentralized algorithm, under quantized-PSP auction mechanism for power electricity sharing games, following which the auction game system converges to the efficient solution.

In the V2G regulation problems, suppose that each of autonomous PEVs deals with the tradeoff between costs and benefits with respect to supplied active and reactive power service rates; while power grid deals with the tradeoff between the benefits with the aggregated service rates provided by the PEV populations and the costs he needs to pay for the services. The players participating in the underlying auction games, including power grid as a single buyer and autonomous PEVs as sellers, do not report a full cost or valuation function but only a multi-dimensional bid signal: the maximum active and reactive power quantities that power grid wants and the maximum per-unit prices he is willing to pay, the maximum active and reactive power quantities that an autonomous PEV can provide and the minimum per-unit prices he asks.

In this chapter the existence of the efficient Nash equilibrium for the V2G auction game is shown, though there may exist other inefficient Nash equilibria. Moreover in order to deal with the V2G coordination problems with large-scale PEVs, it designs a class of V2G coordination auction games with aggregator players each of which represents a collection of individual PEVs. To implement the efficient Nash equilibrium, in this chapter, it adapts and extends the PSP auction mechanism, which is proposed in [22] for double-side auction games, to the underlying V2G auction games with constrained multi-dimension bid strategies. As illustrated with the numerical examples, the game system converges to some Nash equilibrium which is near to the efficient Nash equilibrium.

The organization of this chapter is as follows: In Sect. 8.2, it formulates a class of vehicle-to-grid coordination problems for frequency and voltage regulations. In Sect. 8.3, it proposes a decentralized V2G coordination method under the PSP auction mechanism and shows the efficiency property of the underlying auction game. In Sect. 8.4, it designes V2G coordination auction games with aggregators each of which represents a collection of PEVs. The so-called Q-PSP auction mechanism is extended to the underlying V2G auction games, in Sect. 8.5, and is illustrated with numerical examples. Section 8.6 briefly gives the conclusions of this part.

8.2 Formulation of Vehicle-to-Grid Coordination Problems

In this section, it formulates a class of frequency and voltage regulations for power grid via charging coordination of PEVs, which is initialized in [12], such that an individual PEV can provide frequency regulation service by coordinating its charging/discharging active power and voltage regulation service by properly setting the phase angle difference between its charging current and power grid voltage.

Denote by $\Delta \mathrm{f}$ the deviation of frequency from the nominal value in power grid and by $P(\Delta \mathrm{f})$ the resulting imbalance between supply and demand with respect to $\Delta \mathrm{f}$ respectively; and denote by $\Delta \mathrm{v}$ the deviation of voltage from the nominal value in power grid and by $Q(\Delta \mathrm{v})$ the resulting imbalance between supply and demand with respect to $\Delta \mathrm{v}$ respectively. More specifically, in this chapter it considers that

- $P(\Delta \mathrm{f}) > 0$ ($P(\Delta \mathrm{f}) < 0$ respectively) indicates that the supply is higher (lower respectively) than the demand with $P(\Delta \mathrm{f})$ in power grid;
- $Q(\Delta \mathrm{v}) > 0$ indicates that power grid requires capacitive reactive power compensation;
- $Q(\Delta \mathrm{v}) < 0$ indicates that power grid requires inductive reactive power compensation.

In the work, it is supposed that $(P(\Delta \mathrm{f}), Q(\Delta \mathrm{v}))$ remains fixed over the next short-time interval $\mathscr{T}$. Denote by $\mathscr{N}$ the collection of electric vehicles which plug in power grid and are willing to supply the V2G service. To simply the analysis, it further assumes that, as along as the vehicles agree to supply the service for the interval $\mathscr{T}$, they will keep plugged in power grid over that interval and provide the service as they agreed.

8.2.1 Coordination Capacity of Individual PEVs

8.2.1.1 Apparent Power Constraint

Consider a PEV charger with a power electronics AC/DC inverter which can adjust active and reactive power by coordinating I_n the magnitude of current of PEV n and θ_n the phase difference between the voltage of power grid and the current of PEV n. Let u_{nP} and u_{nQ} denote the active power and reactive power of PEV n; then

$$u_{nP} = V_s \times I_n cos\theta_n, \tag{8.1a}$$

$$u_{nQ} = V_s \times I_n sin\theta_n, \tag{8.1b}$$

where V_s represents the magnitude of voltage in power grid. It considers that

- $u_{nP} > 0$ and $u_{nP} < 0$ represent that PEV n charges and discharges respectively;
- $u_{nQ} > 0$ and $u_{nQ} < 0$ represent that PEV n operates like a capacitive load and an inductive load respectively.

It considers the following nominal apparent power constraint for PEV n as below:

$$u_{nP}^2 + u_{nQ}^2 \le S_n^2, \tag{8.2}$$

where S_n denotes the nominal apparent power of the charger where PEV n plugs, as illustrated in Fig. 8.1, see [12].

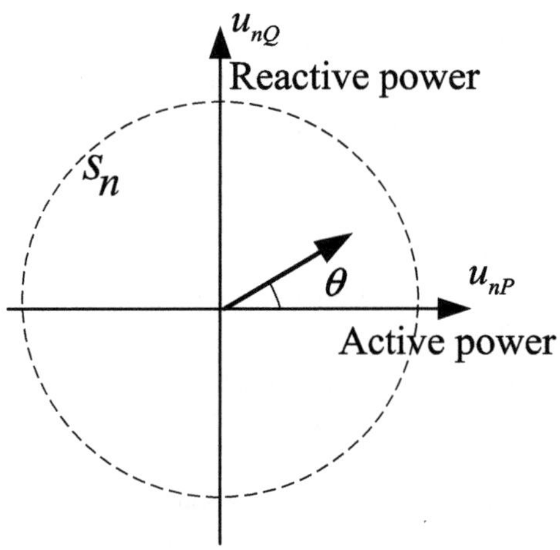

Fig. 8.1 Apparent power constraint for individual PEV n

For analytical simplicity, in this chapter it only studies the V2G coordination in case $P(\Delta\mathrm{f}), Q(\Delta\mathrm{v}) > 0$, i.e. the supply is higher than the demand and power grid needs capacitive reactive power compensation; then it considers

$$u_{nP}, u_{nQ} \geq 0, \tag{8.3}$$

for all $n \in \mathcal{N}$, i.e. each PEV charges and operates like a capacitive load.

8.2.1.2 Active Power Constraint

It considers a maximum charging rate constraint such that

$$u_{nP} \leq u_{nP}^{\max} \triangleq \min\left\{\xi_n, \frac{\Phi_n}{\Delta T}(soc_n^{max} - soc_{n0})\right\}, \tag{8.4}$$

where ξ_n denotes a safe maximum charging rate, soc_n^{min} and soc_n^{max} represent the minimum and maximum state of charge (SoC) of PEV n respectively, soc_{n0} represents the initial SoC value of PEV n, and Φ_n denotes the battery energy capacity of PEV n, and ΔT denotes the length of service interval $\mathcal{T}$.

Note: The specification in (8.4) can guarantee that the SoC of PEV n is in the range of $[soc_n^{min}, soc_n^{max}]$.

In summary, it calls $\boldsymbol{u}_n \equiv (u_{nP}, u_{nQ})$ an admissible active and reactive power coordination of PEV n over the interval $\mathcal{T}$, if $\boldsymbol{u}_n$ satisfies the constraints (8.2)–(8.4).

Denote a local cost function of PEV n by $g_n(\boldsymbol{u}_n; soc_{n0})$, subject to an admissible coordination strategy $\boldsymbol{u}_n \equiv (u_{nP}, u_{nQ})$ as follows:

$$g_n(\boldsymbol{u}_n; soc_{n0}) = g_{nP}(u_{nP}; soc_{n0}) + g_{nQ}(u_{nQ}; soc_{n0}), \tag{8.5}$$

where $g_{nP}(u_{nP}; soc_{n0})$ and $g_{nQ}(u_{nQ}; soc_{n0})$ representing the battery degradation cost subject to charging rate u_{nP}, u_{nQ}, measures the cost related to the decrease of battery energy capacity, due to the battery resistance growth subject to the charging rate, which is analyzed in [25].

For notational simplicity, it may consider $g_n(\boldsymbol{u}_n; soc_{n0})$, $g_{nP}(u_{nP}; soc_{n0})$ and $g_{nQ}(u_{nQ}; soc_{n0})$ as $g_n(\boldsymbol{u}_n)$, $g_{nP}(u_{nP})$ and $g_{nQ}(u_{nQ})$ respectively.

8.2.2 Efficient V2G Behavior for Frequency and Voltage Regulations

It defines a valuation function of power grid, subject to the aggregated V2G coordination strategies $\boldsymbol{u}_s$ with $\boldsymbol{u}_s \equiv (u_{sP}, u_{sQ})$, denoted by $v_s(\boldsymbol{u}_s; P, Q)$, with $P \equiv P(\Delta\mathrm{f})$, $Q \equiv Q(\Delta\mathrm{v})$ for notational simplicity, such that

$$v_s(\boldsymbol{u}_s; P, Q) \triangleq v_{sP}(u_{sP}; P) + v_{sQ}(u_{sQ}; Q). \tag{8.6}$$

As an example, it may consider a quadratic regulation form for valuation function of $v_s(\boldsymbol{u}_s; P, Q)$, such that $v_s(\boldsymbol{u}_s; P, Q) = -\sigma_{sP}(u_{sP} - P)^2 - \sigma_{sQ}(u_{sQ} - Q)^2$, with $\sigma_{sP}, \sigma_{sQ} > 0$.

For notational simplicity, it may consider $v_s(\boldsymbol{u}_s) \equiv v_s(\boldsymbol{u}_s; P, Q)$, $v_{sP}(u_{sP}) \equiv v_{sP}(u_{sP}; P)$ and $v_{sQ}(u_{sQ}) \equiv v_{sQ}(u_{sQ}; Q)$.

Let $\mathscr{U}$ denote a collection of power grid strategies and admissible PEV coordination strategies, denoted by $\boldsymbol{u} \equiv (\boldsymbol{u}_s, \boldsymbol{u}_n; n \in \mathscr{N})$, if $\boldsymbol{u}_n$ satisfies (8.2)–(8.4).

By (8.5) and (8.6), we define the system valuation function, denoted by $J(\boldsymbol{u})$, subject to an admissible coordination strategy $\boldsymbol{u} \in \mathscr{U}$, such that

$$J(\boldsymbol{u}) \triangleq v_s(\boldsymbol{u}_s) - \sum_{n=1}^{N} g_n(\boldsymbol{u}_n), \tag{8.7}$$

such that

$$\boldsymbol{u}_s \equiv (u_{sP}, u_{sQ}) = \left(\sum_{n=1}^{N} u_{nP}, \sum_{n=1}^{N} u_{nQ}\right). \tag{8.8}$$

The objective of the optimal operation of V2G coordination problems is to assign efficient (or socially optimal) allocation, to all PEVs, to maximize the system valuation function (8.7) over the set of admissible strategies $\mathscr{U}$. Let $\boldsymbol{u}^{**}$ denote the efficient allocation, i.e.

$$\boldsymbol{u}^{**} = \underset{\boldsymbol{u} \in \mathscr{U}}{\operatorname{argmax}}\, J(\boldsymbol{u}), \text{ subject to (8.8).} \tag{8.9}$$

The Lagrange function for the optimization problem (8.9) subjected to (8.2)–(8.4) and the equality constraint (8.8) is given below:

$$\begin{aligned} L(\boldsymbol{u}_s, \boldsymbol{u}_n, \boldsymbol{\lambda}, \boldsymbol{\mu}) = {} & v_s(\boldsymbol{u}_s) - \sum_{n=1}^{N} \left\{g_{nP}(u_{nP}) + g_{nQ}(u_{nQ})\right\} \\ & + \lambda_{nP}(u_{nP}^{max} - u_{nP}) + \lambda_{nS}\left(S_n^2 - u_{nP}^2 - u_{nQ}^2\right) \\ & + \mu_P\left(\sum_{n=1}^{N} u_{nP} - u_{sP}\right) + \mu_Q\left(\sum_{n=1}^{N} u_{nQ} - u_{sQ}\right) \end{aligned}$$

with $\boldsymbol{u}_s \equiv (u_{sP}, u_{sQ})$, $\boldsymbol{u}_n \equiv (u_{nP}, u_{nQ}; n \in \mathscr{N})$, $\boldsymbol{\lambda} \equiv (\lambda_{nP}, \lambda_{nS}; n \in \mathscr{N})$ and $\boldsymbol{\mu} \equiv (\mu_P, \mu_Q)$, where λ_{nP} is the Lagrange multiplier corresponding to the active power constraint of PEV n, λ_{nS} is the Lagrange multiplier corresponding to the apparent power constraint, μ_P and μ_Q are the Lagrange multipliers corresponding to the equality constraint related to active and reactive power defined in (8.8) respectively.

The following, together with the equality constraint (8.8), are the KKT conditions for the constrained optimization problems given in (8.9).

$$v'_{sP}(u^{**}_{sP}) - \mu^*_P \leq 0, \qquad u^{**}_{sP} \geq 0, \qquad u^{**}_{sP}(v'_{sP}(u^{**}_{sP}) - \mu^*_P) = 0 \tag{8.10a}$$

$$v'_{sQ}(u^{**}_{sQ}) - \mu^*_Q \leq 0, \qquad u^{**}_{sQ} \geq 0, \qquad u^{**}_{sQ}(v'_{sQ}(u^{**}_{sQ}) - \mu^*_Q) = 0 \tag{8.10b}$$

$$\frac{\partial}{\partial u_{nP}} L(\boldsymbol{u}_s, \boldsymbol{u}_n, \boldsymbol{\lambda}, \boldsymbol{\mu}) \leq 0, \qquad u^{**}_{nP} \geq 0, \qquad u^{**}_{nP} \frac{\partial}{\partial u_{nP}} L(\boldsymbol{u}_s, \boldsymbol{u}_n, \boldsymbol{\lambda}, \boldsymbol{\mu}) = 0 \tag{8.10c}$$

$$- g'_{nQ}(u^{**}_{nQ}) - 2\lambda^*_{nS} u^{**}_{nQ} + \mu^*_Q \leq 0, \qquad u^{**}_{nQ} \geq 0, \qquad u^{**}_{nQ}(-g'_{nQ}(u^{**}_{nQ}) - 2\lambda^*_{nS} u^{**}_{nQ} + \mu^*_Q) = 0 \tag{8.10d}$$

$$u^{**2}_{nP} + u^{**2}_{nQ} \leq S^2_n, \qquad \lambda^*_{nS} \geq 0, \qquad \lambda^*_{nS}(S^2_n - u^{**2}_{nP} - u^{**2}_{nQ}) = 0 \tag{8.10e}$$

$$u^{**}_{nP} \leq u^{max}_{nP}, \qquad \lambda^*_{nP} \geq 0, \qquad \lambda^*_{nP}(u^{max}_{nP} - u^{**}_{nP}) = 0 \tag{8.10f}$$

with $\frac{\partial}{\partial u_{nP}} L(\boldsymbol{u}_s, \boldsymbol{u}_n, \boldsymbol{\lambda}, \boldsymbol{\mu}) = -g'_{nP}(u^{**}_{nP}) - 2\lambda^*_{nS} u^{**}_{nP} - \lambda^*_{nP} + \mu^*_P$.

It considers the following assumptions:

Assumption 8.1

- $v_{sP}(u_{sP})$ and $v_{sQ}(u_{sQ})$ are strictly increasing and concave on u_{sP}, u_{sQ} respectively;
- $g_{nP}(u_{nP})$ and $g_{nQ}(u_{nQ})$, with $n \in \mathcal{N}$, are strictly increasing and convex on u_{nP}, u_{nQ} respectively. ■

Under Assumption 8.1, the underlying regulation problems are convex optimization problems, hence the necessary KKT conditions are also sufficient conditions for the optimality of the optimization problems, see [26]. Thus under Assumption 8.1, there exists a unique solution and it is characterized by the KKT conditions specified in (8.10) together with the equality constraint (8.8).

8.2.3 A Simulation for Frequency and Voltage Regulations

It studies the V2G coordination problems with $N = 1000$, and each of which possesses a battery with capacity of $\Phi_n = 10\,\text{kWh}$. More specifically it is considered that the referred fixed active power $P = 5\,\text{MW}$ and reactive power $Q = 2\,\text{MVar}$ required for power grid.

As studied in the literature, e.g. [27], the distribution of typical travel miles of vehicles approximately satisfies a Gaussian distribution; then the distribution of initial SoC values of PEV batteries, denoted by $\{soc_{n0};\, n \in \mathcal{N}\}$, may approximately satisfy a Gaussian distribution, denoted by $N(\hat{\mu}, \hat{\gamma})$, as well, see [28, 29].

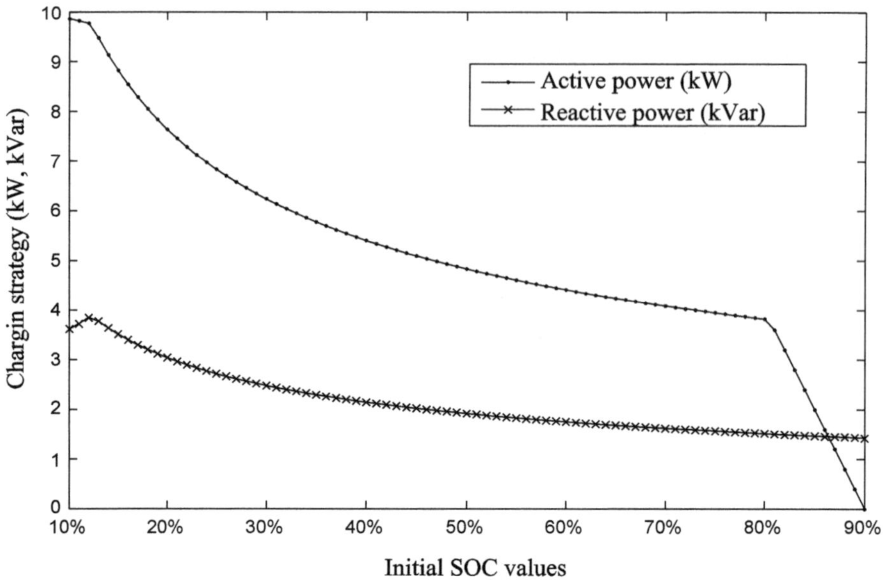

Fig. 8.2 The efficient allocations to individual PEVs with respect to initial SoC values

For purpose of demonstration, in the following simulations, it considers a Gaussian distribution with parameters of $\hat{\mu} = 0.5$ and $\hat{\gamma} = 0.15$, Also it is supposed that $soc_n^{min} = 10\%$ and $soc_n^{max} = 90\%$.

It considers $\xi_n = 10\,\text{kW}$, $S_n = 10.5\,\text{kVA}$ and $\Delta T = 15$ minutes; then based upon the given specifications, The constraints of (8.2)–(8.4) can be specified for each individual PEV. It considers the cost functions of PEV n as $g_{nP}(u_{nP}; soc_{n0}) = 0.1\sqrt{soc_{n0}}u_{nP}^2$ and $g_{nQ}(u_{nQ}; soc_{n0}) = 0.05\sqrt{soc_{n0}}u_{nQ}^2$, which are strictly convex, and the valuation function of power grid as $v_{sP}(u_{sP}; P) = -0.12(u_{sP} - P)^2$ and $v_{sQ}(u_{sQ}; Q) = -0.12(u_{sQ} - Q)^2$ which are strictly concave.

The efficient allocations to all PEVs, denoted by $\boldsymbol{u}^{**}$, are displayed in Fig. 8.2, such that

$$\begin{cases} u_{nP}^{**2} + u_{nQ}^{**2} = S_n^2,\ u_{nP}^{**} < u_{nP}^{max}, & \text{in case } soc_{n0} \in [soc_n^{min}, 12\%] \\ u_{nP}^{**2} + u_{nQ}^{**2} < S_n^2,\ u_{nP}^{**} < u_{nP}^{max}, & \text{in case } soc_{n0} \in (12\%, 81\%) \\ u_{nP}^{**2} + u_{nQ}^{**2} < S_n^2,\ u_{nP}^{**} = u_{nP}^{max} = \frac{\Phi_n}{\Delta T}(soc_n^{max} - soc_{n0}), & \text{in case } soc_{n0} \in [81\%, soc_n^{max}] \end{cases} \tag{8.11}$$

by which it obtains that at the efficient allocations, a PEV n, with $n \in \mathcal{N}$, reaches its apparent power constraint S_n^2 in case its SoC value lower than 12%, and reaches its maximum active power constraint u_{nP}^{max} in case its SoC value higher than 81%. The corresponding aggregated active and reactive regulation service rates $\boldsymbol{u}_s^{**}$ are specified as follows:

$$u_s^{**} \equiv (u_{sP}^{**}, u_{sQ}^{**}) = \left(\sum_{n=1}^{N} u_{nP}^{**}, \sum_{n=1}^{N} u_{nQ}^{**} \right) = (4.997\text{MW}, 1.999\text{MVar}) .$$

8.3 Auction-Based Decentralized Vehicle-to-Grid Coordination Method

The socially optimal V2G coordination problems can be implemented in case that the system controller has complete information and can directly schedule the behaviors of all individual PEVs and power grid. However in practice individual PEVs and power grid may not want to share their private information with others and individual PEVs do not permit the system directly control their batteries. In this situation, decentralized coordination methods can be adapted. It proposes an auction-based decentralized method for the underlying regulation problem, such that each of individual agents, either power grid or an individual PEV, responds with a best bid profile to the bid profiles of the others. The system is at an equilibrium in Nash sense if no single agent can benefit himself by unilaterally deviating from its individual bid profile.

8.3.1 Bid Profiles for Regulation Auction Problems

8.3.1.1 Bid Profile of Power Grid

As a buyer, power grid submits a (4-dimension) bid profile $b_s \equiv (b_{sP}, b_{sQ})$, such that

$$(b_{sP}, b_{sQ}) \equiv \big((\beta_{sP}, r_{sP}), (\beta_{sQ}, r_{sQ})\big) , \tag{8.12}$$

with $r_{sP}, r_{sQ} \geq 0$, which specifies (i) the maximum per unit price β_{sP} of active power that power grid is willing to pay and the demand up to r_{sP} units of the active power; and (ii) the maximum per unit price β_{sQ} of reactive power that power grid is willing to pay and the demand up to r_{sQ} units of the capacitive reactive power.

As a result, the revealed valuation function of power grid with respect to the bid profile given in (8.12) is specified as follows:

$$\widehat{v}_s(u_s) = \beta_{sP} \min(u_{sP}, r_{sP}) + \beta_{sQ} \min(u_{sQ}, r_{sQ}). \tag{8.13}$$

$b_{-s} \equiv \big((\beta_{sP}, 0), (\beta_{sQ}, 0)\big)$ and $\mathscr{B}_s$ are defined as the set of bid profiles of power grid.

8.3.1.2 Bid Profiles of Individual PEVs

As a seller, PEV n specifies a (4-dimension) bid profile $a_n \equiv (a_{nP}, a_{nQ})$, such that

$$(a_{nP}, a_{nQ}) \equiv \big((\alpha_{nP}, s_{nP}), (\alpha_{nQ}, s_{nQ})\big), \tag{8.14}$$

with $0 \le s_{nP} \le u_{nP}^{max}$ and $s_{nP}^2 + s_{nQ}^2 \le S_n^2$, which specifies (i) the maximum per unit price α_{nP} of active power that PEV n asks and the supply up to s_{nP} units of the active power; and (ii) the maximum per unit price α_{nQ} of reactive power that PEV n asks and the supply up to s_{nQ} units of the capacitive reactive power.

As a result, the revealed cost function of PEV n, denoted by $\widehat{g}_n$, with respect to the bid profile given in (8.14) is specified as follows:

$$\widehat{g}_n(\boldsymbol{u}_n) = \alpha_{nP} \min(u_{nP}, s_{nP}) + \alpha_{nQ} \min(u_{nQ}, s_{nQ}). \tag{8.15}$$

8.3.2 Service Allocation Rule Subject to Individual Bid Profiles

Definition 8.1 Considering a collection of bid profiles $\boldsymbol{c} \equiv (b_s, \boldsymbol{a})$, $\boldsymbol{u} \equiv (\boldsymbol{u}_s, \boldsymbol{u}_1, \ldots, \boldsymbol{u}_N)$ is called an admissible allocation with respect to $\boldsymbol{c}$, if the following constraints hold:

$$(0, 0) \le (u_{sP}, u_{sQ}) \le (r_{sP}, r_{sQ}), \tag{8.16a}$$

$$(0, 0) \le (u_{nP}, u_{nQ}) \le (s_{nP}, s_{nQ}), \text{ for all } n \in \mathcal{N}. \tag{8.16b}$$

The set of admissible allocations with respect to $\boldsymbol{c}$ is denoted by $\mathscr{A}(\boldsymbol{c})$. ■

It further defines a function U on an admissible allocation $\boldsymbol{u}$ with respect to a bid profile $\boldsymbol{c}$ as the following:

$$U(\boldsymbol{u}) = (\beta_{sP} u_{sP} + \beta_{sQ} u_{sQ}) - \sum_{n=1}^{N} (\alpha_{nP} u_{nP} + \alpha_{nQ} u_{nQ}), \qquad \text{subject to (8.8).} \tag{8.17}$$

The auctioneer assigns an optimal admissible allocation $\boldsymbol{u}^*$ with respect to a collection of bid profiles $\boldsymbol{c}$, such that

$$\boldsymbol{u}^*(\boldsymbol{c}) = \underset{\boldsymbol{u} \in \mathscr{A}(\boldsymbol{c})}{\operatorname{argmax}} \{U(\boldsymbol{u})\}. \tag{8.18}$$

Let $\widehat{\lambda}_{sP}, \widehat{\lambda}_{sQ}$ denote the Lagrange multipliers corresponding to the quantity constraint of power grid (8.16a), and $\widehat{\lambda}_{nP}, \widehat{\lambda}_{nQ}$ denote the Lagrange multipliers corresponding to the quantity constraint of PEV n (8.16b), and $\widehat{\mu}_P, \widehat{\mu}_Q$ denote the Lagrange multipliers corresponding to the equality constraint (8.8). The associated Lagrange function is given below:

$$\begin{aligned}\widehat{L}(\boldsymbol{u}_s, \boldsymbol{u}_n, \widehat{\boldsymbol{\lambda}}, \widehat{\boldsymbol{\mu}}) &= \widehat{\lambda}_{sP}(r_{sP} - u_{sP}) + \widehat{\lambda}_{sQ}(r_{sQ} - u_{sQ}) \\ &+ \widehat{\lambda}_{nP}(s_{nP} - u_{nP}) + \widehat{\lambda}_{nQ}(s_{nQ} - u_{nQ}) \\ &+ (\beta_{sP}u_{sP} + \beta_{sQ}u_{sQ}) - \sum_{n=1}^{N} \{(\alpha_{nP}u_{nP} + \alpha_{nQ}u_{nQ})\} \\ &+ \widehat{\mu}_P \left(\sum_{n=1}^{N} u_{nP} - u_{sP}\right) + \widehat{\mu}_Q \left(\sum_{n=1}^{N} u_{nQ} - u_{sQ}\right)\end{aligned}$$

with $\boldsymbol{u}_s \equiv (u_{sP}, u_{sQ})$, $\boldsymbol{u}_n \equiv (u_{nP}, u_{nQ}; n \in \mathscr{N})$, $\widehat{\boldsymbol{\lambda}} \equiv \left(\widehat{\lambda}_{sP}, \widehat{\lambda}_{sQ}, \widehat{\lambda}_{nP}, \widehat{\lambda}_{nQ}\right)$ and $\widehat{\boldsymbol{\mu}} \equiv \left(\widehat{\mu}_P, \widehat{\mu}_Q\right)$; then the KKT conditions for the auction optimization problems are listed below:

$$\beta_{sP} - \widehat{\mu}_P^* - \widehat{\lambda}_{sP}^* \le 0, \quad u_{sP}^* \ge 0, \quad u_{sP}^*(\beta_{sP} - \widehat{\mu}_P^* - \widehat{\lambda}_{sP}^*) = 0; \tag{8.19a}$$

$$\beta_{sQ} - \widehat{\mu}_Q^* - \widehat{\lambda}_{sQ}^* \le 0, \quad u_{sQ}^* \ge 0, \quad u_{sQ}^*(\beta_{sQ} - \widehat{\mu}_Q^* - \widehat{\lambda}_{sQ}^*) = 0; \tag{8.19b}$$

$$-\alpha_{nP} - \widehat{\lambda}_{nP}^* + \widehat{\mu}_P^* \le 0, \quad u_{nP}^* \ge 0, \quad u_{nP}^*(-\alpha_{nP} - \widehat{\lambda}_{nP}^* + \widehat{\mu}_P^*) = 0; \tag{8.19c}$$

$$-\alpha_{nQ} - \widehat{\lambda}_{nQ}^* + \widehat{\mu}_Q^* \le 0, \quad u_{nQ}^* \ge 0, \quad u_{nQ}^*(-\alpha_{nQ} - \widehat{\lambda}_{nQ}^* + \widehat{\mu}_Q^*) = 0; \tag{8.19d}$$

$$u_{sP}^* \le r_{sP}, \quad \widehat{\lambda}_{sP}^* \ge 0, \quad \widehat{\lambda}_{sP}^*(r_{sP} - u_{sP}^*) = 0; \tag{8.19e}$$

$$u_{sQ}^* \le r_{sQ}, \quad \widehat{\lambda}_{sQ}^* \ge 0, \quad \widehat{\lambda}_{sQ}^*(r_{sQ} - u_{sQ}^*) = 0; \tag{8.19f}$$

$$u_{nP}^* \le s_{nP}, \quad \widehat{\lambda}_{nP}^* \ge 0, \quad \widehat{\lambda}_{nP}^*(s_{nP} - u_{nP}^*) = 0; \tag{8.19g}$$

$$u_{nQ}^* \le s_{nQ}, \quad \widehat{\lambda}_{nQ}^* \ge 0, \quad \widehat{\lambda}_{nQ}^*(s_{nQ} - u_{nQ}^*) = 0. \tag{8.19h}$$

with $u_{sP}^* = \sum_{n=1}^{N} u_{nP}^*$ and $u_{sQ}^* = \sum_{n=1}^{N} u_{nQ}^*$.

8.3.3 Transfer Money of Agents Subject to Bid Profiles

For each bid profile $\boldsymbol{c}$, it will specify the so-called transfer moneys for power grid and PEV n, denoted by $\tau_s(\boldsymbol{c})$ and $\tau_n(\boldsymbol{c})$ respectively, following the allocation way implemented by the system auctioneer given in (8.18). Essentially, the payment/income of an individual unit can be expressed as, the summation of all users' utility functions when this unit didn't join the auction, minus the summation of the all of other units' utility functions when this unit joined the auction. That is to say, the money transfer τ made by each of related units, is exactly the externality it imposes on others through its participation, just as in the VCG-style mechanism adapted in [21, 30] etc.

8.3.3.1 Money Transfer of Power Grid

Let $\boldsymbol{u}^*(\boldsymbol{c}_{(s)})$ denote the solution to the allocation rule, defined in Sect. 8.3.2, with $\boldsymbol{c}_{(s)} = (b_{-s}, \boldsymbol{a})$ for power grid, i.e.

$$\boldsymbol{u}^*(\boldsymbol{c}_{(s)}) = \underset{\boldsymbol{u}\in\mathscr{A}(\boldsymbol{c}_{(s)})}{\text{argmax}}\{U(\boldsymbol{u})\}.$$

For notational simplicity, $\boldsymbol{u}^*(\boldsymbol{c}_{(s)})$ is rewritten as $\boldsymbol{u}^*_{(s)}$.

The money transfer to be made by power grid with bid profile $\boldsymbol{c}$, denoted by $\tau_s(\boldsymbol{c})$, is given by (8.20) below,

$$\tau_s(\boldsymbol{c}) = U(\boldsymbol{u}^*_{(s)}) - \big(U(\boldsymbol{u}^*) - (\beta_{sP}u^*_{sP} + \beta_{sQ}u^*_{sQ})\big), \tag{8.20}$$

with $\boldsymbol{u}^* \equiv \boldsymbol{u}^*(\boldsymbol{c})$ specified in (8.18).

8.3.3.2 Money Transfer of an Individual PEV

Let $\boldsymbol{u}^*(\boldsymbol{c}_{(n)})$ denote the solution to the allocation rule, defined in Sect. 8.3.2, with $\boldsymbol{c}_{(n)} = (b_s, \boldsymbol{a})$ substituted with $s_{nP} = s_{nQ} = 0$ for PEV n, i.e.

$$\boldsymbol{u}^*(\boldsymbol{c}_{(n)}) = \underset{\boldsymbol{u}\in\mathscr{A}(\boldsymbol{c}_{(n)})}{\text{argmax}}\{U(\boldsymbol{u})\}.$$

For notational simplicity, $\boldsymbol{u}^*(\boldsymbol{c}_{(n)})$ is rewritten as $\boldsymbol{u}^*_{(n)}$.

The money transfer to be made by PEV n with bid profiles $\boldsymbol{c}$, denoted by $\tau_n(\boldsymbol{c})$, is given by (8.21) below,

$$\tau_n(\boldsymbol{c}) = U(\boldsymbol{u}^*_{(n)}) - (U(\boldsymbol{u}^*) + (\alpha_{nP}u^*_{nP} + \alpha_{nQ}u^*_{nQ})). \tag{8.21}$$

8.3.4 *Payoff Functions of Individual Units*

Subject to a collection of bid profiles, the payoff functions of individual units are specified by adapting the auctioneer's optimal resource allocation and the money transfer mechanism.

Power grid has a payoff function, denoted by $f_s(\boldsymbol{c})$, such that

$$f_s(\boldsymbol{c}) \triangleq v_s(u^*_s) - \tau_s(\boldsymbol{c}), \tag{8.22}$$

where $u^*_s \equiv u^*_s(\boldsymbol{c})$ represents the allocation to power grid assigned by auctioneer with the bid profiles $\boldsymbol{c}$, and $\tau_s(\boldsymbol{c})$ is the money payed by power grid defined in (8.20).

Individual PEV n has a payoff function, denoted by $f_n(\boldsymbol{c})$, such that

$$f_n(\boldsymbol{c}) \triangleq -\tau_n(\boldsymbol{c}) - g_n(u_n^*), \tag{8.23}$$

where $u_n^* \equiv u_n^*(\boldsymbol{c})$ represents the allocated generation power for PEV n assigned by auctioneer with $\boldsymbol{c}$, and $\tau_n(\boldsymbol{c})$ is the money transfer of PEV n defined in (8.21).

Definition 8.2 A collection of bid profiles $(b_s^0, a_1^0, ..., a_N^0)$ is a Nash equilibrium for the auction game if the following holds:

$$f_s(b_s^0, \boldsymbol{a}^0) \geq f_s(b_s, \boldsymbol{a}^0), \text{ for all } b_s \in \mathscr{B}_s,$$
$$f_n(b_s^0, a_n^0, \boldsymbol{a}_{-n}^0) \geq f_n(b_s^0, a_n, \boldsymbol{a}_{-n}^0), \text{ for all } a_n \in \mathscr{B}_n,$$

i.e. b_s^0 is a best response of power grid with respect to bid profiles $\boldsymbol{a}^0$ of PEV populations, and a_n^0 represents a best response of PEV n with respect to $(\boldsymbol{a}_{-n}^0, b_s^0)$ with $\boldsymbol{a}_{-n}^0 \equiv (a_1^0, \cdots, a_{n-1}^0, a_{n+1}^0, \cdots, a_N^0)$. ■

8.3.5 *Nash Equilibrium Property of Efficient Bid Profiles*

Denoted by $\boldsymbol{c}^*$ a collection of incentive compatible bid profiles, such that

$$b_s^* = \big((v'_{sP}(u_{sP}^{**}), u_{sP}^{**}), (v'_{sQ}(u_{sQ}^{**}), u_{sQ}^{**})\big), \tag{8.24a}$$
$$a_n^* = \big((g'_{nP}(u_{nP}^{**}), u_{nP}^{**}), (g'_{nQ}(u_{nQ}^{**}), u_{nQ}^{**})\big), \tag{8.24b}$$

for all $n \in \mathscr{N}$, where $\boldsymbol{u}^{**} \equiv (u_{sP}^{**}, u_{sQ}^{**}, u_{nP}^{**}, u_{nQ}^{**}; n \in \mathscr{N})$ represents the efficient allocation specified in (8.9).

8.3.5.1 Existence of an Efficient Nash Equilibrium

It will be first shown that the allocations subject to $\boldsymbol{c}^*$ are efficient in Lemmas 8.1 and 8.2, and show in Theorem 8.1 that $\boldsymbol{c}^*$ is a Nash equilibrium for the underlying auction games.

Lemma 8.1 *Considering bid profile* $\boldsymbol{c}^*$ *specified in (8.24),*

$$\alpha_{nP}^* \begin{cases} \geq \mu_P^*, & \textit{in case } u_{nP}^{**} = 0, e_n^{**} \leq S_n^2 \\ = \mu_P^*, & \textit{in case } u_{nP}^{**} \in (0, u_{nP}^{max}), e_n^{**} < S_n^2 \\ = \mu_P^* - \lambda_{nP}^*, & \textit{in case } u_{nP}^{**} = u_{nP}^{max}, e_n^{**} < S_n^2 \\ = \mu_P^* - 2\lambda_{nS}^* u_{nP}^{**}, & \textit{in case } u_{nP}^{**} \in (0, u_{nP}^{max}), e_n^{**} = S_n^2 \\ = \mu_P^* - 2\lambda_{nS}^* u_{nP}^{**} - \lambda_{nP}^*, & \textit{in case } u_{nP}^{**} = u_{nP}^{max}, e_n^{**} = S_n^2 \end{cases} \tag{8.25a}$$

$$\alpha^*_{nQ}\begin{cases} \geq \mu^*_Q, & \textit{in case } u^{**}_{nQ}=0,\, e^{**}_n \leq S^2_n \\ = \mu^*_Q, & \textit{in case } u^{**}_{nQ}>0,\, e^{**}_n < S^2_n \\ = \mu^*_Q - 2\lambda^*_{nS}u^{**}_{nQ}, & \textit{in case } u^*_{nQ}>0,\, e^{**}_n = S^2_n \end{cases} \tag{8.25b}$$

$$\beta^*_{sP}\begin{cases} = \mu^*_P, & \textit{in case } u^{**}_{sP}>0 \\ \leq \mu^*_P, & \textit{in case } u^{**}_{sP}=0 \end{cases} \tag{8.25c}$$

$$\beta^*_{sQ}\begin{cases} = \mu^*_Q, & \textit{in case } u^{**}_{sQ}>0 \\ \leq \mu^*_Q, & \textit{in case } u^{**}_{sQ}=0 \end{cases} \tag{8.25d}$$

with $e^{**}_n \equiv u^{**2}_{nP}+u^{**2}_{nQ}$, $\boldsymbol{\lambda}^* \equiv (\lambda^*_{nP}, \lambda^*_{nS}; n \in \mathcal{N})$ *and* $\boldsymbol{\mu}^* \equiv (\mu^*_P, \mu^*_Q)$ *specified in (8.10).* ■

Lemma 8.1 can be verified by analyzing the KKT conditions (8.10) and $\boldsymbol{c}^*$ defined in (8.24).

Lemma 8.2 *Suppose* $\boldsymbol{c}^*$ *as the collection of bid profiles specified in (8.24); then* $\boldsymbol{u}^*(\boldsymbol{c}^*) = \boldsymbol{u}^{**}$, *i.e. the allocation* $\boldsymbol{u}^*(\boldsymbol{c}^*)$ *is efficient.* ■

Lemma 8.2 can be verified by substituting (8.24) and (8.25) into the allocation rule (8.18), and then comparing the KKT conditions specified in (8.10) and (8.19) respectively.

Theorem 8.1 *Under Assumption 8.1, the efficient bid profile* $\boldsymbol{c}^*$ *specified in (8.24) is a Nash equilibrium for the underlying auction games.*

Proof It is sufficient to show that the efficient bid profile $\boldsymbol{c}^* \equiv (b^*_s, \boldsymbol{a}^*)$ is a Nash equilibrium if *(1) and (2)* below hold.

(1) To show that b^*_s is a best response of power grid with respect to $\boldsymbol{a}^*$ in *(1.1)* – *(1.2)* below.

(1.1) Suppose that power grid decreases its allocation from u^{**}_{sP} by a $\delta_{sP} \geq 0$ and decreases its allocation from u^{**}_{sQ} by a $\delta_{sQ} \geq 0$, via deviating its bid profile b^*_s to another one $b^\dagger_s$.

Consider $u^{**}_{sP} > 0$ in case $\delta_{sP} > 0$ and $u^{**}_{sQ} > 0$ in case $\delta_{sQ} > 0$, since power grid with $u^{**}_{sP} = 0$ can't decrease its active power allocation and power grid with $u^{**}_{sQ} = 0$ can't decrease its reactive power allocation.

Let $\boldsymbol{u}^\dagger = (u^\dagger_{sP}, u^\dagger_{sQ}, u^\dagger_{nP}, u^\dagger_{nQ}; n \in \mathcal{N})$ denote the allocation with respect to the bid profile $\boldsymbol{c}^\dagger = (b^\dagger_s, \boldsymbol{a}^*)$. By (8.17) and (8.18), it can obtain that the allocations to some PEVs decrease while the allocations to other PEVs are unchanged, in case the allocation to power grid decreases.

Denote by $\Delta\tau_s$ the change of transfer money of power grid by deviating from b^*_s to $b^\dagger_s$; then the following analysis holds:

$$
\begin{aligned}
\Delta\tau_s &= \tau_s(b_s^\dagger, \boldsymbol{a}^*) - \tau_s(b_s^*, \boldsymbol{a}^*) \\
&= \sum_{n=1}^{N} \left(\alpha_{nP}^*(u_{nP}^\dagger - u_{nP}^{**}) + \alpha_{nQ}^*(u_{nQ}^\dagger - u_{nQ}^{**})\right) \\
&\geq \beta_{sP}^* \sum_{n=1}^{N} \left(u_{nP}^\dagger - u_{nP}^{**}\right) + \beta_{sQ}^* \sum_{n=1}^{N} \left(u_{nQ}^\dagger - u_{nQ}^{**}\right) \\
&= -\beta_{sP}^*\delta_{sP} - \beta_{sQ}^*\delta_{sQ},
\end{aligned}
$$

where the 2nd equality is obtained just by taking differences of the payments of power grid, the inequality is obtained by comparing α_{nP}^*, β_{sP}^*, α_{nQ}^* and β_{sQ}^* in (8.25), and the last equality follows by the equality constraint (8.8).

The change in the valuation of power grid, denoted by Δv_s, satisfies the following

$$
\begin{aligned}
\Delta v_s &= v_{sP}(u_{sP}^{**} - \delta_{sP}) - v_{sP}(u_{sP}^{**}) + v_{sQ}(u_{sQ}^{**} - \delta_{sQ}) - v_{sQ}(u_{sQ}^{**}) \\
&< -v_{sP}'(u_{sP}^{**})\delta_{sP} - v_{sQ}'(u_{sQ}^{**})\delta_{sQ} \\
&= -\beta_{sP}^*\delta_{sP} - \beta_{sQ}^*\delta_{sQ},
\end{aligned}
$$

where the 1st equality is the definition of the Δv_s, the inequality is obtained by the 1st part of Assumption 8.1, and the last equality is obtained by the definition of $\boldsymbol{c}^*$ in (8.24). By the analysis given above and (8.22),

$$
f(b_s^\dagger, \boldsymbol{a}^*) - f(b_s^*, \boldsymbol{a}^*) = \Delta v_s - \Delta\tau_s < 0,
$$

i.e. the payoff of power grid decreases.

(1.2) Suppose that the active or reactive power allocation to power grid increases via deviating from b_s^* to $b_s^\dagger$.

The allocation to power grid can't increase any more, since the allocation to each PEV reaches the maximum quantity it can provide under $\boldsymbol{c}^*$.

(2) To show that a_n^* is a best response of PEV n with respect to $\boldsymbol{c}_{-n}^* \equiv (b_s^*, \boldsymbol{a}_{-n}^*)$ in *(2.1)* and *(2.2)* below in case $u_{nP}^{**2} + u_{nQ}^{**2} < S_n^2$ and $u_{nP}^{**2} + u_{nQ}^{**2} = S_n^2$ respectively.

(2.1) In case $u_{nP}^{**2} + u_{nQ}^{**2} < S_n^2$.

(2.1.1) Suppose that PEV n decreases its allocation from u_{nP}^{**} by a $\delta_{nP} \geq 0$ and decreases its allocation from u_{nQ}^{**} by a $\delta_{nQ} \geq 0$, via deviating its bid profile b_n^* to another one $b_n^\dagger$.

Then it considers $u_{nP}^{**} > 0$ when $\delta_{nP} > 0$ and $u_{nQ}^{**} > 0$ when $\delta_{nQ} > 0$, since PEV n with $u_{nP}^{**} = 0$ can't decrease its active power allocation and PEV n with $u_{nQ}^{**} = 0$ can't decrease its reactive power allocation.

Define $\boldsymbol{u}^\dagger = (u_{sP}^\dagger, u_{sQ}^\dagger, u_{nP}^\dagger, u_{nQ}^\dagger; n \in \mathcal{N})$ as the allocation with respect to the bid profile $\boldsymbol{c}^\dagger = (b_s^*, a_n^\dagger, \boldsymbol{a}_{-n}^*)$, then note that $\boldsymbol{u}^\dagger$ satisfies the following analysis by (8.17) and (8.18):

(R1.a) $u_{nP}^\dagger = u_{nP}^{**} - \delta_{nP}$, $u_{mP}^\dagger = u_{mP}^{**}$; $u_{nQ}^\dagger = u_{nQ}^{**} - \delta_{nQ}$, $u_{mQ}^\dagger = u_{mQ}^{**}$, for all $m \neq n$, i.e. the active and reactive power allocations to other PEVs are unchanged.

(R1.b) $u_{sP}^{\dagger} \le u_{sP}^{**}$, the equality holds with $\delta_{nP} = 0$; $u_{sQ}^{\dagger} \le u_{sQ}^{**}$, the equality holds with $\delta_{nQ} = 0$, i.e. power grid buys less, when PEV n sells less. By the equality constraint (8.8), $u_{sP}^{\dagger} = u_{sP}^{**} - \delta_{nP}$, $u_{sQ}^{\dagger} = u_{sQ}^{**} - \delta_{nQ}$.

By (8.21), it gets the change in transfer money of PEV n is

$$\Delta\tau_n = \tau_n(b_s^*, a_n^{\dagger}, \boldsymbol{a}_{-n}^*) - \tau_n(b_s^*, a_n^*, \boldsymbol{a}_{-n}^*) = \beta_{sP}^*\delta_{nP} + \beta_{sQ}^*\delta_{nQ} \ge \alpha_{nP}^*\delta_{nP} + \alpha_{nQ}^*\delta_{nQ},$$

where the 2nd equality holds by (R1.a) and (R1.b), and the inequality holds by comparing α_{nP}^*, β_{sP}^*, α_{nQ}^* and β_{sQ}^* in (8.25).

Under the 2nd part of Assumption 8.1, $g_{nP}(u_{nP})$ and $g_{nQ}(u_{nQ})$ are increasing and strictly convex, then it gets the change in the individual cost of PEV n is

$$\begin{aligned}
\Delta g_n &= g_{nP}(u_{nP}^{**} - \delta_{nP}) - g_{nP}(u_{nP}^{**}) + g_{nQ}(u_{nQ}^{**} - \delta_{nQ}) - g_{nQ}(u_{nQ}^{**}) \\
&> -g_{nP}'(u_{nP}^{**})\delta_{nP} - g_{nQ}'(u_{nQ}^{**})\delta_{nQ} \\
&= -\alpha_{nP}^*\delta_{nP} - \alpha_{nQ}^*\delta_{nQ},
\end{aligned}$$

where the 1st equality is the definition of the Δg_n, the inequality is obtained by the 2nd part of Assumption 8.1 and the 2nd equality is obtained by the definition of $\boldsymbol{c}^*$ in (8.24).

By analysis above and (8.23),

$$f(b_s^*, a_n^{\dagger}, \boldsymbol{a}_{-n}^*) - f(b_s^*, a_n^*, \boldsymbol{a}_{-n}^*) = -\Delta g_n - \Delta\tau_n < 0,$$

i.e. the payoff of PEV n decreases.

(2.1.2) Suppose that PEV n decreases its allocation from u_{nP}^{**} by a $\delta_{nP} \ge 0$ and increases from u_{nQ}^{**} by a $\delta_{nQ} > 0$, via deviating its bid profile b_n^* to another one $b_n^{\dagger}$.

It considers $u_{nP}^{**} > 0$ when $\delta_{nP} > 0$, since PEV n with $u_{nP}^{**} = 0$ can't decrease its active power allocation.

Define $\boldsymbol{u}^{\dagger} = (u_{sP}^{\dagger}, u_{sQ}^{\dagger}, u_{nP}^{\dagger}, u_{nQ}^{\dagger}; n \in \mathscr{N})$ as the allocation with respect to the bid profile $\boldsymbol{c}^{\dagger} = (b_s^*, a_n^{\dagger}, \boldsymbol{a}_{-n}^*)$; then note that $\boldsymbol{u}^{\dagger}$ satisfies the following analysis by (8.17) and (8.18):

(R2.a) $u_{nP}^{\dagger} = u_{nP}^{**} - \delta_{nP}$, $u_{mP}^{\dagger} = u_{mP}^{**}$, for all $m \ne n$, i.e. the active power allocations to other PEVs are unchanged; $u_{nQ}^{\dagger} = u_{nQ}^{**} + \delta_{nQ}$, $u_{mQ}^{\dagger} \le u_{mQ}^{**}$, for all $m \ne n$, such that $\sum_{m \ne n}(u_{mQ}^{**} - u_{mQ}^{\dagger}) = \delta_{nQ}$, i.e. the reactive power allocations to other PEVs decrease or are unchanged.

(R2.b) $u_{sP}^{\dagger} \le u_{sP}^{**}$, the equality holds with $\delta_{nP} = 0$; $u_{sQ}^{\dagger} = u_{sQ}^{**}$ i.e. power grid buys less active power when PEV n sells less and the reactive power allocation to power grid is unchanged. By the equality constraint (8.8), $u_{sP}^{\dagger} = u_{sP}^{**} - \delta_{nP}$.

By (8.21), it gets the change in the transfer money of PEV n is

$$\begin{aligned}\Delta\tau_n &= \tau_n(b_s^*, a_n^\dagger, \boldsymbol{a}_{-n}^*) - \tau_n(b_s^*, a_n^*, \boldsymbol{a}_{-n}^*)\\ &= \beta_{sP}^*\delta_{nP} - \sum_{m\neq n}\alpha_{mQ}^*(u_{mQ}^{**} - u_{mQ}^\dagger)\\ &\geq \alpha_{nP}^*\delta_{nP} - \alpha_{nQ}^*\delta_{nQ},\end{aligned}$$

where the 2nd equality is obtained just by taking differences of the transfer moneys of PEV n and considering (R2.b), the inequality is obtained by comparing the α_{nP}^*, β_{sP}^*, α_{nQ}^* and α_{mQ}^* with $u_{mQ}^{**} > 0$ (since the allocation to PEV m with $u_{mQ}^{**} = 0$ can't decrease) in (8.25) and considering (R2.a).

The change of the individual cost of PEV n, denoted by Δg_n, satisfies the following analysis

$$\begin{aligned}\Delta g_n &= g_{nP}(u_{nP}^{**} - \delta_{nP}) - g_{nP}(u_{nP}^{**}) + g_{nQ}(u_{nQ}^{**} + \delta_{nQ}) - g_{nQ}(u_{nQ}^{**})\\ &> -g_{nP}'(u_{nP}^{**})\delta_{nP} + g_{nQ}'(u_{nQ}^{**})\delta_{nQ}\\ &= -\alpha_{nP}^*\delta_{nP} + \alpha_{nQ}^*\delta_{nQ},\end{aligned}$$

where the inequality is obtained by the 2nd part of Assumption 8.1 and the last equality is obtained by (8.24).

By the analysis given above and (8.23),

$$f(b_s^*, a_n^\dagger, \boldsymbol{a}_{-n}^*) - f(b_s^*, a_n^*, \boldsymbol{a}_{-n}^*) = -\Delta g_n - \Delta\tau_n < 0,$$

i.e. the payoff of PEV n decreases.

(2.1.3) Suppose that PEV n increases its allocation from u_{nP}^{**} by a $\delta_{nP} \geq 0$ and decreases from u_{nQ}^{**} by a $\delta_{nQ} > 0$, via deviating its bid profile b_n^* to another one $b_n^\dagger$.

Then it considers $u_{nQ}^{**} > 0$ with $\delta_{nQ} > 0$, since PEV n with $u_{nQ}^{**} = 0$ can't decrease its reactive power allocation, and $u_{nP} < u_{nP}^{max}$, since PEV n with $u_{nP}^{**} = u_{nP}^{max}$ can't increase its active power allocation.

Define $\boldsymbol{u}^\dagger = (u_{sP}^\dagger, u_{sQ}^\dagger, u_{nP}^\dagger, u_{nQ}^\dagger; n \in \mathcal{N})$ as the allocation with respect to the bid profile $\boldsymbol{c}^\dagger = (b_s^*, a_n^\dagger, \boldsymbol{a}_{-n}^*)$, then note that $\boldsymbol{u}^\dagger$ satisfies the following analysis by (8.17) and (8.18):

(R3.a) $u_{nP}^\dagger = u_{nP}^{**} + \delta_{nP}$, $u_{mP}^\dagger \leq u_{mP}^{**}$, for all $m \neq n$, such that $\sum_{m\neq n}(u_{mP}^{**} - u_{mP}^\dagger) = \delta_{nP}$, i.e. the active power allocations to other PEVs decrease or are unchanged; $u_{nQ}^\dagger = u_{nQ}^{**} - \delta_{nQ}$, $u_{mQ}^\dagger = u_{mQ}^{**}$, for all $m \neq n$, i.e. the reactive power allocations to other PEVs are unchanged.

(R3.b) $u_{sP}^\dagger = u_{sP}^{**}$; $u_{sQ}^\dagger \leq u_{sQ}^{**}$, the equality holds with $\delta_{nQ} = 0$, i.e. the active power allocation to power grid is unchanged and power grid buys less reactive power when PEV n sells less. By the equality constraint (8.8), $u_{sQ}^\dagger = u_{sQ}^{**} - \delta_{nQ}$.

By (8.21), it gets the change in transfer money of PEV n is

$$\begin{aligned}
\Delta\tau_n &= \tau_n(b_s^*, a_n^\dagger, \boldsymbol{a}_{-n}^*) - \tau_n(b_s^*, a_n^*, \boldsymbol{a}_{-n}^*) \\
&= \beta_{sQ}^*\delta_{nQ} - \sum_{m\neq n} \alpha_{mP}^*(u_{mP}^{**} - u_{mP}^\dagger) \\
&\geq \alpha_{nQ}^*\delta_{nQ} - \alpha_{nP}^*\delta_{nP}
\end{aligned}$$

where the 2nd equality is obtained just by taking differences of the transfer moneys of PEV n and considering (R3.b), the inequality is obtained by comparing the α_{nP}^*, α_{mP}^* with $u_{mP}^{**} > 0$ (since the allocation to PEV m with $u_{mP}^{**} = 0$ can't decrease), α_{nQ}^* and β_{sQ}^* in (8.25) and considering (R3.a).

Under the 2nd part of Assumption 8.1, $g_{nP}(u_{nP})$ and $g_{nQ}(u_{nQ})$ are increasing and strictly convex, then it gets the change in the individual cost of PEV n is

$$\begin{aligned}
\Delta g_n &= g_{nP}(u_{nP}^{**} + \delta_{nP}) - g_{nP}(u_{nP}^{**}) + g_{nQ}(u_{nQ}^{**} - \delta_{nQ}) - g_{nQ}(u_{nQ}^{**}) \\
&> g_{nP}'(u_{nP}^{**})\delta_{nP} - g_{nQ}'(u_{nQ}^{**})\delta_{nQ} \\
&= \alpha_{nP}^*\delta_{nP} - \alpha_{nQ}^*\delta_{nQ},
\end{aligned}$$

where the inequality is obtained by the 2nd part of Assumption 8.1 and the 2nd equality is obtained by the definition of $\boldsymbol{c}^*$ in (8.24).

By analysis above and (8.23),

$$f(b_s^*, a_n^\dagger, \boldsymbol{a}_{-n}^*) - f(b_s^*, a_n^*, \boldsymbol{a}_{-n}^*) = -\Delta g_n - \Delta\tau_n < 0,$$

i.e. the payoff of PEV n decreases.

(2.1.4) Suppose that PEV n increases its allocation from u_{nP}^{**} by a $\delta_{nP} \geq 0$ and increases its allocation from u_{nQ}^{**} by a $\delta_{nQ} \geq 0$, via deviating its bid profile b_n^* to another one $b_n^\dagger$.

Then it considers $u_{nP}^{**} < u_{nP}^{max}$, since PEV n with $u_{nP}^{**} = u_{nP}^{max}$ can't increase its active power allocation.

Define $\boldsymbol{u}^\dagger = (u_{sP}^\dagger, u_{sQ}^\dagger, u_{nP}^\dagger, u_{nQ}^\dagger; n \in \mathcal{N})$ as the allocation with respect to the bid profile $\boldsymbol{c}^\dagger = (b_s^*, a_n^\dagger, \boldsymbol{a}_{-n}^*)$, then note that $\boldsymbol{u}^\dagger$ satisfies the following analysis by (8.17) and (8.18):

(R4.a) $u_{nP}^\dagger = u_{nP}^{**} + \delta_{nP}$, $u_{mP}^\dagger \leq u_{mP}^{**}$, for all $m \neq n$, such that $\sum_{m\neq n}(u_{mP}^{**} - u_{mP}^\dagger) = \delta_{nP}$, i.e. the active power allocations to other PEVs decrease or are unchanged with $u_{mP} = 0$; $u_{nQ}^\dagger = u_{nQ}^{**} + \delta_{nQ}$, $u_{mQ}^\dagger \leq u_{mQ}^{**}$, for all $m \neq n$, such that $\sum_{m\neq n} (u_{mQ}^{**} - u_{mQ}^\dagger) = \delta_{nQ}$, i.e. the reactive power allocations to other PEVs decrease or remain unchanged.

(R4.b) $u_{sP}^\dagger = u_{sP}^{**}$; $u_{sQ}^\dagger = u_{sQ}^{**}$ i.e. the active and reactive power allocations to power grid are unchanged.

By (8.21), it gets the change in transfer money of PEV n is

$$\begin{aligned}\Delta\tau_n &= \tau_n(b_s^*, a_n^\dagger, \boldsymbol{a}_{-n}^*) - \tau_n(b_s^*, a_n^*, \boldsymbol{a}_{-n}^*)\\ &= -\sum_{m\neq n}\alpha_{mP}^*(u_{mP}^{**} - u_{mP}^\dagger) - \sum_{m\neq n}\alpha_{mQ}^*(u_{mQ}^{**} - u_{mQ}^\dagger)\\ &\geq -\alpha_{nP}^*\delta_{nP} - \alpha_{nQ}^*\delta_{nQ}\end{aligned}$$

where the 2nd equality is obtained just by taking differences of the transfer moneys of PEV n, the inequality is obtained by comparing the α_{nP}^*, α_{mP}^* with $u_{mP}^{**} > 0$ (since the allocation to PEV m with $u_{mP}^{**} = 0$ can't decrease), α_{nQ}^* and α_{mQ}^* with $u_{mQ}^{**} > 0$ (since the allocation to PEV m with $u_{mQ}^{**} = 0$ can't decrease) in (8.25) and considering (R4.a) and (R4.b).

The change of the individual cost of PEV n satisfies the following:

$$\begin{aligned}\Delta g_n &= g_{nP}(u_{nP}^{**} + \delta_{nP}) - g_{nP}(u_{nP}^{**}) + g_{nQ}(u_{nQ}^{**} + \delta_{nQ}) - g_{nQ}(u_{nQ}^{**})\\ &> g_{nP}'(u_{nP}^{**})\delta_{nP} + g_{nQ}'(u_{nQ}^{**})\delta_{nQ}\\ &= \alpha_{nP}^*\delta_{nP} + \alpha_{nQ}^*\delta_{nQ},\end{aligned}$$

where the inequality is obtained by the 2nd part of Assumption 8.1 and the 2nd equality is obtained by the definition of $\boldsymbol{c}^*$ in (8.24).

By analysis above and (8.23),

$$f(b_s^*, a_n^\dagger, \boldsymbol{a}_{-n}^*) - f(b_s^*, a_n^*, \boldsymbol{a}_{-n}^*) = -\Delta g_n - \Delta\tau_n < 0,$$

i.e. the payoff of PEV n decreases.

(2.2) In case $u_{nP}^{**2} + u_{nQ}^{**2} = S_n^2$.

(2.2.1) Suppose that PEV n decreases its allocation from u_{nP}^{**} by a $\delta_{nP} \geq 0$ and decreases its allocation from u_{nQ}^{**} by a $\delta_{nQ} \geq 0$, via deviating its bid profile b_n^* to another one $b_n^\dagger$.

Following the similar analysis in (2.1.1), it can make a conclusion that the payoff of PEV n decreases in this case.

(2.2.2) Suppose that PEV n increases its allocation from u_{nP}^{**} by a $\delta_{nP} \geq 0$, via deviating its bid profile b_n^* to another one $b_n^\dagger$. Due to the constraint of apparent power in (8.2), the reactive power allocation to PEV n has to be decreased by some positive valued $\delta_{nQ} > 0$, such that

$$(u_{nP}^{**} + \delta_{nP})^2 + (u_{nQ}^{**} - \delta_{nQ})^2 \leq S_n^2.$$

Then it considers $u_{nP}^{**} < u_{nP}^{max}$, since PEV n with $u_{nP}^{**} = u_{nP}^{max}$ can not increase its active power allocation, and $u_{nQ}^{**} > 0$, since PEV n with $u_{nQ}^{**} = 0$ can not decrease its reactive power allocation.

Define $\boldsymbol{u}^\dagger = (u_{sP}^\dagger, u_{sQ}^\dagger, u_{nP}^\dagger, u_{nQ}^\dagger; n \in \mathcal{N})$ as the allocation with respect to the bid profile $\boldsymbol{c}^\dagger = (b_s^*, a_n^\dagger, \boldsymbol{a}_{-n}^*)$, then note that $\boldsymbol{u}^\dagger$ satisfies the following analysis by (8.17) and (8.18):

(R5.a) $u^{\dagger}_{nP} = u^{**}_{nP} + \delta_{nP}$, $u^{\dagger}_{mP} \le u^{**}_{mP}$, for all $m \ne n$, such that $\sum_{m\ne n}(u^{**}_{mP} - u^{\dagger}_{mP}) = \delta_{nP}$, i.e. the active power allocations to other PEVs decrease or are unchanged; $u^{\dagger}_{nQ} = u^{**}_{nQ} - \delta_{nQ}$, $u^{\dagger}_{mQ} = u^{**}_{mQ}$, for all $m \ne n$, i.e. the reactive power allocations to other PEVs are unchanged.

(R5.b) $u^{\dagger}_{sP} = u^{**}_{sP}$, i.e. the active power allocation to power grid is unchanged; $u^{\dagger}_{sQ} < u^{**}_{sQ}$ i.e. the reactive power allocation to power grid decreases. By the equality constraint (8.8), $u^{\dagger}_{sQ} = u^{**}_{sQ} - \delta_{nQ}$.

By (8.21), it gets the change in transfer money of PEV n is

$$\begin{aligned}\varDelta\tau_n &= \tau_n(b^*_s, a^{\dagger}_n, \boldsymbol{a}^*_{-n}) - \tau_n(b^*_s, a^*_n, \boldsymbol{a}^*_{-n}) &(8.26)\\ &= -\sum_{m\ne n}\alpha^*_{mP}(u^{**}_{mP} - u^{\dagger}_{mP}) + \beta^*_{sQ}\delta_{nQ}\\ &\ge -\mu^*_P\delta_{nP} + \mu^*_Q\delta_{nQ}, &(8.27)\end{aligned}$$

where the 2nd equality is obtained just by taking differences of the transfer moneys of PEV n and considering (R5.b), the inequality is obtained by (R5.a) and (8.25) where α^*_{mP}, μ^*_P, β^*_{sQ} and μ^*_Q are specified.

Under the 2nd part of Assumption 8.1, $g_{nP}(u_{nP})$ and $g_{nQ}(u_{nQ})$ are strictly increasing and convex,

$$\varDelta g_n = g_{nP}(u^{**}_{nP} + \delta_{nP}) - g_{nP}(u^{**}_{nP}) + g_{nQ}(u^{**}_{nQ} - \delta_{nQ}) - g_{nQ}(u^{**}_{nQ}) > \alpha^*_{nP}\delta_{nP} - \alpha^*_{nQ}\delta_{nQ},$$

by which together with (8.25), it obtains that

$$\varDelta g_n > (\mu^*_P - 2\lambda^*_{nS}u^{**}_{nP})\delta_{nP} - (\mu^*_Q - 2\lambda^*_{nS}u^{**}_{nQ})\delta_{nQ}. \tag{8.28}$$

By (8.26) and (8.28), it obtain that

$$f(b^*_s, a^{\dagger}_n, \boldsymbol{a}^*_{-n}) - f(b^*_s, a^*_n, \boldsymbol{a}^*_{-n}) = -\varDelta g_n - \varDelta\tau_n < 2\lambda^*_{ns}u^{**}_{nP}\delta_{nP} - 2\lambda^*_{ns}u^{**}_{nQ}\delta_{nQ}. \tag{8.29}$$

By the apparent power constraint of (8.16), it obtains that

$$\begin{aligned}u^{**2}_{nP} + u^{**2}_{nQ} &= S^2_n, &(8.30)\\ (u^{**}_{nP} + \delta_{nP})^2 + (u^{**}_{nQ} - \delta_{nQ})^2 &\le S^2_n. &(8.31)\end{aligned}$$

By (8.30) and (8.31), it obtains that $2u^{**}_{nP}\delta_{nP} - 2u^{**}_{nQ}\delta_{nQ} + \delta^2_{nP} + \delta^2_{nQ} \le 0$; then $2u^{**}_{nP}\delta_{nP} - 2u^{**}_{nQ}\delta_{nQ} < 0$, by which together with (8.29), it obtains that

$$f(b^*_s, a^{\dagger}_n, \boldsymbol{a}^*_{-n}) - f(b^*_s, a^*_n, \boldsymbol{a}^*_{-n}) < 0,$$

i.e. the payoff of PEV n decreases by deviating from a^*_n to any other bid profile $a^{\dagger}_n$.

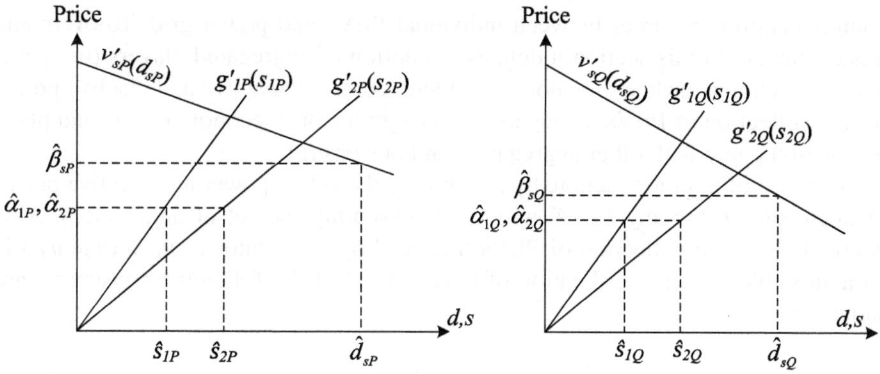

Fig. 8.3 Inefficient Nash equilibrium

(2.2.3) Suppose that PEV n decreases its allocation from u_{nP}^{**} by a $\delta_{nP} \geq 0$ and increases from u_{nQ}^{**} by a $\delta_{nQ} > 0$, via deviating its bid profile b_n^* to another one $b_n^\dagger$.

Following a similar analysis in *(2.2.2)*, it concludes that the payoff of PEV n decreases.

Thus, following the analysis in *(1)* and *(2)*, the efficient bid profile $\boldsymbol{c}^*$ is a Nash equilibrium. ■

In Theorem 8.1, it verifies that there exists the efficient Nash equilibrium for the underlying V2G auction games; however as discussed below, there may exist other inefficient Nash equilibria beside the efficient one.

In Fig. 8.3, it illustrates incentive compatible bid profiles for V2G auction games, such that

$$\widehat{\alpha}_{1P} = \widehat{\alpha}_{2P} < \widehat{\beta}_{sP},\ \widehat{s}_{1P} + \widehat{s}_{2P} = \widehat{d}_{sP},\ \text{and}\ \widehat{\alpha}_{1Q} = \widehat{\alpha}_{2Q} < \widehat{\beta}_{sQ},\ \widehat{s}_{1Q} + \widehat{s}_{2Q} = \widehat{d}_{sQ}. \tag{8.32}$$

Remark For purpose of demonstration, it only considers V2G coordination problems with two PEVs.

According to the KKT conditions, it is easy to check that the allocations to all players subject to bid profiles specified in (8.32) are inefficient, and by applying the technology used in the proof of Theorem 8.1, it can verify that the collection of bid profiles specified in (8.32), is a Nash equilibrium as well.

8.4 V2G Coordination Auction Games with Aggregated Players

Up to now it has studied the auction games for V2G coordination problems. However it is infeasible to implement for the underlying auction games with high-penetration PEV populations in large-scale power grid, since it may require enormous communication resources between individual PEVs and power grid. To overcome this challenge, in this section it defines the notion of aggregated players (or aggregators), such that each aggregator can directly control the active and reactive power for the collection of PEVs managed by this aggregator in economic way, and plays the auction game with other aggregators and power grid.

By abuse of notation, denote by u_{iP} and u_{iQ} the active power and reactive power of aggregator i, for any $i \in \mathscr{I}$ with $\mathscr{I}$ representing the set of aggregators. Also denote by $\mathscr{M}_i$ the collection of PEVs managed by aggregator i.. $\boldsymbol{u}_i \equiv (u_{iP}, u_{iQ})$ is an admissible power coordination of aggregator i, if the following constraints are satisfied:

$$u_{iP} \geq 0, u_{iQ} \geq 0, \tag{8.33a}$$

$$u_{iP} \leq u_{iP}^{\max} \triangleq \sum_{n \in \mathscr{M}_i} u_{nP}^{\max}, \tag{8.33b}$$

$$u_{iP}^2 + u_{iQ}^2 \leq S_i^2 \triangleq \sum_{n \in \mathscr{M}_i} S_n^2, \tag{8.33c}$$

where $u_{iP}^{\max}$ and S_i^2 are the max power charging rate of aggregator i and the nominal apparent power of the power electronic device for aggregator i respectively.

The cost function of aggregator i, denoted by $G_i(\boldsymbol{u}_i)$, subject to an admissible coordination strategy $\boldsymbol{u}_i$ is specified as below:

$$G_i(\boldsymbol{u}_i) = G_{iP}(u_{iP}) + G_{iQ}(u_{iQ}) \tag{8.34}$$

where $G_{iP}(u_{iP})$ and $G_{iQ}(u_{iQ})$ denote the cost functions of aggregator i with respect to active power and reactive power respectively.

Suppose that each aggregator can directly control the active and reactive power for the collection of PEVs managed by this aggregator in economic way; then $G_{iP}(u_{iP})$ shall be defined as the minimal overall cost for all PEVs in $\mathscr{M}_i$ with the total supplied active power of u_{iP}, i.e.

$$G_{iP}(u_{iP}) = \min_{(u_{nP}; n \in \mathscr{M}_i) \in \mathscr{U}_i(u_{iP})} \sum_{n \in \mathscr{M}_i} g_{nP}(u_{nP}) \tag{8.35}$$

with $\mathscr{U}_i(u_{iP})$ specified in the following:

$$\mathscr{U}_i(u_{iP}) \triangleq \left\{ (u_{nP}; n \in \mathscr{M}_i) : \text{ s.t. } u_{iP} = \sum_{n \in \mathscr{M}_i} u_{nP}, \text{ and } u_{nP} \leq u_{nP}^{\max} \right\}. \tag{8.36}$$

Similarly, $G_{iQ}(u_{iQ})$ shall be defined as the minimal overall cost for all PEVs in $\mathscr{M}_i$ with the total supplied reactive power of u_{iP}, i.e.

$$G_{iQ}(u_{iQ}) = \min_{u_{iQ}=\Sigma_{n\in\mathscr{M}_i} u_{nQ}} \sum_{n\in\mathscr{M}_i} g_{nQ}(u_{nQ}). \tag{8.37}$$

Lemma 8.3 *Under the 2nd part of Assumption 8.1, $G_{iP}(u_{iP})$ and $G_{iQ}(u_{iQ})$ are strictly increasing and convex.* ■

Proof of Lemma 8.3 *is given in Appendix* 8.7.1.

By the specifications of V2G coordination problems with collection of aggregators given in this section and Lemma 8.3, it can obtain that the analysis studied in Sect. 8.3 for auction games for individual PEVs still hold for the auction games for aggregators. As a consequence, the existence of efficient Nash equilibrium is verified for the aggregated auction games as stated in Corollary 8.1.

Corollary 8.1 *Under Assumption 8.1, the efficient bid profile for the aggregated V2G coordination problems is a Nash equilibrium for the aggregated auction games.* ■

8.5 Implementation of Nash Equilibrium

In this section, it will discuss the implementation of efficient Nash equilibrium for the underlying V2G coordination auction games by extending a so-called dynamical quantized-PSP (Q-PSP) mechanism, proposed in [22] for double-side auction games, under which the game system can converge to a Nash equilibrium close to the efficient Nash equilibrium and show that under the Q-PSP mechanism the system converges to a Nash equilibrium, in finite steps, which approximate the efficient Nash equilibrium up to a quantized level.

Essentially the double-side auctions are formulated as two single-side auctions, say buyer-side and seller-sided auctions, coupled by a so-called "joint potential quantity".

The Q-PSP mechanism is extended to the V2G auction games as specified in Steps (1)–(4) below:

(0) Initialize a collection of bid profiles for PEVs and power grid;
(1) Set accumulated demand as the bid quantity of power grid, and the accumulated supply as the sum of the bid quantities of those PEVs whose bid price is less than the price of power grid;
(2) Set potential active (reactive respectively) power quantity as

- Larger value of accumulated active (reactive respectively) power demand and supply, to encourages to supply or to demand more electricity in the market, when the active (reactive respectively) power selling price of any players whose constraint is not reached, is not equal to the buying price;
- Smaller value of accumulated demand and supply, otherwise.

(3) PEVs and power grid update their bid profiles w.r.t. potential quantity simultaneously as follows:

   – Each PEV updates its bid to maximize its own utility; then it bids a smaller value of the price of power grid in last iteration and higher quantized price than its bid price in case accumulated supply is less than potential quantity, and bids a smaller value of the price of power grid in last iteration and lower quantized price than his bid price, otherwise.
   – Power grid updates its bid to maximize its own utility; then it bids a larger value of the matched selling price and lower quantized price than his bid price in case accumulated demand is less than potential quantity; and bids a larger value of the matched selling price and higher quantized price than his bid price, otherwise.

(4) Go to Step (1) until joint potential quantity does not update any more. ■

Remark The matched selling price, appeared in Step (3) above, represents the highest price of PEV whose price is not higher than price of power grid.

In the following, it studies the numerical simulations for the V2G coordination games with four aggregators under the Q-PSP mechanism. For purpose of demonstration it is supposed that each aggregator possesses a common active power constraint of 9MW and an apparent power constraint of 10 MVA, and consider the referred active and reactive powers as $P = 30\,\text{MW}$ and $Q = 15\,\text{Mvar}$ respectively.

By Lemma 8.3, the cost function of each aggregator is strictly convex under convexity assumption of cost functions of individual players. Here it is supposed a simplified quadratic form for cost functions of aggregators, such that $G_{iP}(u_{iP}) = \kappa_{iP} u_{iP}^2$ and $G_{iQ}(u_{iQ}) = \kappa_{iQ} u_{iQ}^2$ with parameters given in Table 8.1. It considers that the (concave) valuation function of power grid has a common negative quadratic form as $v_{sP} = -1200(u_{sP} - P)^2$ and $v_{sQ} = -1200(u_{sQ} - Q)^2$.

The updates of players' bid profiles and potential quantity under Q-PSP auction mechanism are displayed in Figs. 8.4 and 8.5 respectively, and the evolution of constraints of players is illustrated in Table 8.2 where 1 and 0 are used to represent the case that players reach and don't reach at constraint respectively. From Figs. 8.4, 8.5 and Table 8.2, it can be observed that the system converges to a Nash equilibrium close to the efficient one.

Table 8.1 Cost function parameters for aggregated players

Parameter of aggregators	Player 1	Player 2	Player 3	Player 4
κ_{iP}	98	135	156	172
κ_{iQ}	50	83	93	105

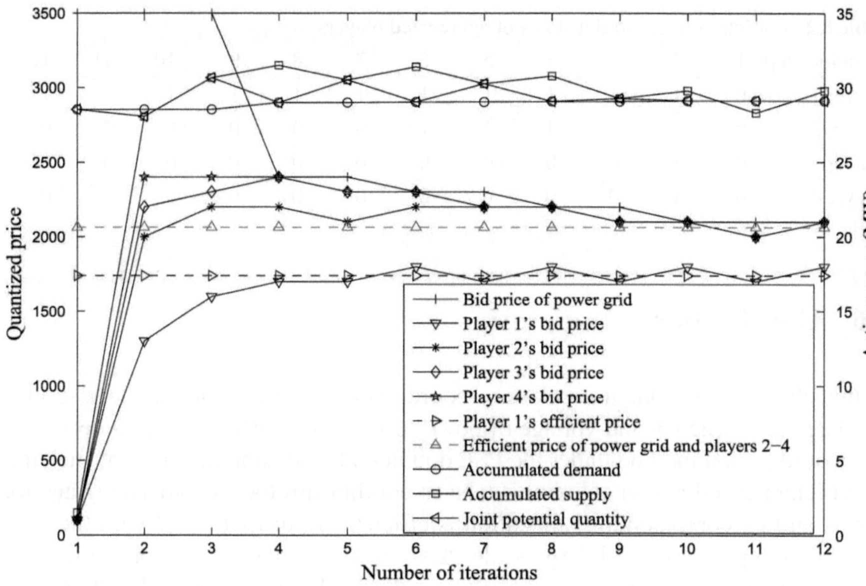

Fig. 8.4 Updates of players' bid profiles and potential quantity of active power under Q-PSP auction mechanism

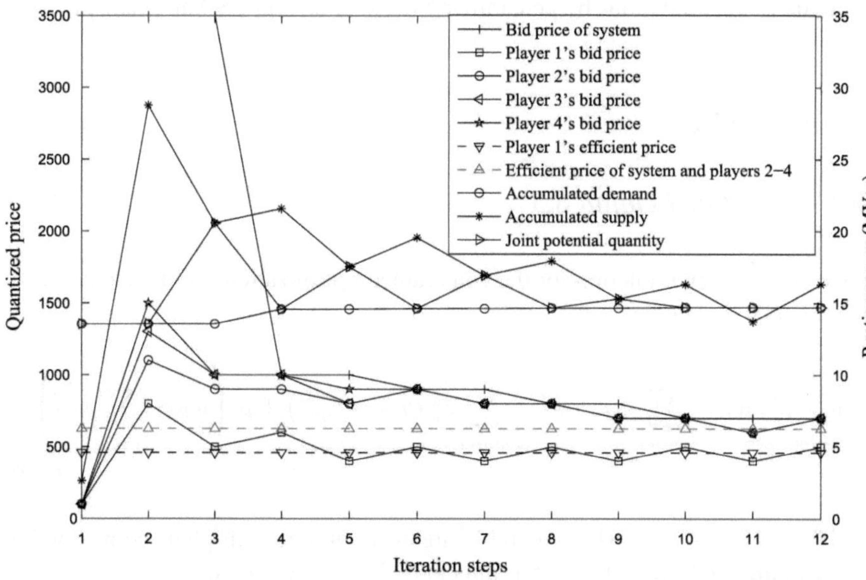

Fig. 8.5 Updates of players' bid profiles and potential quantity of reactive power under Q-PSP auction mechanism

Table 8.2 Evolution of constraint status of aggregated players

Iteration step	1	2	3	4	5	6	7	8	9	10	11	12
Player 1	0	1	1	1	1	1	1	1	1	1	1	1
Player 2	0	1	1	1	1	1	0	0	0	0	0	0
Player 3	0	1	0	0	0	0	0	0	0	0	0	0
Player 4	0	1	0	0	0	0	0	0	0	0	0	0

8.6 Conclusions

In this chapter it formulates a class of coordination problems of vehicle-to-grid to regulate the frequency and voltage of power grid, and further develops a decentralized coordination method under the PSP double-sided auction mechanism such that the efficient coordination solution is a Nash equilibrium for the underlying auction games and its corresponding allocation is efficient. In order to deal with the V2G coordination problems with large-scale PEVs, it formulates the auction games with aggregators each of which can centrally coordinate the power rates of collection of individual PEVs.

The efficient Nash equilibrium for the underlying auction games is implemented with numerical simulations by generalizing the so-called Q-PSP mechanism.

8.7 Appendices

8.7.1 *Proof of Lemma 8.3*

Firstly the Lagrange function for the constrained optimization problem appeared in (8.35) is specified below:

$$L(u_{nP}, \boldsymbol{\lambda}^p, \mu) = \sum_{n \in \mathcal{M}_i} g_{nP}(u_{nP}) + \sum_{n \in \mathcal{M}_i} \lambda_n^p \left(u_{nP} - u_{nP}^{\max}\right) + \mu \left(u_{iP} - \sum_{n \in \mathcal{M}_i} u_{nP} \right), \tag{8.38}$$

with $\boldsymbol{\lambda}^p \equiv (\lambda_n^p, n \in \mathcal{M}_i)$, where μ and λ_n^p are the Lagrange multipliers corresponding to the equality and inequality constraints in (8.36), respectively.

(8.39) listed below, together with the equality constraint in (8.36), are the KKT conditions for the constrained optimization problems appeared in (8.35).

$$g'_{nP}(u^{\diamond}_{nP}) + \lambda^{p*}_{nP} - \mu^* \geq 0, \quad u^{\diamond}_{nP} \geq 0, \quad u^{\diamond}_{P}\left(g'_{nP}(u^{\diamond}_{nP}) + \lambda^{p*}_{n} - \mu^*\right) = 0; \quad (8.39a)$$

$$u^{\diamond}_{nP} - u^{\max}_{nP} \leq 0, \qquad \lambda^{p*}_{n} \geq 0, \qquad \lambda^{p*}_{n}\left(u^{\diamond}_{nP} - u^{\max}_{nP}\right) = 0. \quad (8.39b)$$

Under the 2nd part of Assumption 8.1, the optimizations specified in (8.35) are constrained convex optimization problems; then there exist a unique solution for (8.35) with an aggregated active power u_{iP}, denoted by $\left\{u^{\diamond}_{nP}(u_{iP}); n \in \mathscr{M}_i\right\}$, that is to say,

$$\left\{u^{\diamond}_{nP}(u_{iP}); n \in \mathscr{M}_i\right\} = \underset{(u_{nP};n\in\mathscr{M}_i)\in\mathscr{U}_i(u_{iP})}{\operatorname{argmin}} \sum_{n\in\mathscr{M}_i} g_{nP}(u_{nP}).$$

In (1) and (2) below, it will show strictly increasing and convexity properties of $G_{iP}(u_{iP})$ respectively.

(1) *To show the strictly increasing property of $G_{iP}(u_{iP})$.*
Consider a pair of distinct valued $(u^{\dagger}_{iP}, u^{\ddagger}_{iP})$, such that $u^{\dagger}_{iP} < u^{\ddagger}_{iP}$; then it is straightforward to check that $u^{\diamond}_{nP}(u^{\dagger}_{iP}) \leq u^{\diamond}_{nP}(u^{\ddagger}_{iP})$ for all $n \in \mathscr{M}_i$, and moreover there at least exists a PEV m such that $u^{\diamond}_{mP}(u^{\dagger}_{iP}) < u^{\diamond}_{mP}(u^{\ddagger}_{iP})$, since $\sum_{n\in\mathscr{M}_i} u^{\diamond}_{nP}(u^{\dagger}_{iP}) = u^{\dagger}_{iP} < u^{\ddagger}_{iP} = \sum_{n\in\mathscr{M}_i} u^{\diamond}_{nP}(u^{\ddagger}_{iP})$; then by these together with the 2nd part of Assumption 8.1, it obtains that

$$G_{iP}(u^{\dagger}_{iP}) = \sum_{n\in\mathscr{M}_i} g_{nP}\left(u^{\diamond}_{nP}(u^{\dagger}_{iP})\right) < \sum_{n\in\mathscr{M}_i} g_{nP}\left(u^{\diamond}_{nP}(u^{\ddagger}_{iP})\right) = G_{iP}(u^{\ddagger}_{iP}),$$

in case $u^{\dagger}_{iP} < u^{\ddagger}_{iP}$.

(2) *To show the strictly convex property of $G_{iP}(u_{iP})$.*
By calculate the second derivative of $G_{iP}(u_iP)$, it obtains that

$$G''_{iP}(u_{iP}) = \sum_{n\in\mathscr{M}_i}\left\{g''_{nP}(u^{\diamond}_{nP})\left(\frac{du^{\diamond}_{nP}}{du_{iP}}\right)^2 + g'_{nP}\left(u^{\diamond}_{nP}\right)\frac{d^2u^{\diamond}_{nP}}{du^2_{iP}}\right\}, \quad (8.40)$$

with $u^{\diamond}_{nP} \equiv u^{\diamond}_{nP}(u_{iP})$.

By the KKT conditions of (8.39), it can obtain that $\dfrac{du^{\diamond}_{nP}}{du_{iP}} = 0$ in case $g'_{nP}\left(u^{\diamond}_{nP}\right) \neq \mu^*$; then $\sum_{n\in\mathscr{M}_i} g'_{nP}\left(u^{\diamond}_{nP}\right)\dfrac{d^2u^{\diamond}_{nP}}{du^2_{iP}} = \mu^* \sum_{n\in\mathscr{M}_i}\dfrac{d^2u^{\diamond}_{nP}}{du^2_{iP}}$, by which together with (8.40), the following holds

$$
\begin{aligned}
G''_{iP}(u_{iP}) &= \sum_{n \in \mathcal{M}_i} \left\{ g''_{nP}(u^{\diamond}_{nP}) \left(\frac{du^{\diamond}_{nP}}{du_{iP}} \right)^2 \right\} + \mu^* \sum_{n \in \mathcal{M}_i} \frac{d^2 u^{\diamond}_{nP}}{du^2_{iP}} \\
&= \sum_{n \in \mathcal{M}_i} \left\{ g''_{nP}(u^{\diamond}_{nP}) \left(\frac{du^{\diamond}_{nP}}{du_{iP}} \right)^2 \right\},
\end{aligned}
$$

where the last equality holds because $\sum_{n \in \mathcal{M}_i} \frac{du^{\diamond}_{nP}}{du_{iP}} = 1$.

In summary under the 2nd part of Assumption 8.1 and the fact there exists a PEV n with $\frac{du^{\diamond}_{nP}}{du_{iP}} > 0$, $G''_{iP}(u_{iP}) > 0$, i.e. $G_{iP}(u_{iP})$ is strict convex.

Following the same technology above, it can show that $G_{iQ}(u_{iQ})$ is strictly increasing and convex as well. ■

References

1. I. Cvetkovic, T. Thacker, Future home uninterruptible renewable energy system with vehicle-to-grid technology, in *2009 IEEE Energy Conversion Congress and Exposition, 2009. ECCE 2009* (2009), pp. 2675–2681
2. M. Erol-Kantarci, H. Mouftah, Management of PHEV batteries in the smart grid: towards a cyber-physical power infrastructure, in *2011 7th International Wireless Communications and Mobile Computing Conference (IWCMC)* (2011), pp. 795–800
3. C. Wu, H. Mohsenian-rad, J. Huang, Vehicle-to-aggregator interaction game. IEEE Trans. Smart Grid (2011)
4. D. Infield, J. Short, C. Home, L. Frerism, Potential for domestic dynamic demand-side management in the uk. *2007 IEEE Power Engineering Society General Meeting* (2007), pp. 1–6
5. S. Jang, Optimal decision on contract size for V2G aggregator regarding frequency regulation, in *Optimization of Electrical and Electronic Equipment (OPTIM)* (2010), pp. 54–62
6. H. Sekyung, H. Soohee, K. Sezaki, Development of an optimal vehicle-to-grid aggregator for frequency regulation. IEEE Trans. Smart Grid **1**(1) (2010)
7. C. Hutson, G. Venayagamoorthy, K. Corzine, Intelligent scheduling of hybrid and electric vehicle storage capacity in a parking lot for profit maximization in grid power transactions, in *Energy 2030 Conference* (2008), pp. 1–8
8. M. Baran, F. Wu, Optimal sizing of capacitors placed on a radial distribution system. IEEE Trans. Power Deliv. **4**(1), 735–743 (1989)
9. R. Gallego, A. Monticelli, R. Romero, Optimal capacitor placement in radial distribution networks. IEEE Trans. Power Syst. **16**(4), 630–637 (2001)
10. S. Sundhararajan, A. Pahwa, Optimal selection of capacitors for radial distribution systems using a genetic algorithm. IEEE Trans. Power Syst. **9**(3), 1499–1507 (1994)
11. Y.C. Huang, H.T. Yang, C.L. Huang, Solving the capacitor placement problem in a radial distribution system using tabu search approach. IEEE Trans. Power Syst. **11**(4), 1868–1873 (1996)
12. C. Wu, H. Mohsenian-Rad, J. Huang, J. Jatskevich, PEV-based combined frequency and voltage regulation for smart grid, in *Innovative Smart Grid Technologies (ISGT)* (2012), pp. 1–6

13. A. Lazar, N. Semret, The progressive second price auction mechanism for network resource sharing, in *8th International Symposium on Dynamic Games* (1998)
14. A.A. Lazar, N. Semret, Design and analysis of the progressive second price auction for network bandwidth sharing, in *Telecommunication Systems - Special issue on Network Economics* (1999)
15. W. Vickrey, Counterspeculation, auctions, and competitive sealed tenders. J. Financ. **16**(1), 8–37 (1961)
16. E.H. Clark, Multipart pricing of public goods. Public Choice **11**(1), 19–33 (1971)
17. T. Groves, Incentives in teams. Econometrica **41**(4), 617–631 (1973)
18. L. Ausubel, P. Cramton, Demand reduction and inefficiency in multi-unit auctions. Working papers, University of Maryland (2002)
19. G. Federico, D. Rahman, Bidding in an electricity pay-as-bid auction. J. Regul. Econ. **24**(2), 175–211 (2003)
20. F. Wen, A. Kumar David, Optimal bidding strategies and modeling of imperfect information among competitive generators. IEEE Trans. Power Syst. **16**(1), 15–21 (2001)
21. Rahul Jain, Jean Walrand, An efficient nash-implementation mechanism for network resource allocation. Automatica **46**, 1276–1283 (2010)
22. P. Jia, P. Caines, Analysis of decentralized decision processes in competitive markets: quantized single and double-sided auctions, in *49th IEEE Conference on Decision and Control* (2010), pp. 237–243
23. P. Jia, P. Caines, Analysis of decentralized quantized auctions on cooperative networks. IEEE Trans. Autom. Control **58**, 529–534 (2013)
24. P. Jia, P. Caines, Analysis of quantized double auctions with application to competitive electricity markets. INFOR: Inf. Syst. Oper. Res. **48**(4), 239–250 (2010)
25. S. Bashash, S. Moura, J. Forman, H. Fathy, Plug-in hybrid electric vehicle charge pattern optimization for energy cost and battery longevity. J. Power Sources **196**(196), 541–549 (2011)
26. S. Boyd, L. Vandenberghe, *Convex Optimization* (Cambridge University Press, Cambridge, 2004)
27. J. Orr, A. Emanuel, Current harmonics generated by a cluster of electric vehicle battery chargers. IEEE Trans. Power Appar. Syst. **101**(3), 691–700 (1982)
28. Z. Luo, Z. Hu, Y. Song, Z. Xu, H. Lu, Optimal coordination of plug-in electric vehicles in power grids with cost-benefit analysis — part i: Enabling techniques, *To appear in IEEE Transactions on Power Systems* (2013)
29. Z. Luo, Z. Hu, Y. Song, Z. Xu, H. Lu, Optimal coordination of plug-in electric vehicles in power grids with cost-benefit analysis — part ii: a case study in china, *To appear in IEEE Transactions on Power Systems* (2013)
30. S. Yang, E.B. Hajek, VCG-Kelly mechanisms for allocation of divisible goods: adapting VCG mechanisms to one-dimensional signals. IEEE J. Sel. Areas Commun. **25**(6), 1237–1243 (2007)

Chapter 9
Conclusions and Future Work

9.1 Conclusions

This book focuses on the designs of decentralized methods to implement optimal charging strategies for large-population PEVs in power systems. This book studies several classes of charging coordination problems of large-scale PEVs by considering distinct characteristics of PEV populations and electrical power systems to implement the optimal/nearly optimal charging strategies in effective ways.

- A class of PEV charging coordination problems is formulated as large-population games on a finite charging interval. Following the decentralized method, it shows that, under certain mild conditions, the large-population charging games converges to a unique Nash equilibrium which is either globally optimal for homogeneous case or nearly globally optimal for the heterogeneous case.
  Some examples demonstrate that the convergence to the Nash equilibrium occurs very quickly over a broad range of parameters. The results in this book will be important when PEV market penetration becomes sufficiently large that electricity demand patterns change significantly with PEV charging, since the algorithm allows users to choose their own locally optimal charging pattern while still achieving near-optimal global conditions. The strategy may improve PEV market penetration, especially relative to centralized strategies that could deter consumers who wish to independently determine their charging strategy.
  To improve the performance of the optimality of proposed method, a deviation cost from the average value of the homogeneous subpopulation is used, and each of homogeneous populations updates their strategies successively. By adopting the revised update mechanism, the system converges to the exact valley-fill equilibrium and the artificially introduced deviation cost vanishes at the equilibrium.
- Some key characteristics of chemical batteries, like the state of health, the growth of resistance, and the cycle life, are effected by the charging behaviors. In this book, the cost function underpinning the strategy establishes a tradeoff between the cost of energy and costs associated with battery degradation. It also introduces

Z. Ma, *Decentralized Charging Coordination of Large-scale Plug-in Electric Vehicles in Power Systems*, https://doi.org/10.1007/978-981-13-7652-8_9

a charge that penalizes high demand, thereby mitigating occurrences of high coincident charging on local distribution grids.

A price-based strategy is formulated to coordinate the charging strategies of large-scale PEVs in a decentralized way with the consideration of the battery degradation cost. The optimal charging strategy avoids the damage caused by the excessive charging power of the valley filling strategy to the grid and the battery of PEVs. A decentralized scheme is proposed where all PEVs simultaneously update their optimal charging strategies with respect to a common price profile, and then the price profile is updated using these latest proposed charging strategies. It establishes sufficient conditions that ensure this iterative process converges to the unique efficient collection of charging strategies. At convergence, the price profile coincides with the generator marginal cost.

- To mitigate the significant management, computation and communication capabilities of the charging coordination problems with large-scale PEVs, in this work, it constructs a hierarchical model for the PEV charging coordination problems, and proposes an off-line decentralized method for the constructed hierarchical optimization problems. Under the decentralized method, each PEV implements its best behavior with respect to a given local charging price curve set by its agent. Each agent submits the collected aggregated charging behaviors under this agent to the system operator who then updates the electricity generation price and broadcasts it to PEVs via agents. It is verified that, under certain conditions, the proposed dynamical procedure converges to the socially optimal solution. The proposed method under the hierarchical structure presents the advantage of the autonomy of the individual PEVs and the low computation and communication capability requirements on the system.
- The charging & discharging coordination problem has been studied in this book in which PEVs serve as ESSs and reduce the fluctuation of the RES. Firstly the underlying coordination problem is formulated as an optimization problem, such that the system deals with the tradeoff between the generation cost and the aggregated battery degradation cost over a multi-time period. The optimal solution possesses the valley-fill and peak-shift properties on each pair of adjacent time instants. Moreover, a decentralized method has been presented based on the gradient projection method such that PEVs' strategies were updated locally and simultaneously. It is shown that, under certain mild conditions, the system is guaranteed to converge to the optimal solution without imposing any considerations on the shape of the demand curve. To address the forecast errors in the predictions of PEV populations, RES productions and inelastic load profiles, a receding horizon based method is designed such that the prediction proceeds at each iteration.
- It proposes a novel decentralized pricing mechanism for DR based upon the auction rules to optimize large-scale PEVs' charging coordination over a multi-time interval. Each PEVs performs energetically in the auction game so that it can achieve a private maximum benefit, and at the same time the system is shown to converge to the efficient solution under the specified sufficient condition. Through simulation examples, it can be observed that the proposed dynamic pricing mechanism improves the performance of the system by encouraging individual PEVs

to reduce their power consumption and shift their charging demands to off-peak hours.

- It formulates a class of PEV charging coordination problems in distribution networks with capacity constrained feeder lines. A decentralized method was proposed such that the strategies of all of the PEVs are updated simultaneously by applying the gradient projection method. To avoid the overloading of the aggregated behaviors of PEV populations on the feeder lines, we adjust the step size of the gradient projection method. It is shown that, under certain mild conditions, the system converges to a coordination behavior which satisfies the capacity constraints, and the system cost is minimized over a converged space by applying the proposed method.
- It formulates a class of coordination problems of vehicle-to-grid to regulate the frequency and voltage of power grid, and further develops a distributed coordination method under the PSP double-sided auction mechanism such that the efficient coordination solution is a Nash equilibrium for the underlying auction games and its corresponding allocation is efficient. In order to deal with the V2G coordination problems with large-scale PEVs, it formulates the auction games with aggregators each of which can centrally coordinate the power rates of collection of individual PEVs.
  The efficient Nash equilibrium for the underlying auction games is implemented with numerical simulations by generalizing the so-called Q-PSP mechanism.

9.2 Future Work

It lists a few of possible future research topics in the following:

- In many parts of this book, it has been considered that the supply and non-PEV demand are deterministic and completely predictable. This is clearly not the case in practice, as demand is stochastic, conventional generators experience forced outages, and wind and solar generation is variable and difficult to predict. Therefore a natural extension to the associated optimization methods would incorporate uncertain supply and demand forecasting models into the optimization process. Moreover, it is necessary to implement the feasible and optimal solutions in case of the hard constraints on the systems, e.g. the capacity constraints on the transformers in the transmission systems and the feeder lines of the distribution networks, and the feasible range of the nodes of the distribution network, etc.
- In the current work, it considers a class of PEV charging coordination problems that deal with the tradeoff between the total generation cost and the accumulated battery degradation cost of PEV populations. However the battery degradation cost model only fits for the the LiFePO4 battery, which is a lithium-ion type battery. It is interesting to establish the battery degradation cost with respect to the charging/discharging rate for other types of batteries, like Lithium Cobalt Oxide adopted in Tesla, Nickel metal hydride type adopted in Honda Civic Hybrid, and

Ford Ranger PEVs, and to design the associated decentralized optimal charging coordination methods for these types of PEV populations.

- The pricing mechanism developed in this book can be extended in several directions. For example, a system with multiple energy providers can be considered, as well as a system with on-site distributed generators and storage units, and hierarchical system structure with microgrids as the medium. As an interesting direction, it may be also worth to explore the effect of malicious individual PEVs in the future.
- An efficient auction-based decentralized method has been designed for V2G coordination problems with a single-step time horizon in this book, it may be promising to extend the method to the PEV charging and V2G coordination problems during a multi-step time horizon. The multi-time PEV charging coordination problems belong to the allocation of the multi-type divisible resources among players including users and providers. Moreover, the model underpinning the charging coordination scheme assumes that the PEV population and the background demand are known with certainty prior to the charging period. The resulting auction game is static and the auction process can be undertaken off-line ahead of actual charging. In reality, the required information may be difficult to predict accurately, for example PEVs may come and go without any advanced warning. Consequently, the efficient Nash equilibrium determined ahead of the charging interval may be suboptimal relative to a solution that considers disturbances. A practical approach to addressing this deficiency is to adopt a receding horizon strategy.

MIX
Papier aus verantwortungsvollen Quellen
Paper from responsible sources
FSC® C105338

If you have any concerns about our products,
you can contact us on
ProductSafety@springernature.com

In case Publisher is established outside the EU,
the EU authorized representative is:
Springer Nature Customer Service Center GmbH
Europaplatz 3, 69115 Heidelberg, Germany

Printed by Libri Plureos GmbH
in Hamburg, Germany